Big Data and Electric Mobility

This book details how to assess electric mobility characteristics within electric vehicles, discussing energy management methods, automated systems, and the enormous potential of data resources mined from software, navigation systems, and connectivity.

Big Data and Electric Mobility presents methods to mine data specifically for electric vehicles, to comprehend their performance and to present opportunities to develop data-driven technological advancements. Including contributions from experts across the world, the book will look at topics such as human mobile behavior, battery charging and health, powertrain simulation, energy management, and multiphysics-constrained optimal charging.

The book will be key reading for researchers and engineers in the fields of automotive engineering, electrical engineering, and data mining.

Big Data and Electric Mobility

Edited by Quan Zhou and Haoran Zhang

CRC Press
Taylor & Francis Group
Boca Raton London New York

CRC Press is an imprint of the
Taylor & Francis Group, an **informa** business

Designed cover image: Getty Images

First edition published 2025
by CRC Press
2385 NW Executive Center Drive, Suite 320, Boca Raton FL 33431

and by CRC Press
4 Park Square, Milton Park, Abingdon, Oxon, OX14 4RN

CRC Press is an imprint of Taylor & Francis Group, LLC

ISBN: 978-1-032-29939-6 (hbk)
ISBN: 978-1-032-29940-2 (pbk)
ISBN: 978-1-003-30282-7 (ebk)

DOI: 10.1201/9781003302827

Typeset in Times LT Std
by Apex CoVantage, LLC

Contents

PART II Design, Management, and Control of the Powertrain of Electric Vehicles

PART III Prediction and Planning of Electric Vehicles in Real-World Driving

Preface

In the past 20 years, the decarbonization target has promoted the rapid development of electrified mobility (eMobility) systems. At the same time, as a landmark of the fourth industrial revolution, the Internet of Things appeared as a great potential platform to have everything connected, with Big Data accessible in many areas, including the transport and automotive sectors. Today, two powerful forces, that is, Big Data and eMobility, have demonstrated the great potential to reshape our industry, society, and, in fact, our life.

In the realm of future mobility systems, we expect significant transformations as governments worldwide prioritize the reduction of carbon emissions and the transition to cleaner and more sustainable energy sources. The development of connected, autonomous, shared, and electric (CASE) vehicle technology is making contributions towards the carbon-neutral target. Connected mobility models and mobility-as-a-service (MaaS) frameworks will gain further prominence, emphasizing the utilization of shared transportation options and integrated mobility services to enhance transport efficiency and sustainability. Intelligent transportation systems (ITS) will also play a vital role in optimizing energy efficiency and minimizing congestion. Electric vehicles (EVs) will continue their ascent as an advantageous mode of transportation, decreasing dependence on fossil fuels. The utilization of biofuels and newly merged electric fuels (eFuel) in internal combustion engine (ICE) powered vehicles will lead to lower or zero-carbon emissions. Hydrogen fuel cells are promising in the longer run, particularly for heavy-duty and long-haul transportation applications. The transition to the hydrogen era is attractive to all vehicles because hydrogen allows the highest energy efficiency in the life cycle efficiency and offers the most convenient way for energy storage.

Big Data and artificial intelligence will play key roles in the investigation and integration of renewable energy sources, electrified powertrains, intelligent transport systems, and shared mobility models, crucial in advancing eMobility systems. Artificial intelligence will exhibit capabilities in enhancement of the efficiency and performance of eMobility systems at four levels (unit, powertrain, vehicle, and platoon), whereas the conventional design-of-experiments (DoE) method is insufficient to meet the growing demands for cost-effective vehicle product development. Digitalization in new vehicle design is rapidly evolving, with the technological development of digital twins, the Internet of Things (IoT), and artificial intelligence (AI), for example, in the ability to update vehicle control software through over-the-air (OTA) services. Automotive original equipment manufacturers (OEMs) now have the opportunity to use vast amounts of data to develop digital twin models in vehicle design. The deep integration of digital twin technology and artificial intelligence in eMobility systems will drive further advancements in the transport sector.

Bringing the expertise of researchers from global leading universities, most worked at the University of Birmingham before, this book explores the intersection of Big Data and eMobility, delving into the profound impact they have on present and future opportunities in vehicle development. It aims to foster a platform for discussing the outcome, challenges, and opportunities identified in the research carried out by the authors. It is a comprehensive guide, offering state-of-the-art techniques with examples and demonstrations that offer an improved understanding of how to make the eMobility system more sustainable, efficient, and intelligent by using the technologies discussed.

This book is organized into twelve chapters, which are categorized into three parts. Followed by an introductory chapter, Part I lays the foundations for the design, optimization, and management of the power sources in electric vehicles. Part II explores the design, management, and control of the electric vehicle at the powertrain level. In Part III, the impact of real-world driving uncertainties, including driver behaviors, traffic lights, and inter-vehicle interactions on the performance of electric vehicles is discussed.

We hope that this book will provide a thought-provoking and enlightening journey, offering valuable insights, inspiring ideas, and a vision for a future where Big Data and eMobility intertwine to create a smarter, greener, and more efficient world.

Prof Hongming Xu, CEng, FIMechE, FHEA, FSAE
Chair Professor in Energy and Automotive Engineering,
University of Birmingham

Editor Biographies

Dr. Haoran Zhang is a researcher with an interdisciplinary background, holding dual bachelor's degrees in engineering and economics, complemented by two PhD degrees in oil and gas storage and transportation engineering and environmental science. His research is at the forefront of sustainable development, focusing on clean energy supply chains, green transportation, and smart urban energy systems. His innovative research has garnered prestigious accolades, including the Energy Globe Award, the R&D 100 Award, the John Tiratsoo Award for Young Achievement, and the Smart 50 Award among others.

Quan Zhou is Professor of Automotive Engineering in the School of Automotive Studies at Tongji University, Shanghai, China. He received his PhD in mechanical engineering from the University of Birmingham, Birmingham, UK, in 2019. Before joining Tongji University with the NSFC International Fellowship, Zhou was Assistant Professor and Research Fellow at the University of Birmingham. His research interests include evolutionary computation, fuzzy logic, reinforcement learning, and their application in automotive systems. He is associate editor for *IEEE Transactions on Transportation Electrification* and *IET Journal of Intelligent Transportation.*

Contributors

Haoxuan Dong
Southeast University
Nanjing, China

Yinglong He
University of Surrey
Surrey, United Kingdom

Ruicheng Jiang
Zhejiang University
Hangzhou, Zhejiang, China

Fei Ju
Southeast University
Nanjing, China

Yan Kong
Jilin University
Changchun, Jilin, China

Ji Li
University of Birmingham
Birmingham, United Kingdom

Yanfei Li
Tsinghua University
Beijing, China

Zhi Li
Zhejiang University
Hangzhou, Zhejiang, China

Haoye Liu
Tianjin University
Tianjin, China

Qiao Liu
Jilin University
Changchun, Jilin, China

Qingyang Liu
Tianjin University
Tianjin, China

Yan Sui
Jilin University
Changchun, Jilin, China

Yanhong Wu
Tianjin University
Tianjin, China

Fuguo Xu
Chiba University
Chiba, Japan

Nan Xu
Jilin University
Changchun, Jilin, China

Xiaoli Yu
Zhejiang University
Hangzhou, Zhejiang, China

Fenglai Yue
National Innovation Center of Intelligent and Connected Vehicles
Beijing, China

Cetengfei Zhang
Tongji University
Shanghai, China

Haoran Zhang
Peking University
Beijing, China

Yuanjian Zhang
Loughborough University
Loughborough, United Kingdom

Di Zhao
Jilin University
Changchun, Jilin, China

Quan Zhou
Tongji University
Shanghai, China

Weichao Zhuang
Southeast University
Nanjing, China

1 Introduction

Quan Zhou and Haoran Zhang

1.1 BACKGROUND

Decarbonization requires global actions from almost all industry sectors. The figure from the International Energy Agency (IEA) shows that the transport sector contributed to one-fifth of the total carbon emissions; therefore, decarbonization in the transportation sector is in urgent demand, and it requires collaborations from multiple disciplines. Chemical engineers are working on new ways of producing zero-emission fuels and advanced energy storage devices. Civil engineers are improving transport networks, making them smarter, safer, more efficient, and greater integration with the power grid. Automotive engineers are optimizing the design of vehicles, making them transit from fossil fuel-based propulsion to electrified propulsion.

In the era of Industry 4.0, the power and scope of the Internet are becoming all-encompassing connecting people, vehicles, and infrastructure. Ultimately, all vehicles will be connected and work as intelligent agents to benefit society. The technology roadmap of the UK advanced propulsion center (APC) highlights that the electrified vehicle is more suitable for the Internet of Vehicles and autonomous driving. And the connected vehicles would be mobile platforms collecting Big Data for advanced control and management, and also for the evolution of vehicle products.

Therefore, incorporating Big Data and electric mobility systems would generate huge impacts on people's life. This will not only contribute to environmental protection in terms of reducing CO2 and other harmful emissions but also provide more energy-efficient and cheaper solutions for the transportation of people and goods.

1.2 BIG DATA: DEFINITION, HISTORY, AND RECENT DEVELOPMENT

Big Data is made up of larger, complex datasets that originate from increasingly new sources of data. These datasets are so massive that the abilities of traditional data processing software are insufficient to manage them. Nevertheless, these massive volumes of data can be used to address business problems that would not have been tackled before.

A simple definition of Big Data organized around Big Data's three Vs could be large volumes of data that contain immense variety and are generated at increasing velocity. Recently, six other Vs and one C have been added to define the truthfulness and meaningfulness of data as veracity, infrequency as validity, extracting value from the collected data, replicability in the form of visualizations,

DOI: 10.1201/9781003302827-1

processability in a virtual cloud platform, data variability, and complexity, purely in computation form.

Though the concept of big data is still emerging, the origins of large datasets go back to the 1970s when the world of organized data in the first data centers was emerging. In addition, the development of the relational database further consolidated the concept. Around 2005, researchers identified just how much data users generated through SNS, video streaming services, and other online services. In parallel to these developments, data organization and storage systems were developed as well. Hadoop, an open-source framework created specifically to store and analyze large datasets, was developed around the same year. This development of open-source frameworks such as Hadoop, Spark, and others was essential for the growth of Big Data because they make data storage cheaper and enhance the ease of engaging with Big Data sets.

Nevertheless, the volume of Big Data has been at an all-time high, as is the dependency on it for decision-making. Users of the system are still generating significant amounts of data. However, it is not just humans that are engaging with systems. Evolution in smartphone and sensor technology now allows devices to communicate through the Internet of Things (IoT) network. Besides, advancements in devices and communication network systems have eased data gathering for various indicators and drawing performance insights. Moreover, the emergence of artificial intelligence and machine learning has churned out diversity and cross-operability across various platforms.

Furthermore, cloud computing has expanded Big Data possibilities in truly elastic scalability. Such scalability is of critical importance in analyzing and supporting demands in services. We focus on this aspect of Big Data to solve challenges that hinder future development, particularly in understanding the emerging area of the Internet of Things.

1.3 EMOBILITY: THE FUTURE OF TRANSPORT SYSTEM

Electrification is a global action to achieve carbon neutrality. Apart from railway vehicles, road vehicles are now largely dependent on electrified powertrain systems. According to the IEA's prediction, there would be over 3 billion electrified vehicles on road by 2050. The electrified vehicles include battery electric vehicles, fuel cell vehicles, and plug-in hybrid vehicles, which are mainly driven by electric motors so that the propulsion system can operate with much higher energy efficiency compared to the conventional internal combustion (IC) engine-driven vehicles. However, the energy supply systems of electrified vehicles are more complex than conventional vehicles as they normally have more than one energy source. Therefore, the selection of onboard energy sources and the design and control of the energy management systems are critical for electrified vehicles.

Vehicle electrification provides more degrees of freedom for the selection of energy resources. Distinguished from the conventional IC engine vehicles that are fuelled with fossil fuels, for example, gasoline and diesel, the electrified vehicles are capable of being driven by low-carbon or zero-carbon fuels. Electro-fuel

(E-fuel) is an emerging topic and is normally produced by storing energy from renewable sources in the chemical bonds of liquid or gaseous fuels in a synthesis reactor. The most popular E-fuel is hydrogen, and there are two mainstreams to utilize hydrogen fuel for vehicle energy supply. The world's first hydrogen-powered car, Toyota MIRAI, has been produced in 2014, which is powered by a 114 kw fuel cell stack. The volume power density of Toyota's fuel cell is 3.1 kW/L. Nevertheless, fuel cell vehicles are still suffering some underlying challenges, for example, high cost. Hydrogen engines are also under rapid development as they can be based on the existing IC engine technology. The hydrogen engine is cheaper since it does require high-purity hydrogen, but the design of high durability hydrogen engine is a challenging task. Other carbon-containing E-fuels, methane, methanol, DME, etc. can be produced if the renewable hydrogen gas is reacted with CO_2 captured from the air or the exhaust gas of industrial burners. The power units for those E-fuels are also being developed as an alternative solution to hydrogen.

The design and control of the onboard energy system for the electrified vehicle are challenging because engineers need to balance multiple design targets through the optimization of hundreds and thousands of design/control parameters. Most of the optimization tasks are highly nonlinear and involve unpredictable factors, such as optimization of the energy management control. Advanced optimization and control algorithms are in urgent demand so that the vehicle can achieve maximum energy efficiency in real-world driving. Recent research highlights that energy consumption can be reduced by up to 20% if the powertrain control is incorporated with the information from the intelligent transportation system, such as traffic flow and traffic light signals. The optimization algorithms need to be fast enough to enable online optimization in real-time control so that the control strategy can be adapted to the dramatical change of the driving environments. Conventional design of experiment (DoE)–based R&D of vehicle powertrains is transiting from model-based design into artificial intelligence-empowered design and optimization. Therefore, future vehicle and transport systems would be benefitted from the rapid development of Big Data and the Internet of Things (Vehicles).

1.4 BIG DATA FOR EMOBILITY

eMobility offers the ideal opportunity to introduce renewable energy sources more widely into the transport sector. Electric vehicles are a key player in transport-oriented smart cities powered by smart grids, as they help make these cities greener by reducing vehicle emissions and carbon footprints. The number of electric passenger vehicles has the potential to increase from 2 million in 2016 to 200 million by 2030.

From dashboard apps to battery charge rates or position and acceleration levels, electric vehicles generate, store, and collate a huge amount of data every minute. The vast amount of data generated by electric vehicles could be beneficial to engineers and developers of these models. A comprehensive summary of current advanced data mining efforts for electric vehicles can provide a true understanding of their performance and pave the way for widespread commercial applications.

Big Data and eMobility are at the forefront of information technology and transpiration, respectively. Incorporating state-of-the-art techniques in these fields would generate huge impacts on IT and transportation and other related industry, making our lives greener, more harmonic, and more intelligent. However, current knowledge of Big Data and eMobility is fragmented. Big Data–based electric mobility development is an emerging topic both in academic and industrial aspects. Currently, all studies about Big Data in electric mobility are fragmented. No work has summarized the systemic knowledge in this field. Specifically, there is no book focusing on introducing how to screen and process the potential value from the deluge of unverified, noisy, and sometimes incomplete information for electric mobility development. However, this knowledge is significant for stakeholders, such as researchers, engineers, operators, company administrators, and policymakers in related fields, to comprehensively understand current technologies' infra-knowledge structure and limitations. Therefore, we planned to write a book aiming to establish a systematic knowledge of Big Data methods for eMobility and thus make a timely contribution to this new area.

This book aims to help audiences systematically understand

1) How to define and reinvent data-driven mobility models by studying urban mobility, transportation behavior, energy system management, and electrification potential.
2) Within the regular electric mobility research framework, how can we characterize the nature of data-enabled electric mobility research direction, and what are the similarities and differences?
3) The existing positive and successful electric mobility technologies that can be identified in the studied domain, and how can they be best applied for practical success.

The contributions of our book to the literature and dialogue are

1) In addition to the factors of policy and market, the most significant development bottleneck is from technical limitations. This book mainly wants to systematically summarize the current fundamental technologies of electric mobility to help to promote the development of the electric mobility industry.
2) The introduction from the technical aspect can help people to understand the potential of electric mobility in the future city more intuitively than the conceptual introduction.
3) Currently, topics of AI and Big Data are very hot and popular. Our book can help the audiences to understand electric mobility technologies from the view of AI researchers and data scientists.

Cornerstone technologies in the sphere of Big Data and eMoblity will be introduced in the chapters of this book. Chapters are written in the form of summaries of the frontier technologies applicable in eMobility harnessing the emerging IoT and Big Data. This book will introduce the key technologies from carbon-neutral fuels to onboard power and energy systems and their synergies. The utilization of Big Data and advanced artificial intelligence techniques and their application in the design, control, and operation management of the eMobility systems also will be introduced. Within the scope of Big Data and mobility systems, the book will offer demonstrated answers to the following 11 questions:

1) How are carbon-neutral fuels produced, and how can they power the primary movers to mobility systems?
2) How can the primary and secondary energy resources be stored in the vehicle systems, and what is the most energy-efficient method to convert the energy into propulsion power?
3) How to maximize the use of used energy storage devices, and how to recycle them with the minimum environmental impacts.
4) How to maximize the vehicle's energy efficiency through the multi-objective optimization of powertrain topology and component size?
5) What are the impacts of vehicle electrification on vehicle R&D? How to develop world-leading electrified vehicles with the minimum cost and time?
6) How does vehicle electrification impact the design of future thermal propulsion systems, and what is the future of internal combustion engines?
7) How much waste energy can be recovered from the vehicle powertrain system, and what are the main solutions to recover the thermal energy into electricity for automotive applications?
8) What are the impacts of driver behaviors on vehicle energy performance? How to regulate vehicle control strategy for different end-users?
9) What is the theoretical optimal energy performance over a given driving cycle, and how to optimize the vehicle energy through advanced control?
10) How does the traffic condition impact the vehicle energy performance, and how can the models of traffic scenarios contribute to vehicle energy optimization?
11) What is the impact of vehicle automation on vehicle energy performance, and how to incorporate advanced driving control with vehicle energy optimization?

The rest of the book is organized into 11 chapters to address the preceding 11 questions.

Chapter 2 will address the first challenge by introducing the next-generation fuels for automotive, electric fuels (e-fuels). It will provide a brief introduction to the definition, significance, and classification of e-fuels. The implementation of e-fuels on internal combustion engines and how they will make a timely contribution to carbon neutrality will also be demonstrated in Chapter 2.

The second research question will be answered in Chapter 3, which provides a comprehensive review of current energy storage devices and discusses the principle for selecting energy storage devices for automotive applications. Chapter 3 will also demonstrate data acquisition methods for the design of driving cycles and artificial intelligence techniques for the design and control of energy storage devices.

In Chapter 4, health identification algorithms and second-life applications for vehicle energy storage devices will be introduced to address the third challenge. This chapter will start with the performance parameters of common powertrain components such as fuel cells, batteries, and supercapacitors, and explore the prediction method of component remaining service life and the way to improve second-life value from the degradation mechanism and influencing factors.

Chapter 5 of the book will be primarily centered around the fourth challenge, delving into the obstacles encountered within the automotive sector when adopting model-based product development. The chapter will introduce the notion of automotive digital twins (DT) as a potential solution, outlining a comprehensive system comprised of four core components: the physical characteristics of the vehicle, virtual systems, data interface, and connected intelligence. Furthermore, a detailed case study will be presented, illustrating the practical applications of DT in the realm of vehicle control. To conclude the chapter, an overview of forthcoming technology trends and potential avenues for further research will be provided.

Chapter 6 provides the answers to the fifth question. It covers the topology and component size design of electrified powertrains. Following a classification of electrified vehicles based on their degree of hybridization, two types of pure electric powertrains and four types of hybrid powertrains are introduced and analyzed. Component sizing methods, including conventional design of experiment methods and artificial intelligence-driven methods are introduced.

To address the sixth challenge, Chapter 7 introduces the principles and methods for the design of dedicated thermal propulsion systems for electrified vehicles. This chapter provides a preliminary overview of the advanced technologies used and to be used for electrified passenger cars in recent years. Some potential techniques will also be presented and discussed, including gasoline compression ignition engines, hydrogen, and fuel/engine co-optimization.

Chapter 8 will introduce thermal management methods for eMobility and thus make a timely contribution to challenge 7. Following the analysis of the potential of heat energy recovery, this chapter introduces the main thermal management methods and heat recovery techniques. Based on the summarization and discussions on the development and thermal management of HPAC, future potential development directions for HPAC for EVs are pointed out.

In Chapter 9, the answers to Challenge 8 will be discussed. It serves to introduce the concept of driving behavior modeling and its integral role in vehicle supervisory control. Followed by a comprehensive overview of driving behavior, the significance of driver behavior within both industrial and academic realms is explicated, particularly within the context of VSC. The mainstream driver-oriented control methods are introduced while demonstrations are provided for control of hybrid and electric vehicles. Finally, outlook and future technology development trends are provided.

In Chapter 10, the answers to Challenge 9 will be provided. This chapter introduces the global optimization methods for control of the electrified vehicle over a given driving cycle. The main challenges for global optimization will be discussed in terms of standardization, real-time application, control accuracy, and drivability satisfaction. State-of-the-art techniques to address these four challenges are discussed, and a prospect of future research directions is provided.

Chapter 11 evaluates the impact of the traffic environment on vehicle energy performance and discusses the methods to mitigate vehicle energy consumption based on traffic information. In this chapter, a traffic simulation platform is introduced to simulate main traffic scenarios in real-world driving. Following the modeling work of the vehicle powertrain, optimization methods considering traffic impact is introduced. An outlook and prospects for traffic-informed vehicle energy optimization are given in this chapter.

Chapter 12 aims to address the last challenge through driving control and traffic predictive modeling for electrified vehicles with autonomous driving functions. The requirement for the development and assessment of electrified vehicles will be discussed and mathematical models will be developed for the multiple objective optimization problems. Emerging methods for driving and traffic control and their cooperation for energy optimization will be introduced. Outlook for future intelligent transportation systems and eMobility systems will be discussed.

1.5 SUMMARY

The power and scope of the Internet are becoming all-encompassing connecting people, vehicles, and infrastructure. Ultimately, all vehicles will be connected and work as intelligent agents to benefit society. Decarbonization requires global actions from almost all industry sectors. Therefore, incorporating Big Data and electric mobility systems would generate huge impacts on people's life. This book aims to establish a systematic knowledge of Big Data methods for eMobility and thus make a timely contribution to this new area. This book has 11 individual chapters to address the grand challenges in the field of eMobility harnessing Big Data and artificial intelligent methods. State-of-the-art technologies are introduced to improve energy efficiency and reduce carbon emissions from the perspectives of energy carrier selection, component design, powertrain system integration, operation control, and cooperative optimization harnessing intelligent transportation systems. The power of Big Data for eMobility has been demonstrated in different scenarios, and the outlook of future research directions is given in each chapter.

Part I

Design, Optimization, and Management of the Power Sources in Electric Vehicles

2 E-Fuels and Their Implementation in Zero-Emission Propulsion Systems

Haoye Liu and Qingyang Liu

State	Abbreviation
carbon dioxide	CO_2
dimethyl ether	DME
polyoxymethylene dimethyl ether	PODE
spark ignition	SI
compression ignition	CI

2.1 WHAT IS E-FUEL?

E-fuels, also known as electrofuels, power-to-liquids, power-to-gas, e-gas, air-to-fuels, or CO_2-based synthetic fuels are produced by storing energy from renewable sources in the chemical bonds of liquid or gaseous fuels in a synthesis reactor. Hydrogen gas can be produced through electrolysis using electricity from wind turbines, solar cells, geothermal, or hydroelectricity, eliminating carbon emissions entirely at the point of production.[1] Ammonia can also be considered as an e-fuel if it is produced by nitrogen (from air) and hydrogen gas produced from renewable energy sources. Other carbon-containing e-fuels, methane, methanol, and DME, etc., can be produced if the renewable hydrogen gas is reacted with CO_2 captured from air or from exhaust gas of industrial burners.[2]

Sometimes in the literature, metal fuel is also called electrofuels; metal fuels are produced through electrochemical reduction reactions of their oxides, if the energy source for the reaction was electricity.[3] In this chapter, e-fuel only refers to liquid and gaseous fuels synthesized by using renewable energy sources.

DOI: 10.1201/9781003302827-3

2.2 WHY E-FUEL?

Carbon dioxide (CO_2) is the most important greenhouse gas on earth. Its concentration has reached 413.2 parts per million in 2020, 149% of the preindustrial level. Even the economic slowdown caused by COVID-19 has no significant impact on the level of greenhouse gases and its growth rate. If no measures are taken to inhibit the continuous rise of CO_2 concentration, global temperatures will continue to rise. Given the long life of CO_2, even if CO_2 emissions were rapidly reduced to net zero, the observed temperature levels will continue for decades. While the temperature rises, it also means that extreme weather events will occur more frequently, including elevated temperature and heavy rainfall, melting ice and snow, rising sea level, and ocean acidification, with far-reaching socio-economic impacts.

Today, about half of the CO_2 emitted by human activities remains in the atmosphere. The other half is absorbed by marine and terrestrial ecosystems. The communique points out that it is worried that the capacity of terrestrial ecosystems and oceans as "sinks" may become less effective in the future, thereby reducing their ability to absorb carbon dioxide and cushion greater temperature rise.[4]

Two-thirds of global CO_2 emissions come from the energy system, and fossil fuels account for 80% of the current global primary energy demand. Fossil fuel were formed by ancient plants and animals through pressure, temperature, and changes of tens of millions to hundreds of millions of years. Fossil fuels make us no longer rely on direct sunlight today, but use concentrated solar energy that has been stored for millions of years. This also has a negative impact. Considering the timeliness of the carbon cycle, burning fossil fuels is constantly releasing carbon dioxide and promoting global warming.[5]

Transportation sector consumed about 1/4 total world energy. The main energy carrier in the transportation sector is gaseous or liquid fuel due to the high energy density. If fuel production is to capture CO_2 in the atmosphere and release CO_2 in the subsequent combustion process, it can achieve life-cycle net zero carbon emission. Using such fuel in the transportation sector will have a significant impact on the level of greenhouse gases and its growth rate.

To achieve life-cycle net zero carbon emission, fuels can be produced in two ways, biofuel and e-fuel.

Plants can seal CO_2 in itself from the atmosphere through photosynthesis. Therefore, biofuels made from plants naturally have the attribute of zero carbon emission. Biofuels are by now a well-established component in the liquid fuels market. For example, biofuels are commonly added to vehicle fuel in EU countries. According to the estimation of fuel use data of all EU member states of Eurostat, from 2015 to 2019, the total mixing proportion of ethanol in gasoline fuel increased from 5.09% to 5.92%, while the total mixing proportion of biodiesel in diesel fuel increased from 5.92% to 6.50%.[6]

To date, all commercial approaches to biofuels involve photosynthetic capture of solar radiation and conversion to reduced carbon; however, there are some limitations by using biofuel. With the development of using biofuels as vehicle fuels, there is a growing voice about the food crisis caused by biofuels. People artificially use corn and cereals to make biofuels which are supposed as food, reducing the global food supply. To solve these problems, it was proposed to use non-food crops to produce biofuels. This avoids direct competition between biofuel and grain crops. However, it must be recognized that producing non-food crops still need farmers and appropriate farming conditions. Due to the limited number of farmers and cultivated land, if part of the resources is allocated to produce plants for biofuel manufacturing, the output of food crops will be reduced, which will also push up the price of food crops. The second-generation biofuels use lignocellulosic raw materials and the third-generation biofuels from algae use nonedible raw material sources that can be used to produce biodiesel and bioethanol. However, the low efficiency presents significant challenges to scaling the new biofuel production.

The CO_2 obtained by air carbon capture can be sealed in e-fuel through the electric energy generated by renewable energy power generation (solar, wind, hydropower, and other renewable energy). At present, renewable energy, such as wind energy and solar energy, is developing very fast. The first reason is that the cost is reduced due to technological improvement, and the second reason is that it has received strong government support. However, the instability of wind and solar energy makes it difficult to match the stable power demand. With the proportion of renewable energy such as wind energy and solar energy rising, reasonable measures must be taken to solve this problem. E-fuel is a good carrier for the electric energy from renewable energy sources, which helps address the problem of intermittency that confronts efforts to increase the amount of renewable power on the electric grid. Compared to battery and biomass, the mass energy density of e-fuel is much higher.[8]

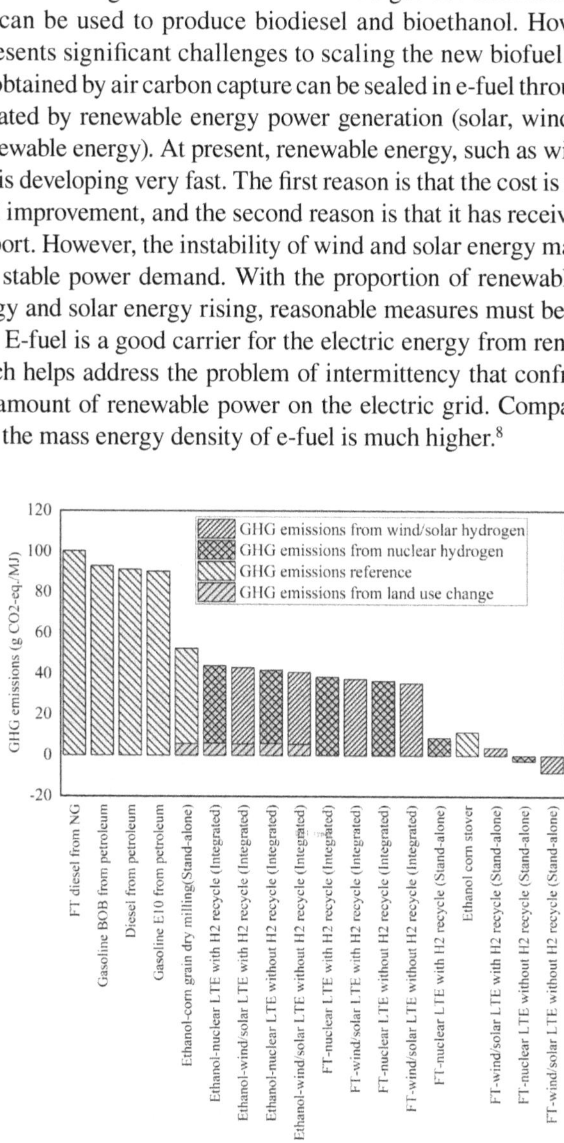

FIGURE 2.1 WTW GHG emissions comparison of different fuel production pathways.[7]

In the meantime, the energy loss during storage and transportation is low for e-fuel compared to that with batteries. In addition, because renewable energy such as solar energy and wind energy do not need additional land and farmers, the contradiction between e-fuel production and crop planting is not significant.[9]

2.3 CLASSIFICATION OF E-FUEL

Conventional fossil fuels, gasoline and diesel, have great advantages as vehicle engine fuels in the following aspects:

1) They are liquid fuel with a high energy density both in mass and volume, so that they can provide a relatively long vehicle driving range.
2) They have moderate viscosity and volatility, so that they can mix with air in the cylinder in a reasonable duration for the subsequent combustion event.
3) The ignitability of gasoline and diesel are just the opposite.

The combustion organization of modern gasoline and diesel engines are designed for the specific fuel ignitability, and only those fuel with ignitability in the requirement of gasoline/diesel engines can be used. E-fuel includes a variety of fuels. Most of the e-fuels are very common substances in our life. E-fuels, which have been mentioned in the literature, are listed in Table 2.1.[10] Alcohols, including methanol, ethanol, and butanol have quite similar properties compared with gasoline. Other e-fuels, however, differ from gasoline and diesel in terms of one or more properties. Therefore, based on different properties, e-fuel can be classified in different ways. These properties which are highly related to engine operation, are the fuel ignitability, the state under normal temperature and pressure, the elemental composition, and the fuel component.

The first and most important property is fuel ignitability, which determines the combustion organization of the engine. Fuels can be divided into gasoline-like fuel with low ignitability scaled by fuel octane number and diesel-like fuel with high ignitability scaled by cetane number. Luckily, e-fuel is either gasoline-like fuels or diesel-like fuel, and no e-fuel exhibits ignitability between the gasoline and diesel ignitability requirement (not suitable for either gasoline engines or diesel engines). Gasoline-like include synthetic gasoline, hydrogen gas, ammonia, alcohol, and methane, while diesel-like fuels include synthetic diesel and ether fuels.

According to the state under normal temperature and pressure, e-fuels can be divided into gaseous fuel and liquid fuel. The state under normal temperature and pressure determines the design of fuel supply system. The fuel supply system of conventional gasoline and diesel engines is designed for liquid fuel, and the system requires modification if gaseous fuel is applied. Gaseous fuels include hydrogen gas, ammonia, methane, and dimethyl ether, while liquid fuels include synthetic gasoline, synthetic diesel, alcohol, etc.

The element composition and component of fuel influence more detailed properties. If the fuel contains oxygen or nitrogen, the mass energy density is low, and the engine exhaust gas contains more by-products which is not common in

TABLE 2.1
Fuel Properties of E-Fuels

Property	Formula	Density in liquid	Boling point	Latent heat of evaporation	Autoignition temperature	Laminar flame speed	Octane number	Cetane number	Lower heating value	Flammability limit
Unit		Kg/m^3	°C	kJ/kg	K	m/s			kJ/kg	%
Hydrogen gas	H_2	39	–252	–	773–850	3.51	130	–	120	4–75
Ammonia	NH_3	600	–33	–	930	0.07	130	–	18.8	15–28
Methanol	CH_3OH	791	65	1109	742	0.52	110	–	19.7	6.7–36
Ethanol	C_2H_5OH	690	78	904	690	0.54	110	–	26.8	4.3–19
Butanol	C_4H_9OH	809	~118	584	614	0.5	96	–	33.1	1.4–11.2
DME	CH_3OCH_3	660	–29.5	460	506	0.45	–	55	28.8	3.4–17
Methane	CH_4	420	–161.5	–	650	0.355	130	–	55	5–13.9
Gasoline	C5-C12	700–750	25–215	310–320	575–675	0.58	90–100	–	44	1–6
Diesel	C10-C26	800–860	180–360	251–270	520	–	–	>50	42.5	1.4–7.6

conventional engines. Different components have different laminar flame speed and flammability limit. According to the element composition, e-fuel can be divided into carbon free fuel, hydrocarbon fuel, and oxygenated hydrocarbon fuel. Carbon free fuels include hydrogen gas and ammonia; hydrocarbon fuels include methane, synthetic gasoline, and synthetic diesel; and oxygenated hydrocarbon fuels include alcohol and ether fuels. According to the components, e-fuel can be divided into hydrogen, ammonia fuel, alcohol, hydrocarbon, and ether fuels.

2.4 COMPARISON OF E-FUELS IN FINANCIAL COST AND WELL-TO-WHEEL ANALYSIS

2.4.1 Manufacturing Cost

At present, large-scale e-fuel production is immature. The financial costs of e-fuel manufacturing are still very high. Figure 2.2 shows the comparison of the manufacturing cost of the various synthesis routes towards alcohols, ethers, and hydrocarbons. The energy demand presented is the energetic equivalent of 1 L of diesel, the values being independent of the size of the plant. For all e-fuel, the first step of manufacturing is hydrogen gas production, which accounts for most of the cost. Due to the difference of conversion efficiency, the hydrogen gas production cost of other fuels except H_2 is also increased accordingly. In addition, CO_2 capture, high-pressure steam, and other processes also increase costs. Therefore, regarding the cost of manufacturing, hydrogen gas is significantly lower than other e-fuels, and methanol, DME, and synthetic gasoline (MTG) are better than other fuels.

2.4.2 Well-to-Wheel Analysis

Beside financial cost, the energy consumption is another important aspect for e-fuel industrial chain. From electric energy to vehicle wheel driving power, energy goes through a series of processes, including production, storage, transportation, as

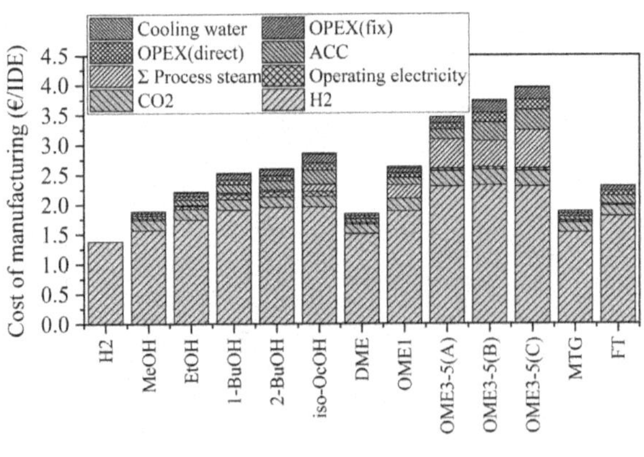

FIGURE 2.2 Comparison of base case costs for the synthesis of 1 l_{DE} e-fuel.[11]

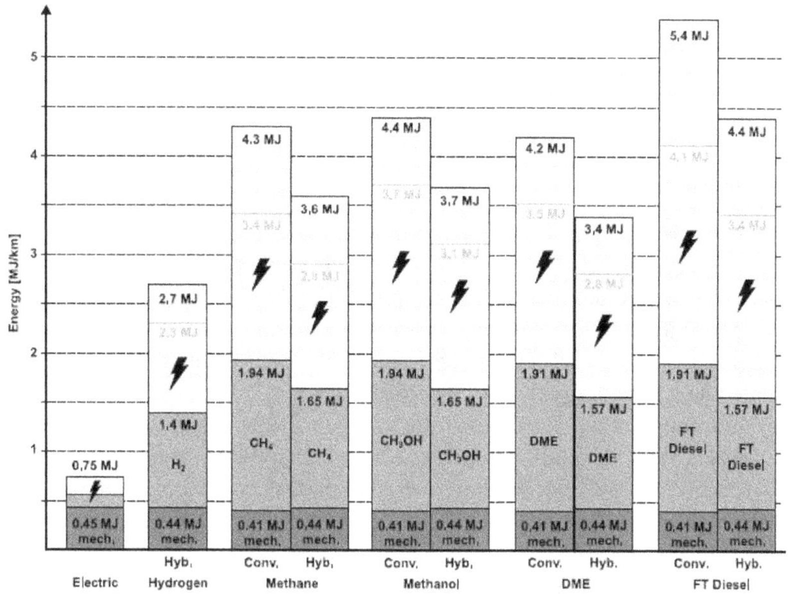

FIGURE 2.3 Well-to-miles analysis results for the five e-fuel fuels.[12]

well as the conversion inside the vehicle powertrain, and each process consumes energy and reduces the well-to-wheel efficiency. Figure 2.3 show well-to-wheel analysis results for the five e-fuels. In addition, an electric vehicle using electric energy directly is included for comparison. Here, hydrogen gas is considered as using in fuel-cells, while other fuels as using in internal combustion engines both with or without a hybrid powertrain system which is close to the present scenario. H2 still has advantages in energy conversion rate in Figure 2.3, compared with other fuels. On one hand, the manufacturing energy loss is lower, and on the other hand, the brake thermal efficiency of fuel-cells is higher than that of internal combustion engines.

2.5 HYDROGEN

2.5.1 Resources

Hydrogen widely exists in nature. Water contains 11% hydrogen, and oil, natural gas and animals and plants also contain hydrogen. However, the amount of hydrogen gas in nature is quite small. As an e-fuel, hydrogen gas can be produced directly by electrolyzing water. It is the e-fuel with the lowest production cost and highest energy efficiency.

$$H_2O \rightarrow H_2 + O_2$$

At present, the main sources of hydrogen gas are still natural gas reforming and gasification reaction. Electricity—from the grid or from renewable sources such as wind, solar, geothermal, or biomass—is also currently used to produce hydrogen gas, but it has not been widely used yet.[13]

2.5.2 Properties

Hydrogen gas is the lightest gas with no color and taste. Hydrogen gas is a typical gasoline-like fuel because it has high auto-ignition temperature and high octane number. The octane number of hydrogen is 130, which is higher than the lower limit of gasoline fuel (90). But some properties are quite different between hydrogen gas and gasoline fuel. Hydrogen is gaseous under normal temperature and pressure. The mass energy density of hydrogen is very high, which is three times that of gasoline, but the density of hydrogen gas is low, which leads to low volume energy density. Liquid hydrogen needs to be kept at a very low temperature (about—252.8 °C), and does not have a very high density. Hydrogen exhibits wide flammability limits (4%–75%), extremely high diffusion speed and flame propagation rate. However, the minimum ignition energy (0.02 mJ) is one order of magnitude lower than that of conventional fossil fuels.

2.5.3 Benefits

One of the important benefits of using hydrogen as vehicle fuel is that the only product of complete combustion of hydrogen is water, so the emission is cleaner than ordinary hydrocarbon fuels. Hydrogen gas has good potential to achieve high thermal efficiency in internal combustion engines. First, the flammability limits of hydrogen are wide, which is suitable for lean combustion. Secondly, hydrogen gas exhibits a high diffusion rate, and the laminar burning velocity of hydrogen is over 3 m/s, which is much higher than other fuels, so the combustion duration of hydrogen is short. For internal combustion engines, thermal efficiency is highly related to combustion duration. Finally, the irreversible loss of hydrogen combustion is low, providing more exhaust energy to be recovered if exhaust energy recovery system is used.

2.5.4 Shortages

Hydrogen gas has many shortages in terms of storage, transportation, and volumetric energy density. The density of gaseous hydrogen is very low, leading to very low volumetric energy. The volumetric energy density can be increased through liquification or compression. Liquid hydrogen needs to be kept at a very low temperature (about—252.8 °C), and the density is only 71 g/L, which is one order of magnitude lower than gasoline fuel. The density of hydrogen is 28 g/L and 40 g/L with the storage pressures of 300 bar and 700 bar, respectively. Liquification of hydrogen consumes 30%-40% total energy of hydrogen. Hydrogen also has embrittlement effects, which makes the container leak after storing hydrogen for a period of time, which has great potential safety hazards during storage and transportation.

In terms of engine combustion, although hydrogen has high auto-ignition temperature and high octane number, its minimum ignition energy is one order of magnitude lower than that of conventional hydrocarbon fuel. Due to the low minimum ignition energy, hydrogen is easy to be ignited by hot spots and even residual gas in the cylinder, resulting in strong knocking during engine operation and engine damage. The quenching distance of hydrogen is short, and the temperature gradient near the combustion chamber wall is large, resulting in high heat transfer loss and high wall temperature. In addition to the extremely high flame propagation speed of hydrogen, it is easy to produce inlet manifold backfire. Hydrogen combustion only generates water, so the humidity of the burned gas is very high and easy to emulsify and denature the lubricating oil of internal combustion engine, reducing the durability of internal combustion engines. High-humidity exhaust gas is also harmful to the durability of the aftertreatment system.[14]

2.6 AMMONIA

2.6.1 Resources

Ammonia is the second largest chemical product in the world, mainly used as fertilizer and refrigerant. At present, ammonia fuel is mainly synthesized by N_2 and H_2 through the Haber-Bosch process.

$$N_2 + H_2 \rightarrow NH_3$$

N_2 can be easily separated from the air, and H_2 comes from either the conversion of fossil fuels or electrolyzing water. China produced 31.9% total ammonia in the world, followed by Russia with 8.7%, India with 7.5%, and the United States with 7.1%.

2.6.2 Properties

Ammonia molecule has a trigonal pyramidal shape with three hydrogen atoms and an unshared pair of electrons attached to a nitrogen atom. Ammonia is gaseous under normal temperature and pressure. Ammonia is a chemical substance with strong lipophilicity, hydrophilicity, permeability, and corrosivity. At 20 °C, 56 g ammonia can be dissolved in 100 g water. With high auto-ignition temperature and high octane number, ammonia is also a typical gasoline-like fuel. The mass energy density of ammonia is about 45% of that of gasoline. Compared to hydrogen, ammonia liquification is much easier. Usually, liquid ammonia is stored in a liquid ammonia bottle with a pressure of around 10 bar. The volume energy density of liquid ammonia is more than one third of that of gasoline. The latent heat of vaporization of ammonia is 1371 kJ/kg, which is about 4 times that of gasoline. It exhibits relatively narrow flammability limits (16%-25%), and extremely low laminar flame speed (7 cm/s). The minimum ignition energy (8 mJ) is more than one order of magnitude higher than that of conventional fossil fuels.[15]

2.6.3 Benefits

Ammonia is also a carbon-free fuel. Ideally, the only product of complete combustion of ammonia is water and nitrogen, so the emission is cleaner than hydrocarbon fuels. Compared to hydrogen, the infrastructure for the production, storage, and transportation of ammonia fuel is relatively mature, and the energy and cost of ammonia liquification are far lower than that of hydrogen. The energy of liquid ammonia is over five times of compressed hydrogen gas.

2.6.4 Shortages

The low activity of ammonia is the biggest obstacle to its application in internal combustion engine. The extremely high ignition temperature and minimum ignition energy make it difficult for ammonia to achieve stable ignition in the internal combustion engine, resulting in the high combustion variation and even misfire with a large amount of unburned ammonia at the end of the cycle. In addition, the laminar burning velocity of ammonia is the lowest in all the e-fuels, and its application in the engine will greatly extend the combustion duration, reducing the engine thermal efficiency. Therefore, to use ammonia as fuel in an internal combustion engine, it is necessary to use combustion assistance technology, such as hydrogen blending, pre-chamber jet ignition to improve the combustion quality. However, all these technologies will increase the complexity and cost of the engine system.

Regarding exhaust emissions, because ammonia itself contains nitrogen, a large amount of fuel NOx will inevitably be produced in the combustion process, but other nitrogen-free fuel will not have this problem, so the NOx emission of the engine fuelled with ammonia fuel is generally much higher, and it is difficult to suppress it by simply reducing the combustion temperature. In addition, there is a considerable proportion of N_2O in NOx produced by ammonia combustion, which is also caused by the unique reaction path of ammonia combustion. The greenhouse effect of N_2O is more than 200 times that of CO_2. Therefore, the aftertreatment system of ammonia engines could be more expensive than others.

2.7 METHANE

2.7.1 Resources

Methane is the main component of natural gas, and biogas, so it is widely distributed in nature. Natural gas and biogas can be used directly as engine fuels. Methane can also be synthesized from hydrogen and CO_2.

$$C_2O + H_2 \rightarrow CH_4 + H_2O$$

Similar to ammonia, the source of hydrogen directly determines whether synthetic methane is a renewable fuel or not.

2.7.2 Properties

Methane is a chemical compound of one carbon atom bonded to four hydrogen atoms, which is the simplest organic matter and the hydrocarbon with the lowest carbon content (the highest hydrogen content). Methane is also a gasoline-like fuel with high auto-ignition temperature and high octane number. Methane is gaseous under normal temperature and pressure. The mass energy density of methane is higher than that of gasoline and diesel. Methane can also be stored by liquification or compression with much less energy and cost compared with hydrogen. The minimum ignition energy (0.29 mJ) is slightly higher than that of conventional fossil fuels.

2.7.3 Benefits

Compared with hydrogen and ammonia, methane is obviously more suitable as an engine fuel. Methane has a high octane number, and its anti-knock performance is much better than that of hydrogen. The minimum ignition energy of methane is not as high as that of ammonia, and the laminar burning velocity is at the same level as that of gasoline. Because the anti-knock performance is better than gasoline, methane can be used in spark ignition engines with higher compression ratio to achieve higher thermal efficiency. In addition, methane molecule contains only one carbon atom, so its CO_2 emission in same energy release is about 10% lower than that of gasoline. With the same reason, the emission of polycyclic aromatic hydrocarbons and soot produced in the process of engine combustion is much lower for methane, so it is a relatively cleaner fuel.

2.7.4 Shortages

The main disadvantage of methane is still that it is a gaseous fuel, so it is still troublesome in the process of methane storage and transportation. Another problem with methane is that it itself is a typical greenhouse gas. Therefore, the leakage of methane and the high natural gas emissions caused by incomplete combustion of the engine will increase the greenhouse effect.

2.8 METHANOL

2.8.1 Resources

Methanol can be produced from a wide range of sources, including abundant fossil fuels (such as natural gas, coal, oil shale, oil sand, etc.), as well as agricultural products and municipal waste, garbage, wood, and various biomass. Methanol in most European countries and US is produced from natural gas, while in China it is produced from coal.

Methanal can also be synthesized from hydrogen and CO_2.

$$CO_2 + H_2 \rightarrow CH_3OH + H_2O$$

The source of hydrogen directly determines whether synthetic methanol is a renewable e-fuel or not.

2.8.2 Properties

Methanol is the simplest alcohol, and the molecular formula of methanol is CH3OH. With high auto-ignition temperature and high octane number, methanol is also a typical gasoline-like fuel. In fact, methanol is an excellent fuel for gasoline engines. Methanol is liquid at normal temperature and pressure, and its density is close to gasoline. The volume and mass calorific value of methanol is about half that of gasoline. The latent heat of vaporization of methanol is 1100 kJ/kg, close to three times that of gasoline. Methanol can be miscible in water in any proportion. It is easy to dissolve in gasoline, but difficult to dissolve in diesel. Methanol exhibits wide flammability limits (6.7%-36%), and comparable laminar burning velocity to gasoline. The minimum ignition energy (0.14 mJ) is also comparable to that of gasoline fuel (0.2 mJ).

2.8.3 Benefits

Methanol has high octane number, good anti-knock performance, and the vaporization process of methanol can reduce charge temperature, so earlier ignition time can be adopted when methanol is used in spark ignition engines and engines obtain higher thermal efficiency. The molecule of methanol has only one carbon atom, and the molecule contains oxygen, so the potential to generate soot is extremely low, which can achieve cleaner combustion compared with gasoline.

2.8.4 Shortages

The volumetric energy density of methanol is only half that of gasoline, so the direct application of methanol in the production gasoline engine will reduce the engine power. Methanol has corrosivity. Methanol has strong water absorption, and aqueous methanol is easy to oxidize and produce acid, which further increases the corrosivity and is easy to cause corrosion and damage to the fuel system. Therefore, engines using methanol as fuel need to adopt a special fuel supply system to increase the manufacturing cost.

Methanol is a toxic substance to human beings. Methanol combustion is also easy to produce irregular emissions such as formaldehyde, and the toxicity of formaldehyde and methanol mixture to organisms is further enhanced. The humidity of burned gas of methanol is also relatively large, which can make the lubricating oil emulsified and denatured. In addition, the high humidity and the high concentrations of irregular emissions raise higher requirements for the durability and reliability of the aftertreatment system, which also raises the cost of the aftertreatment system.

2.8.5 Higher Alcohol

Besides methanol, higher alcohol, including ethanol, butanol, and octanol can also be used as internal combustion engine fuel. Ethanol can be produced using dimethyl ether, hydrogen, and CO_2 as raw materials. Butanol can be produced from ethanol, and octanol can be produced by butanol. With the increase of carbon number, the ignitability of alcohol fuel increases and the octane number decreases, and the solubility in diesel increases.[16]

2.9 DIMETHYL ETHER (DME)

2.9.1 Resources

At present, dimethyl ether (DME) can be produced from methanol through dehydration and condensation, so its source mainly depends on the source of methanol. The energy consumption and cost of this process are not high. Because methanol is a typical e-fuel, dimethyl ether can also be considered as a potential e-fuel.

2.9.2 Properties

DME is the simplest ether fuel. DME is a colorless, non-toxic gas with a slight ether smell. It contains 34.8% oxygen and has no C-C bond. The lower heating value of DME is 27.6 MJ/kg, while the ignition temperature is only 512 K. Dimethyl ether has good ignitability and cetane number as high as 55 (cetane number of diesel generally needs to be more than 50), so it is a typical diesel-like fuel.

2.9.3 Benefits

DME can be used as a diesel alternative fuel because of its high cetane number. Because DME contains oxygen in its molecule and there is no carbon-carbon bond, the soot formation during diffusion combustion is much lower than that of diesel. If only combustion process is considered, DME is a very ideal clean diesel fuel.

2.9.4 Shortages

Similar to other gaseous fuels, the main problem of DME is the requirement for a specially-designed fuel supply system for gaseous fuels on the engine, which will increase the manufacturing cost of the engine. Because DME is suitable for diesel engines, the engine needs to equip an in-cylinder high-pressure injection system for DME considering the diffusion combustion of diesel engines. The lubricity and viscosity of DME are far lower than that of diesel, which is a disadvantage to the establishment of high injection pressure in the fuel system, and also reduces the durability of the high-pressure fuel system. In the meantime, DME is corrosive to the seal materials of rubber and other fuel systems, so it is easy to cause seal damage.

2.9.5 Polyoxymethylene Dimethyl Ethers (PODE)

DME is a gaseous fuel under the standard condition with disadvantages when used as a diesel fuel in compression ignition engines. Polyoxymethylene dimethyl ethers (PODE) stand for the ethers with the formula of $CH_3O(CH_2O)_n\ CH_3$ where n is the polymerization degree, usually ranging from 1 to 7. Compared with DME, PODEs are liquid fuel and generally have higher oxygen content (>40%) and cetane number. PODEs are a promising diesel alternative fuel or a green fuel additive for diesel. PODEs can also be synthesized from methanol, so it can be regarded as a potential e-fuel.

PODEs have different polymers with a degree of polymerization from 1 to 7. As the polymerization degree increases, the cetane number, oxygen content, flash point, density, viscosity, and surface tension of PODE become larger, as well as the melting point and boiling point. When the polymerization degree is higher than 2, the cetane numbers of PODEs are all higher than 63, which is higher than the requirement as a diesel engine fuel. With the polymerization degree of 2 and 3, PODEs have flash points within the range of the normal atmospheric temperature (between 15 and 25 °C), not fulfilling the safety criterion. If polymerization degree is higher than 5, melting points of PODEs are higher than 18.5 °C. Therefore, the upper limit of PODE polymerization degree should be at least no higher than 5 in order to avoid the precipitation in the fuel supply system especially at low temperature conditions. Meanwhile, a polymerization degree higher than 5 is also not recommended considering the cost of the feedstocks.[17]

The ideal polymerization degree range for PODEs should be 4–5 considering all the aspects discussed before. However, for the present technologies, it is difficult to exactly control the polymerization range in industrial production of PODEs. PODEs are still produced in the form of mixture with wider polymerization degree range rather than 4–5.

2.10 APPLICATION OF E-FUEL IN VEHICLE INTERNAL COMBUSTION ENGINES

The application of e-fuel mainly depends on its ignitability, that is whether it is a gasoline-like fuel or diesel-like fuel. Gasoline-like fuels are commonly used in spark ignition (SI) engines, while diesel-like fuels are commonly used in compression ignition (CI) engines. For SI engines, the thermal efficiency is relatively low and the power range is limited. However, most e-fuels, especially carbon-free fuels are gasoline-like fuels. To break the limit of using gasoline-like fuels in SI engines, the dual fuel compression ignition mode, which were firstly proposed to apply gasoline in high efficiency compression ignition mode, could also be a solution for the application of gasoline-like e-fuels in compression ignition mode.[18]

2.10.1 SI Mode Engines

The SI mode is also known as the conventional gasoline engine mode. In SI mode engines, fuel is prepared either by low-pressure fuel injection in the intake port or by early in-cylinder high-pressure fuel injection, so fuel is fully premixed with air before combustion occurs. A spark plug supplies external energy to trigger the ignition of the air-fuel mixture, and the mixture is consumed by flame propagation from the spark plug to the edge of the combustion chamber.

As a suitable fuel for SI mode engines, the fuel should have gasoline-like ignitability (high octane number), and the fuel should be gaseous fuel or liquid fuel with high volatility. Until now, in gasoline-like e-fuels, methanol, ethanol, and methane have been applied in production SI engines. Hydrogen, ammonia, and other alcohols are still in engine lab test stage.

2.10.2 CI Mode Engines

The CI mode is also known as the conventional diesel engine mode. In CI mode engines, fuel is injected into the cylinder very close to top dead center, when the temperature and pressure in the cylinder is high. In addition to the high ignitability of the diesel-like fuel, the fuel ignites quickly after being injected into the cylinder, and the injected fuel is consumed by diffusion combustion.

As a suitable fuel for CI mode engines, the fuel should have diesel-like ignitability (high cetane number), and the fuel should have a good lubricity to avoid the wear of the high-pressure fuel pump. Until now, no diesel-like e-fuels have been applied in production CI engines.

2.10.3 Dual-Fuel CI Mode Engines

The main obstacle of using gasoline-like e-fuel in CI mode engine is the low ignitability. Dual fuel CI mode solves this issue by applying diesel-like fuel as the ignition sources. In the dual fuel CI mode, gasoline-like e-fuels are injected through low-pressure injection in the intake port, while diesel-like fuel (normally diesel) is injected in-cylinder with high injection pressures.[19]

Until now, diesel-methanol dual fuel CI mode engines and diesel-natural gas dual fuel CI mode engines have been applied on commercial vehicles in practical use.

2.11 SUMMARY AND OUTLOOK

Based on the world energy report 2021 from the International Energy Agency, although electrification is a central element in the carbon economy, it is not possible to electrify everything. Liquid, gaseous, and solid fuels of various types will continue to make a major contribution to the global energy mix through to 2050. There is a growing role for alternative, low emissions fuels such as e-fuels in all scenarios. These play a key role in the achievement of net zero targets, especially in sectors where direct electrification is most challenging. Today, 17 governments have published low-carbon hydrogen strategies and more than 20 countries are developing them.[20]

However, with the present technology, e-fuels are energy inefficient, with approximately 50% of available fuel energy "lost" during the fuel production process, compared with about 10% lost when generating and using electricity to drive an electric vehicle. The production cost is likely to remain high.[21] However, as an energy storage medium, e-fuel has advantages in energy density, storage, and transportation. If the cost of renewable electric energy continues to decrease in the future, the disadvantages of e-fuel in energy efficiency and cost will be gradually diluted.

For a long time, the evolution of fuel has played a leading role in the development of internal combustion engine technology. In the future, with the increasing share of e-fuel in the daily-used fuel, the internal combustion engine will inevitably need technological innovation to adapt to this change. A foreseeable change is that the market share of dual fuel compression ignition internal combustion engines will increase. At present, in dual fuel compression ignition, the gasoline fuel inducing method is still dominated by the port fuel injection, but the port fuel injection brings about problems such as limited substitution rate of gasoline-like fuel, decreased engine power, and increased unburned fuel. Considering that most of e-fuels are gasoline-like fuels, it is unfavorable to the improvement of e-fuel substitution rate. Therefore, in the future, gasoline-like e-fuels may also be introduced into the cylinder through in-cylinder direct injection to solve these problems. Regarding this, marine engines have taken the lead in technology development. MAN and Wartsila have successively launched dual fuel compression ignition marine engines (Figure 2.4) that can inject both diesel and e-fuel through in-cylinder direct injection (dual direct injection technology).[22] However, currently the cost of this technology is still high, and whether the technology can be applied on a large scale in vehicle engines in the future depends on the policy and the iteration of relevant technologies.

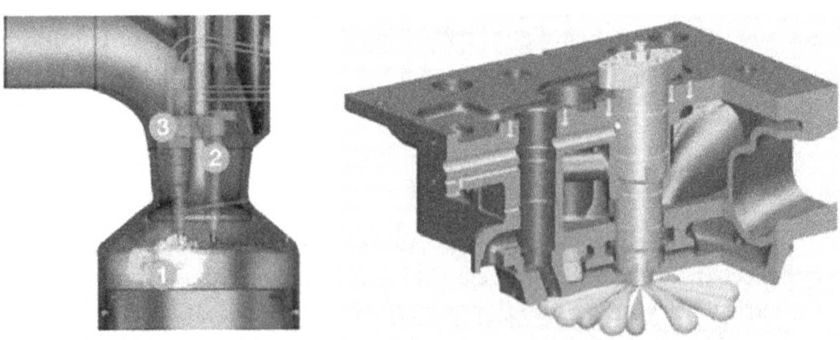

FIGURE 2.4 Injection and combustion system of dual fuel marine engine, MAN (left),[23] Wärtsilä (right).[24]

NOTES

1 Davidson FT, Nagasawa K, Webber ME. Electrofuel. *Mechanical Engineering*, 2017: 30–35.
2 Searle S, Christensen A. *Decarbonization potential of electrofuels in the European Union*. The International Council on Clean Transportation, 2018. www. theicct.org/publications/decarbonization-potential-electrofuels-eu.
3 Ababneh H, Hameed BH. Electrofuels as emerging new green alternative fuel: A review of recent literature. *Energy Conversion and Management*, 2022, 254: 115213.
4 WMO Greenhouse Gas Bulletin (GHG Bulletin)—No. 17: The State of Greenhouse Gases in the Atmosphere Based on Global Observations Through 2020.
5 https://wisburg.com/articles/730482.
6 Biofuels Annual. *EU biofuels annual 2019*. https://apps.fas.usda.gov/newgainapi/api/report/downloadreportbyfilename?filename=Biofuels%20Annual_The%20Hague_EU-28_7-15-2019.pdf.
7 Zang G, Sun P, Elgowainy A, et al. Life cycle analysis of electrofuels: Fischer-Tropsch fuel production from hydrogen and corn ethanol byproduct CO2. *Environmental Science & Technology*, 2021, 55(6): 3888–3897.
8 Bergthorson JM, Goroshin S, Soo MJ, et al. Direct combustion of recyclable metal fuels for zero-carbon heat and power. *Applied Energy*, 2015, 160: 368–382.
9 Galimova T, Ram M, Bogdanov D, et al. Global trading of renewable electricity-based fuels and chemicals to enhance the energy transition across all sectors towards sustainability. *Renewable and Sustainable Energy Reviews*, 2023, 183: 113420.
10 Schemme S, Breuer JL, Köller M, et al. H2-based synthetic fuels: A techno-economic comparison of alcohol, ether and hydrocarbon production. *International Journal of Hydrogen Energy*, 2020, 45(8): 5395–5414.
11 Schemme S, Breuer JL, Köller M, et al. H2-based synthetic fuels: A techno-economic comparison of alcohol, ether and hydrocarbon production. *International Journal of Hydrogen Energy*, 2020, 45(8): 5395–5414.
12 Hänggi S, Elbert P, Bütler T, et al. A review of synthetic fuels for passenger vehicles. *Energy Reports*, 2019, 5: 555–569.
13 Dimitriou P, Tsujimura T. A review of hydrogen as a compression ignition engine fuel. *International Journal of Hydrogen Energy*, 2017, 42(38): 24470–24486.
14 Yip HL, Srna A, Yuen ACY, et al. A review of hydrogen direct injection for internal combustion engines: Towards carbon-free combustion. *Applied Sciences*, 2019, 9(22): 4842.
15 Dimitriou P, Javaid R. A review of ammonia as a compression ignition engine fuel. *International Journal of Hydrogen Energy*, 2020, 45(11): 7098–7118.
16 Kumar BR, Saravanan S. Use of higher alcohol biofuels in diesel engines: A review[J]. *Renewable and Sustainable Energy Reviews*, 2016, 60: 84–115.
17 Liu H, Wang Z, Li Y, et al. Recent progress in the application in compression ignition engines and the synthesis technologies of polyoxymethylene dimethyl ethers. *Applied Energy*, 2019, 233: 599–611.
18 Sjoberg CMG. *Light-duty multimode: Current status and planning for the future*. U.S Department of Energy, 2018.
19 Song H, Liu C, Li Y, et al. An exploration of utilizing low-pressure diesel injection for natural gas dual-fuel low-temperature combustion. *Energy*, 2018, 153: 248–255.
20 IEA. *World energy outlook*. IEA, 2021.
21 Mock P, Díaz S. Pathways to decarbonization: The European passenger car market in the years 2021–2035. *Communications*, 2021, 49: 847129–848102.

22 Andersson K, Salazar CM. *Methanol as a marine fuel report.* Methanol Institute, 2015.
23 Aabo K. Ammonia-fuelled MAN B&W 2-stroke dual-fuel engines. *Journal of the JIME*, 2020, 55, 6: 737–744.
24 Haraldson L. *Methanol as fuel.* Methanol as Fuel & Energy Storage Workshop. Lund, Sweden, 2015.

3 Design and Integration of Energy Storage Devices for Automotive Applications

Yuanjian Zhang and Di Zhao

3.1 DRIVING CYCLES

As a vehicle fuel economy and driving range certification standardization reference, driving cycles can truly reflect the real environment of driving conditions, which help the design of fuel cell vehicle efficient power system and energy storage device, so as to meet the vehicle performance requirements, reduce component degradation, and evaluate and verify the economy and life cycle cost of vehicles. This chapter will analyze the differences and connections between driving cycles from the category of driving cycles to study the driving cycles suitable for running fuel cell vehicles, and discuss the guiding significance of driving cycles for the design of fuel cell vehicles.

3.1.1 Classification of Driving Cycles

Driving cycles for vehicle can be divided into two categories: real driving cycles and standard driving cycles.

3.1.1.1 Real Driving Cycles

Real driving cycles can effectively reflect the actual driving environment of a specific city or region. In order to meet the comprehensive performance of vehicles in a specific driving area, it is necessary to generate representative vehicle driving cycles with regional characteristics. Generating method of real driving cycles is roughly divided into two categories, one is the Markov chain theory applied to driving cycles design, the basic idea of which is that the next speed and acceleration state only depends on the current state, regardless of the previous state. By collecting data under the real condition and getting the corresponding state transfer matrix, combined with Monte Carlo method or simulated annealing, the purpose of driving cycles generation can be achieved. The other is to apply machine learning to driving cycles generation, such as K-mean clustering, support vector

DOI: 10.1201/9781003302827-4

machine, decision tree regression, etc., general process of which is a) test route selection, b) data acquisition and filtering processing, c) motion segment division and dimensionality reduction, d) motion segment classification, and e) motion segment combination. In order to effectively improve the fuel economy and component life of fuel cell vehicles in the specific operating environment and market, it is of great significance to study the driving cycles generation technology and guide the design of the performance parameters and energy management strategies of fuel cell vehicle components.

3.1.1.2 Standard Driving Cycles

The so-called standard driving cycles represents driving cycles promulgated by the authority, with general evaluation standards and regulations. According to the driving area and traffic conditions, standard driving cycles can be divided into urban driving cycles, suburban driving cycles, high speed driving cycles, aggressive driving cycles and mixed driving cycles, which are characterized by maximum speed, average speed, acceleration, deceleration, proportion of acceleration, proportion of deceleration, proportion of uniform speed, and proportion of idle.

1) Urban driving cycles
   In the urban traffic environment, traffic congestion, traffic lights, pedestrians, and other factors will cause the slow or even frequent parking speed of the vehicles. Therefore, the maximum speed and average speed of the urban driving cycles are relatively low, while the acceleration and deceleration processes account for a relatively large and more radical role. Typical urban driving cycles are New York City Cycle (NYCC), WVUCITY, and MANHATTAN. The New York City Cycle (NYCC) features low speed stop-and-go traffic conditions, which has a maximum and average speed of 44.58km/h and 11.41km/h, with a more aggressive maximum acceleration and deceleration of 2.682m/s^2 and 2.28m/s^2, respectively. The top and average vehicle speed of the ECE15 is slightly higher than the NYCC, and the acceleration and deceleration is less intense, and the J10 has characteristic data like the ECE15.
2) Suburban driving cycles
   Suburban driving cycles consist of low-speed urban driving and medium-high speed suburban driving, often including low-speed parking and high-speed parking. Typical suburban driving cycles include UDDS, FTP-75, NEDC, J10–15, JC08, WLTP, etc. Their speed and acceleration time course are shown in Figure 3.1, and the characteristic parameters are shown in Table 3.1.

   The FTP-75 (Federal Test Procedures 75) is similarly characterized as an extended version of the UDDS (Urban Dynamometer Driving Schedule). The FTP-75's maximum deceleration and deceleration is only 1.475 m/s^2, while the acceleration and deceleration process is relatively gentle and does not cover high speed aggressive driving and air conditioning on. In order to compensate for these shortcomings, the

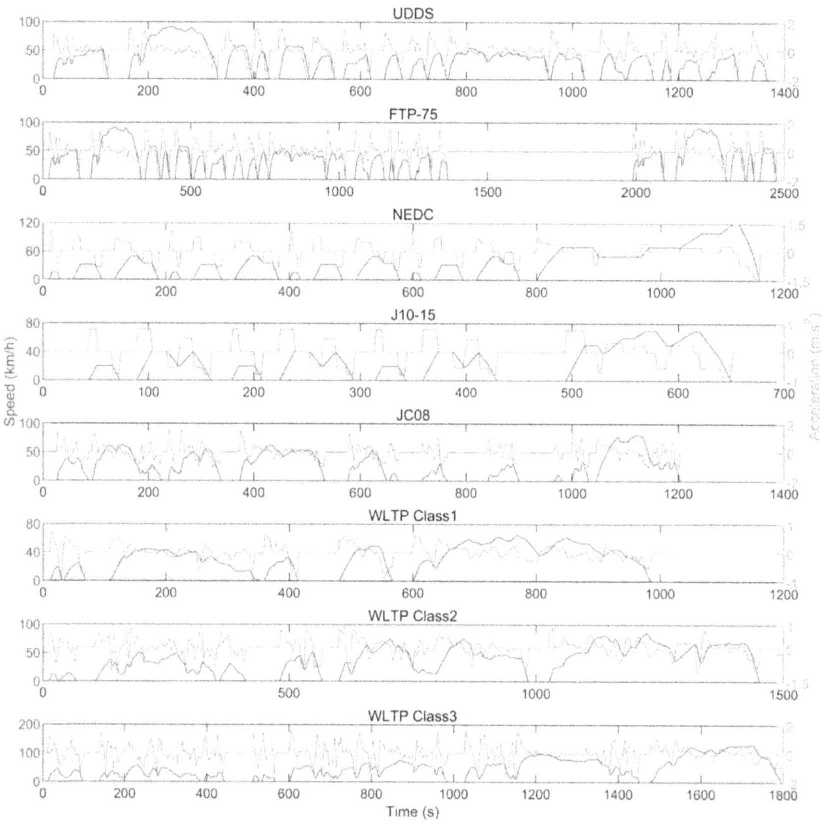

FIGURE 3.1 Speed and acceleration time course of suburban driving cycles.

FTP-75 has added two supplementary driving cycles, SFTP (Supplement Federal Test Procedures): US06 for high-speed aggressive driving and SC03 for air conditioning on. The NEDC (New European Driving Cycle) is officially used in Europe for emission certification and fuel economy testing in light vehicles, is slightly faster but more moderate than the FTP-75. Similar to NEDC, J10–15 (Japanese 10–15 Mode Cycle) changes less in velocity in a short time, and is more moderate than NEDC. Both urban and suburban, the JC08 (Japanese Industrial Standards Committee 08 test cycle) has a higher top speed than the J10–15 and travels shorter at a constant speed, replaced by frequent acceleration and deceleration processes. In addition, the average vehicle speed of the vehicle is kept at 50 km/h before full braking, which also simulates the traffic lights such as urban traffic. JC08 was more aggressive than FTP-75 during acceleration, but more moderate during deceleration. WLTP (Worldwide harmonized Light vehicles Test Procedure) is divided into

TABLE 3.1
Technical Specifications of the Suburban Driving Cycles

							WLTP		
Characteristics	**M.U.**	**UDDS**	**FTP-75**	**NEDC**	**J10–15**	**JC08**	**Class1**	**Class2**	**Class3**
Maximum speed	km/h	91.25	91.25	120.00	70.00	81.60	64.40	85.20	131.30
Average speed	km/h	31.51	25.86	33.60	22.72	24.41	28.47	35.72	46.50
Maximum acceleration	m/s^2	1.475	1.475	1.042	0.806	1.528	0.764	0.958	1.583
Maximum deceleration	m/s^2	1.475	1.475	1.389	0.833	1.125	1.000	1.111	1.486
Proportion of acceleration	%	32.8	32.4	23.8	25.9	27.8			29.4
Proportion of deceleration	%	29.3	28.2	17.6	26.4	25.9			27.8
Proportion of uniform speed	%	20.9	21.2	34.8	22.3	17.4			30.3
Proportion of idle	%	18.0	18.2	23.8	25.4	28.9			12.5

three categories of Class1, Class2, and Class3, different categories of low speed, medium speed, high speed, and super high speed four speed interval combination. The vehicle cycle category depends on the vehicle power quality than power to mass ratio (PMR). The greater the PMR is, the higher the categories, the maximum speed, and the average speed will be, and the more radical the driving mode will be. Overall, WLTP Class3 is the most radical of the suburban driving cycles.

3) High-speed driving cycles

High-speed driving cycles can simulate and evaluate the fuel economy of the vehicle when driving at high speed, and the representative driving cycle is Highway Fuel Economy Driving Schedule (HWFET). HWFET simulates highway driving cycles below 96.4 km/h (60mph) with an average speed of 77.58 km/h. The vehicle includes only a single start-stop process in a single cycle, and the absolute values of maximum acceleration and deceleration are mostly less than 0.45 m/s^2, with a very low average. The mountain driving cycle NREL2VAIL runs for 86.8 miles in time period of 5692 sec from the city of NREL Golden to the city of VAIL. It is also one of the high-speed driving cycles, with the mountain slope information as well.

4) Aggressive driving cycles

As a supplement to the FTP-75, the US06 has more aggressive speed fluctuations and high-speed driving characteristics, thus dividing the US06 into aggressive driving cycles. The maximum speed and average vehicle speed of the US06 are 129.2km/h and 77.2 km/h, respectively, and the maximum acceleration and deceleration even reach 3.241 m/s^2 and 2.816 m/s^2. Compared with the highest and average speed, the FTP-75 driving cycle is increased, reaching 108.1 km/h and 39.6 km/h, and the maximum acceleration and deceleration increase significantly, with 2.816 m/s^2 and 3.755 m/s^2, respectively. Therefore, the LA92 is also considered to be an aggressive driving cycle. LA92 has far more acceleration and deceleration than US06, and the deceleration process is more aggressive, but US06 contains a long high-speed driving range, so the average speed of US06 is higher than LA92.

5) Mixed driving cycle
In the process of vehicle design and verification, it may occur that the existing standard driving cycles cannot fully meet the requirements. Therefore, several sections of standard driving cycles are split, spliced, and recombined to obtain mixed driving cycles. Typical mixed driving cycles include NEDC composed of ECE15 and EUDC and the J10–15 composed of J10 and J15.

The preceding standard driving cycles are more for light vehicles, while World Transient Vehicle Cycle (WTVC) is specifically for heavy commercial vehicles. Furthermore, China announced in October 2019 and implemented CLTC and CHTC in May 2020. In addition to the standard conditions classified, there are some transient non-official cycles, such as Artemis Driving Cycle on urban, rural, and highways, and Hyzem, developed specifically for the evaluation of hybrid vehicles, which are complementary to vehicle design and evaluation.

3.1.1.3 Fuel Cell Vehicle Design Considering Driving Cycles

As mentioned before, the operating environment of fuel cell vehicles directly affects vehicle performance, such as fuel economy and component life. Studying the driving cycles suitable for fuel cell vehicles can guide the design of fuel cell vehicle component performance parameters, energy management strategies, and fleet operation strategies.

First, the driving cycles of fuel cell vehicles will affect the fuel economy, low speed mild driving cycles like ECE15 is beneficial to improve the fuel cell vehicle fuel economy. This is because different driving cycles created different vehicle running state; the average speed is higher or frequent acceleration and acceleration of large driving cycles obviously cause greater energy consumption. In addition, the urban driving cycles and suburban driving cycles with relatively large deceleration processes bring greater braking energy recovery potential, and the braking energy recovery of vehicles will improve the fuel economy.

In addition, the driving cycles of the fuel cell vehicles will affect the component life. Thanks to braking energy recovery, frequent start and stop cities like NYCC may be good for vehicle fuel economy, but this may not be good for vehicle component life. From the perspective of life, fuel cell system should work under stable load driving cycles. Namely the frequent changes in power demand and start-stop may lead to insufficient reactants and seriously affect life, instead of the vehicle fuel economy and acceleration performance; therefore, urban driving cycles are not conducive to life cycles of fuel cell vehicles, while high speed driving cycles are more suitable for extending the service life of fuel cell vehicles.

From the perspective of component performance parameters, the power of the fuel cell system should meet the average requirements of real driving cycles or standard driving cycles to satisfy the demands of vehicle acceleration performance. If the vehicle is only running in urban driving cycles, frequent acceleration and deceleration strengthens the role of the auxiliary energy unit. This can lead to the fuel cell system being small because of the size of the auxiliary energy unit, which not only prolongs the fuel cell life increasing the braking energy but also slightly compensates for the low efficiency caused by the low load of the fuel cell system, thereby improving fuel economy. For high-speed driving cycles, fuel cell size can be appropriately increased and auxiliary energy unit size can be reduced to minimize weight to improve fuel economy. For the aggressive driving cycles with large speed changes, the performance and cost of the vehicle should comprehensively be taken into account, and the size of the fuel cell and auxiliary energy unit should be appropriately increased to minimize the fuel consumption and life degradation and reduce the cost on the premise of meeting the performance requirements.

From the perspective of the energy management strategy, the vehicle power requirements determined by the driving cycles are further allocated, and the distribution results will affect the vehicle performance, fuel economy, and component life. The degradation of the fuel cell life is related to four bad driving cycles: load change, start-stop cycle, idle speed, and high load. In order to extend the life of the fuel cell, the areas with little speed change in the driving cycles should be borne by the fuel cell, and the other areas should be assisted by auxiliary energy units to extend the life of the fuel cell. In addition, the best fuel economy and component life can be obtained by using different energy management strategy parameters in different driving cycles, which requires a combined driving cycle identification technology.

3.2 TOPOLOGY

The power system of the fuel cell vehicle is mainly composed of the fuel cell system, auxiliary energy unit, DCDC, DCAC (Inverter), electric motor, and reducer. The auxiliary energy unit can be divided into photovoltaic and other energy generation units and the energy storage units of battery, supercapacitor, flywheel, superconducting magnetic energy storage (SMES), and other storage units.

According to the different combined types (used energy units) of energy supply units and energy storage units, fuel cell vehicles can be divided into the following types: a) Full FCEV; b) PEMFC+Batt FCEV; c) PEMFC+SC FCEV; d) PEMFC+Batt+SC FCEV; e) PEMFC+Batt+PV FCEV; f) PEMFC+FW FCEV; and g) PEMFC+SMES FCEV, whose advantages and disadvantages are shown in Table 3.2.

TABLE 3.2
The Classification of FCEVs

Type	Advantages	Disadvantages
Full FCEV	• Less system units and simple structure	• Fuel cells require high power • The dynamic response and reliability of fuel cells cannot meet the vehicle requirements • No brake energy recovery
PEMFC+Batt FCEV	• Dynamic response performance is relatively good • Fuel cells can work in good condition • Cold start performance is good • Brake energy recovery can be achieved • Manufacturing cost and use cost are low	• The battery has the lowest power density • There is an energy loss in the battery charge and discharge process • The battery life cycle is short and it is affected by the temperature • The battery is limited to the maximum charge and discharge power
PEMFC+SC FCEV	• Supercapacitors have the highest power density • Supercapacitors can achieve rapid charge and discharge • Supercapacitors have a wide operating temperature range • Supercapacitor charge and discharge life cycle is long	• Supercapacitors have the lowest energy density • The manufacturing cost is high
PEMFC+Batt+SC FCEV	• It combines the advantages of batteries and supercapacitors • It can fully ensure the dynamic response performance in cold environment • Regenerative braking can be borne by supercapacitors to reduce the number of battery charge and discharge cycles	• The system is complex, and the system control and overall layout are more difficult • The manufacturing cost is high
PEMFC+Batt+PV FCEV	• Clean and silent	• Intermittent power output • Large in size, the body platform needs to be redesigned • The manufacturing cost is high
PEMFC+FW FCEV	• High-speed charging capability • High power rating	• Charging time is long • The energy density is low • Big and heavy • High manufacturing cost and use cost
PEMFC+SMES FCEV	• High power density • Long life cycle	• The energy density is low • Short-term energy storage • Ultracold environments are required to ensure coil superconductivity • High manufacturing cost and use cost

3.2.1 Full FCEV

The power system of full FCEV (Figure 3.2) includes fuel cell stack, hydrogen storage bottle, one-way DCDC, inverter, drive motor, and reducer, etc., which only has the fuel cell as one energy source, and all the power requirements of the vehicle are borne by the fuel cell, which requires the dynamic response performance of the fuel cell system. However, due to factors such as delayed reaction gas supply, the dynamic response performance of fuel cells is not excellent, which also causes limited acceleration performance such as vehicle acceleration time and climbing degree. But the fuel cell can continue to generate electricity with the reaction gas supply, so a single fuel cell type also shows a satisfactory driving range. The prospect of a single fuel cell type vehicle is not broad until the key technical issues of the cost, reliability, durability, and dynamic performance of the fuel cell system are significantly improved.

3.2.2 PEMFC+Batt FCEV

PEMFC+Batt FCEV (Figure 3.2 and Figure 3.3) adds a set of cells to a full FCEV. If the battery end voltage matches the DC bus voltage, the bidirectional DCDC at the output of the battery is not required. The fuel cell of this type is the main power source. The dynamic response performance advantage of the battery greatly improves the acceleration performance of the vehicle and retains the advantage of the fuel cell driving range. However, the matching and control of

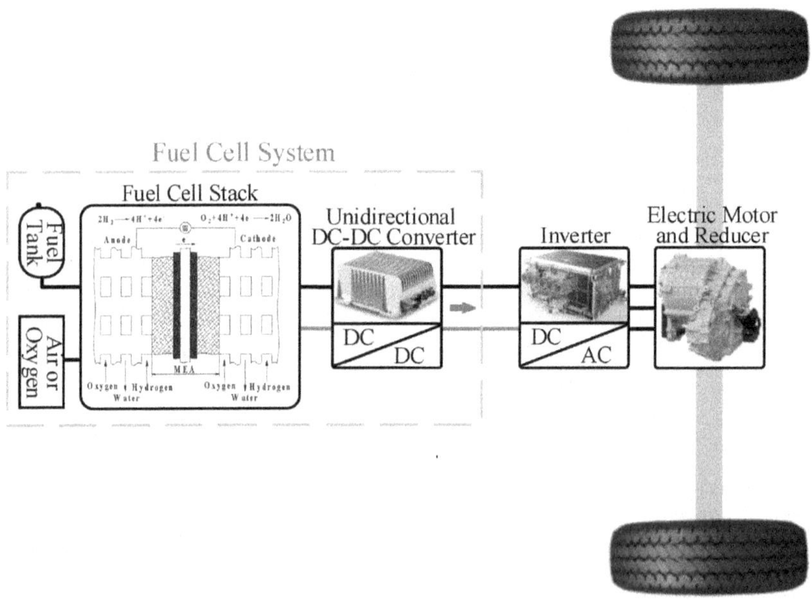

FIGURE 3.2 Power system of full FCEV.

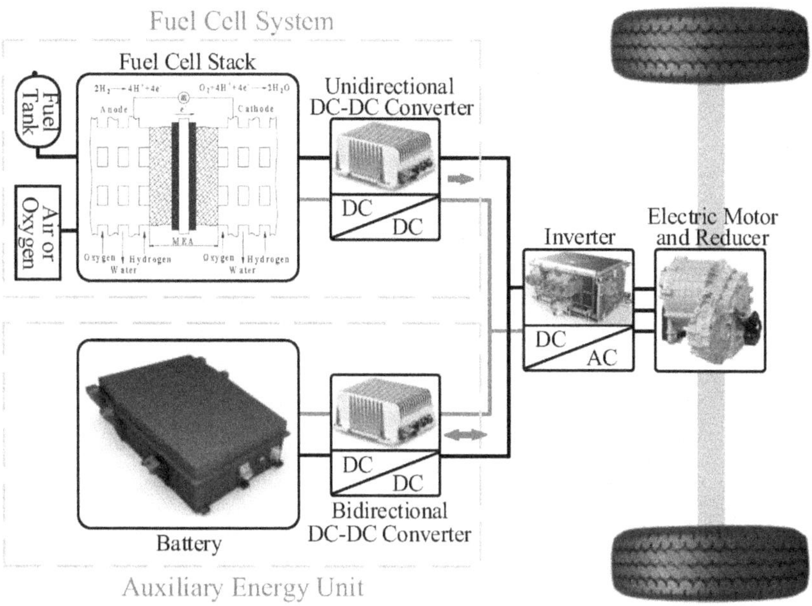

FIGURE 3.3 Power system of PEMFC+Batt FCEV(Non-plug-in).

multiple power sources has become the difficulty and focus of vehicle design, including the manufacturing, use and operation, and loss cost caused by component performance parameters and energy management strategies. The following other hybrid structures also face the same problems. Taking the component performance parameter design as an example, the battery capacity of the PEMFC+Batt FCEV(Non-plug-in) (Figure 3.2) is small, and the battery has limited auxiliary power of the vehicle demand. Therefore, this type is more suitable for intercity buses/trucks. And because PEMFC+Batt FCEV(Plug-in) battery capacity is larger, the battery needs external charging to get a complete charge and discharge cycle, with longer charge and discharge time, but a larger capacity battery provides more confidence in urban cycles, and power consumption ratio reduces the use cost. Thus, the type is more suitable for cost-sensitive city bus. For passenger cars, the PEMFC+Batt FCEV can also be fully competent.

3.2.3 PEMFC+SC FCEV

PEMFC+SC FCEV (Figure 3.4) is similar to PEMFC+Batt FCEV. Compared with the battery, the supercapacitor can charge and discharge quickly, with high power density, and high charge and discharge efficiency, so the supercapacitor can be used to efficiently and quickly recover the braking energy. The vehicle acceleration performance is superior, and the supercapacitor can work in a wide temperature range, greatly improving the cold start performance of the vehicle. But

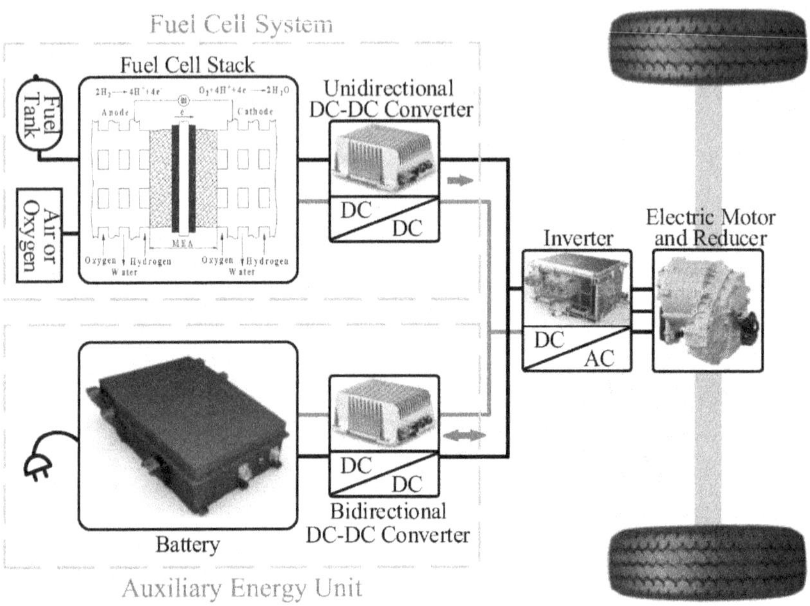

FIGURE 3.4 Power system of PEMFC+Batt FCEV(Plug-in).

supercapacitors store a limited energy density, which is lower specific energy and more expensive than batteries. This type is also generally applicable to all types of passenger vehicles and commercial vehicles.

3.2.4 PEMFC+Batt+SC FCEV

PEMFC + Batt + SC FCEV (Figure 3.5) with battery and supercapacitor coupled, can give full play to the advantages of battery and supercapacitor, for which the energy demand changes of low frequency and high frequency part are respectively borne by both, so the energy output is relatively flat. Parts aging and cost loss can be effectively reduced at the same time to make up for the use of super capacitor driving range loss. Because the type is more complex, the system control and the overall layout is more difficult, and the manufacturing cost may be affected, which is also the key point of the type design. Considering various factors comprehensively, perhaps this type is more suitable for commercial vehicles with low difficulty in space layout.

3.2.5 Others

PEMFC + Batt + PV FCEV (Figure 3.6) couples PEMFC + Batt FCEV to the photovoltaic system, which is also more conducive to the economy of the vehicle. Different from auxiliary energy units, the photovoltaic system is an energy supply

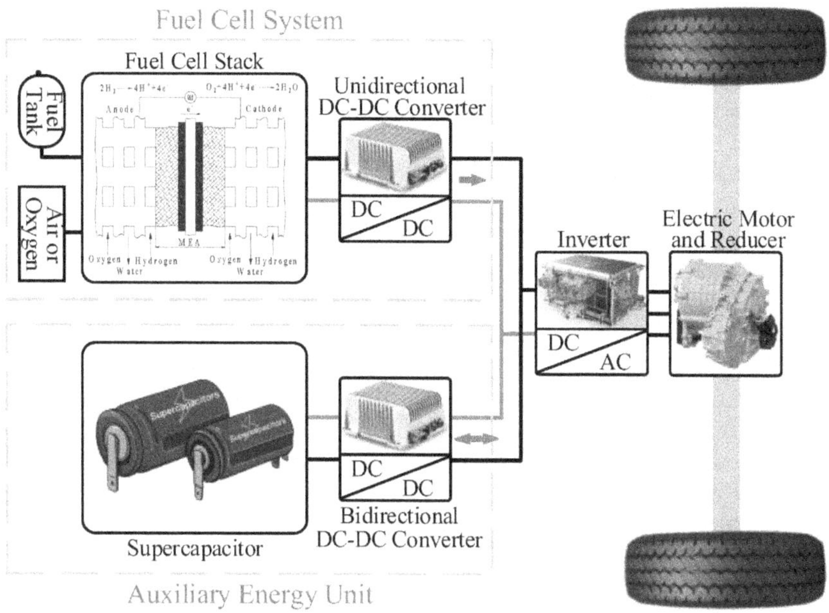

FIGURE 3.5 Power system of PEMFC+SC FCEV.

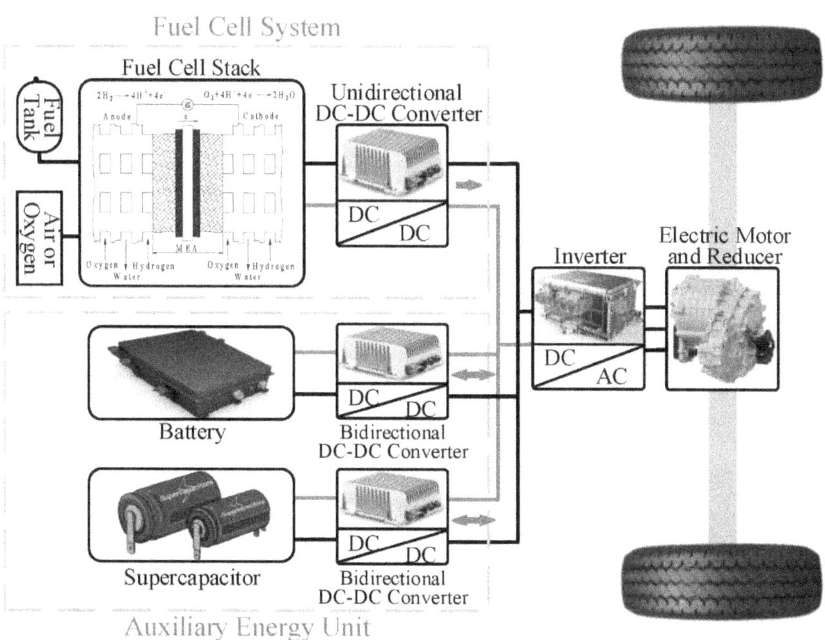

FIGURE 3.6 Power system of PEMFC+Batt+SC FCEV.

device, whose power generation is related to the solar radiation intensity, temperature, and direction, so its output power is intermittent, which is not conducive to the vehicle power performance and increases the difficulty of energy management. At the same time, the volume of the photovoltaic system is large, and the resulting redesign of the body platform is not conducive to the manufacturing cost. Therefore, this type is more suitable for good photovoltaic conditions and easy layout of commercial vehicles.

The flywheels in PEMFC + FW FCEV (Figure 3.7) use high-speed rotation for energy storage, but the correct packaging of the high-speed flywheels is crucial to store the equipment in case of fracture. In addition, it is difficult to design reasonable energy management strategies, which together constitute the key points of design. Because the flywheel uses mechanical devices to store energy, it has a high-speed charging capability, high power rating, and is not sensitive to the temperature, which is conducive to the performance of the vehicle performance. However, along with the mutual conversion process of mechanical energy and electric energy, the energy loss is relatively large, and the flywheel has a long charging time, a large volume and weight, and the manufacturing cost and use cost are high. Therefore, commercial vehicles are more suitable for this type.

PEMFC + SMES FCEV (Figure 3.8) has not yet been commercially used in fuel cell vehicles, but this type is a future development direction. Superconducting energy storage uses direct current to generate a magnetic field to store electricity

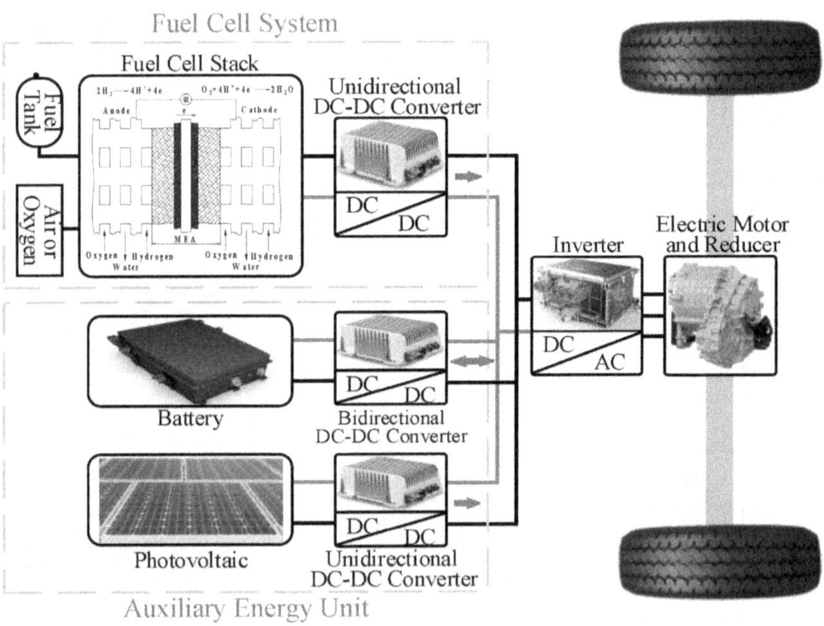

FIGURE 3.7 Power system of PEMFC+Batt+PV FCEV.

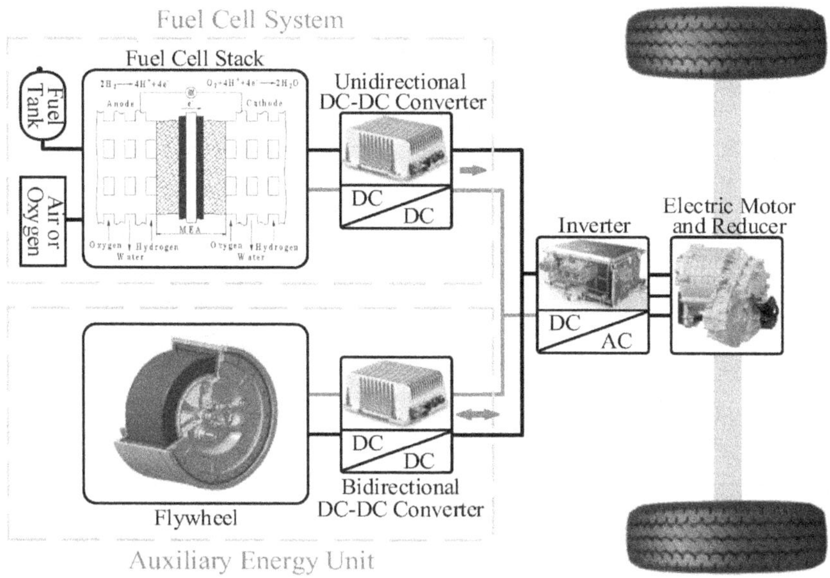

FIGURE 3.8 Power system of PEMFC+FW FCEV.

in the superconducting coil. In order to keep the coil in a superconducting state, cryogenic cooling at very low temperature is required, but its superconducting properties make it almost no resistance in the conductive process, increasing the storage capacity. Superconducting energy storage has super power response characteristics, high power density and a long life cycle, which is beneficial to vehicle performance, but the lower energy density and higher manufacturing and use cost limit its application in fuel cell vehicles. At present, the geometry optimization and reliability of the superconducting coil are still the design difficulties.

3.3 ENERGY MANAGEMENT STRATEGY AND USING COST

Since there are multiple energy units in the fuel cell vehicle power system, in order to achieve the coordinated control of multi-energy units, it is necessary to have less fuel consumption and better vehicle performance under the power that the driver needs for the vehicle. Appropriate energy management strategies cannot only reduce the use cost by reducing energy consumption, but also extend the service life of components by combining the characteristics of rapid response requirements, especially in the start and stop frequency, dynamic power fluctuation, working range, etc. The EMS of fuel cell vehicles can be divided into general energy management strategies and energy management strategies considering component characteristics. Among them, the general energy management strategies can be divided into rule-based, optimization-based, and learning-based energy management strategies.

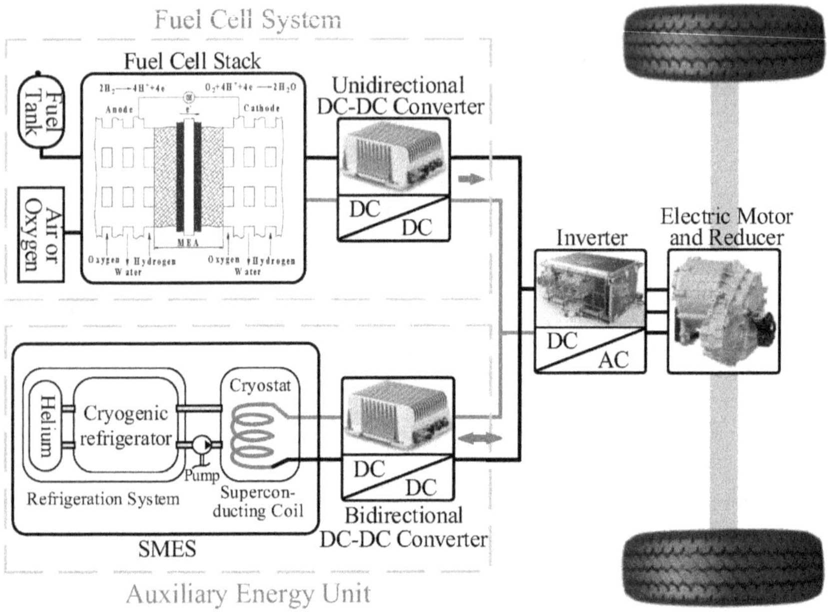

FIGURE 3.9 Power system of PEMFC+SMES FCEV.

3.3.1 Rule-Based

In the fuel cell vehicle energy management strategy, the rule-based energy management strategy is the most common energy management method that can realize real-time control. Its core content is to use human knowledge, inspiration, mathematical model to form a rule set, which can be a fuzzy rule, state table, or state flow chart. Under the transient input, the rule set determines the real-time control of the energy unit to meet the vehicle requirements in the most effective way. Therefore, the rule-based energy management strategy without prior conditions can realize online application, but the rule set is easily affected by heuristic design criteria and experiences arbitrariness and poor adaptability to driving cycles. It is difficult to achieve optimal control that mainly considers the dynamic response performance of the system, the use of the system cost optimization, and limited global optimization. The rule-based energy management strategies can be divided into two major categories: deterministic rule-based and fuzzy rule-based.

3.3.1.1 Deterministic Rule-Based

The deterministic rule-based uses the determined state table or state flow chart as the basis for control to achieve the purpose of coordinated control of multi-energy units. The control strategy method based on deterministic rules is simple and easy to implement. Its operation speed is fast, with low manufacturing cost. So it is often used in commercial fuel cell vehicles, such as state machine, thermostat, and power following control. In the application of the fuel cell vehicle energy management strategy, each state of the state machine is divided according to the SOC level (Figure 3.10a), that is, the state variable is often the SOC of the auxiliary energy unit, and the control variable is often the required power of the fuel cell system and the auxiliary energy unit. Different from the state machine, the control variables controlled by the thermostat are mostly the start-stop state of multiple energy units and do not control the performance parameters such as the power of the energy unit. The state variable of power is often the SOC of the vehicle demand power and the auxiliary energy unit, and the two-dimensional table formed by both is used as the control rule (Figure 3.10b).

3.3.1.2 Fuzzy Rule-Based

The energy management strategy of fuzzy rule-based fuel cell vehicle mainly represents the fuzzy logic control (FLC). Fuzzy logic control is an intelligent control method that imitates human's fuzzy reasoning and decision process behaviorally. It includes three basic processes: fuzzy, fuzzy reasoning, and fuzzy solution. Among them, the formulation of membership function and fuzzy rules will directly affect the control amount of fuzzy logic control. In fuel cell vehicle applications, the

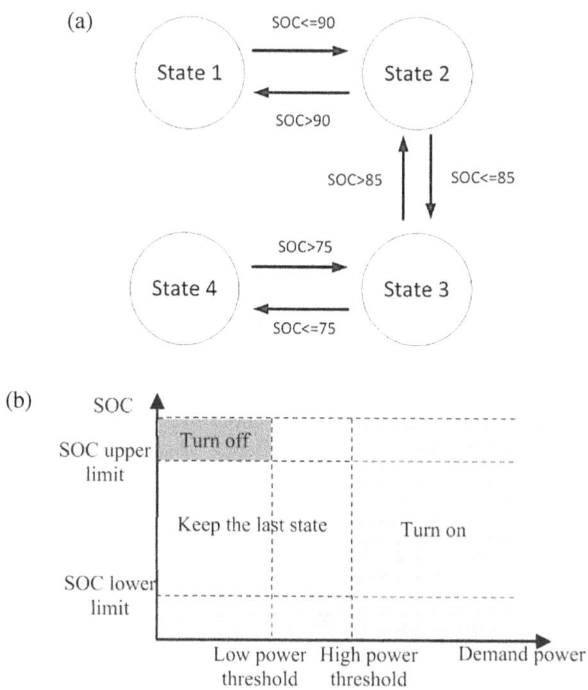

FIGURE 3.10 Deterministic rule-based EMS: (a) state machine control; (b) power following control.

input of fuzzy logic control is often the component SOC and the required power of the vehicle, and the output is often the power control amount of the component, as shown in Figure 3.11. The fuzzy rules are more insensitive to model uncertainty than the deterministic rules, but the fuzzy control process requires faster micro-controllers and a larger memory.

3.3.2 Optimization-Based

The basic idea of the optimization-based energy management strategy is to select reasonable power control to minimize the cost function, which can be divided into global optimization and instantaneous optimization. Global optimization requires battery SOC, driving cycles, etc., so it is not suitable for real-time optimization; instead, by introducing a cost function that only depends on the system parameters in the current state, which is also more suitable for real-time optimization. The cost function of the optimization-based energy management strategy usually considers the energy consumption and the use cost, and the component dynamic response performance is less considered than the rule-based EMS, and the computational amount and complexity is large, and the manufacturing cost is higher.

3.3.2.1 Global Optimization

1) Dynamic programming (DP)
   Dynamic programming is a numerical method to solve multi-stage decision problems, where DP is based on the Bellman equation and uses a recursive way to reduce the optimization problem to a multi-step decision process, as shown in Figure 3.12. Therefore, the DP process requires prior knowledge to be conducted. DP has a wide range of applications, but with the increase of optimization objectives, DP falls into a dimensional disaster, and then limits its application in complex systems. Using DP to the fuel cell and battery power in PEMFC + BAT type, the state variable is battery SOC, the control variable is fuel cell power, cost function for the energy consumption of hydrogen consumption, constraints include the range of fuel cell and battery power range, battery SOC and terminate SOC and battery SOC single step change dynamic performance, in order to achieve better economy.

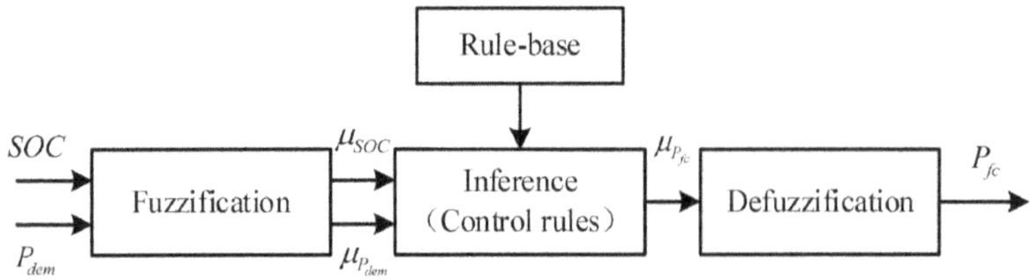

FIGURE 3.11 Fuzzy logic control.

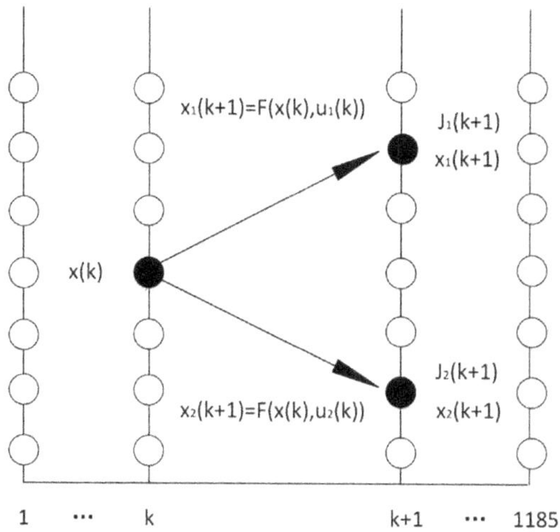

FIGURE 3.12 State transfer process of DP.

2) Pontryagin's minimum principle (PMP)
   PMP determines the optimal control trajectory by iterative search for state transfer in the presence of constraints on some state or input control. In the application of fuel cell vehicle energy management, the basic process is: a) initialization variable; b) define and randomly initialize the control variable; c) calculate power demand and discretization; d) determine the optimal power distribution control amount using Hamiltonian equation; e) repeat c) ~ d) until the termination condition is reached; f) evaluate the termination SOC, if the constraint condition, b) as the optimal control amount, otherwise repeat a) ~f). PMP-based control requires less computational time to obtain the optimal trajectory, and in fuel cell vehicles, PMP is an indirect optimization method to obtain the forced global problem. That is, under certain assumptions, the result of PMP can be regarded as the global optimal solution.
3) Meta-heuristic algorithm
   Meta-heuristic algorithm is a stochastic search method based on iterative mechanisms to solve optimal or satisfactory solutions to complex optimization problems through the understanding of relevant behaviors, functions, experience, rules, and mechanism of action in biological, physical, chemical, social, and other fields. The meta-heuristic algorithm has the advantages of simplicity, strong applicability, no formula derivation, and avoiding local optimal, but it is not universal.

The application of Ant colony optimization in the energy management of fuel cell vehicles is shown in Figure 3.13. The search process adopts the distributed computing method. Robustness is strong, and Ant colony optimization is easily combined

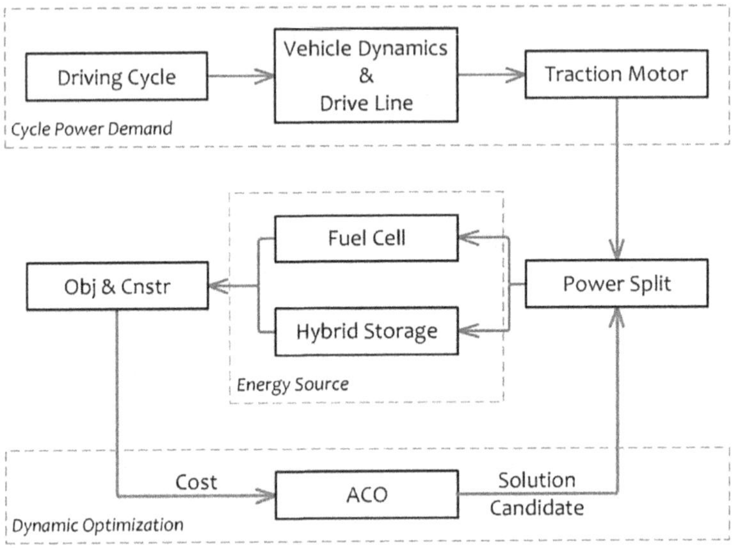

FIGURE 3.13 Application of Ant colony optimization in energy management.

with other algorithms, but it is slower to converge, easy to fall into a local optimum (local optimal). We solved the problem of power distribution of PEMFC + SC vehicles using Ant colony optimization. To maintain the supercapacitor SOC at the same level, its cost function is set to the weighted SOC variation quantity. Genetic algorithm is also commonly used in fuel cell vehicles, genetic algorithm from the random solution, through the adaptive function to evaluate each candidate solution, and using selection, crossover, variation, and other operations to create the best solution for a technical problem, can effectively explore the target parameters, but it is very time-consuming to optimize the parameter value in the rule table, such as the membership function in fuzzy logic control. Design and the use of the offline rule banks of the GA blur energy management controllers to reduce fuel consumption and meet the constraints by minimizing the current disturbances present in the FC (Figure 3.14). Similar to genetic algorithms, particle swarm optimization also does not rely on the assumptions of the optimization problem, and can search for candidate solutions in large space, but particle swarm optimization is more suitable for continuous nonlinear functions and can solve a wide range of constrained optimization problems with small computation time and memory. Grey Wolf Optimizer's optimization process is similar to the genetic algorithm, which also uses the fitness function to judge the position of individuals and to find the optimal solution for the whole iterative process through the location update. The optimal FC reference current for the PEMFC+SC fuel cell vehicles was modified using the Grey Wolf Optimizer, as an input for the negative feedback regulation of the fuel cell current.

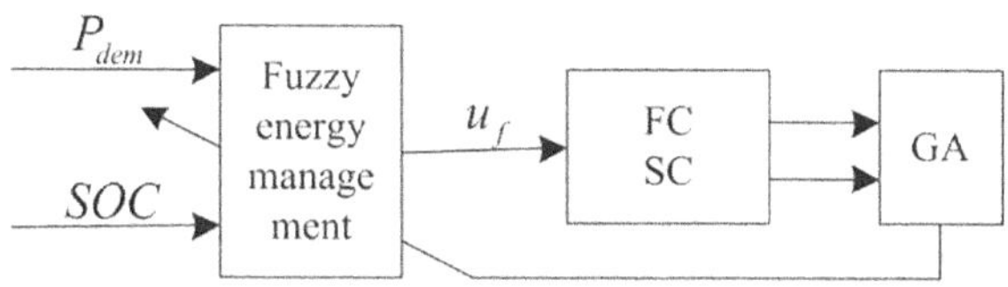

FIGURE 3.14 Fuzzy energy management controller based on genetic algorithm.

3.3.2.2 Instantaneous Optimization

1) (Equivalent consumption minimization strategy (ECMS)
   Equivalent consumption minimization strategy (ECMS) is originally developed based on the heuristic concept, through equivalent factors to auxiliary energy units into equivalent hydrogen consumption. And according to the set energy unit energy priority introduced penalty coefficient and common cost function, equivalent hydrogen consumption at a moment to control the operation of parts is to be seen. Since the equivalent factor and penalty function are initially set, ECMS achieves satisfactory results without prior knowledge of driving cycles, with realizable ones, but cannot guarantee the sustainability of charging components. Due to the initial equivalent factor of SOC reference track has a direct impact on the hydrogen consumption of vehicles, in order to improve the near optimality of ECMS and charging sustainability, which can be real-time optimization adjustment equivalent factor, put forward a method of using DP optimization equivalent factor, while maintaining the auxiliary energy unit SOC, to achieve close to the best fuel economic performance.
2) (Model predictive control (MPC)
   Model predictive control can predict future output trajectories and calculate control sequences based on certain model past and current values to minimize the cost function or error, and MPC therefore depends heavily on higher model accuracy, as well as prior knowledge of the reference trajectories. MPC mainly includes future prediction, rolling optimization and feedback compensation. More accurately, in the application of fuel cell vehicle energy management strategy, the general steps are (Figure 3.15): a) estimate the future power according to the current and historical power; b) calculate the optimal control track; c) apply the control amount and feedback the system state; and d) repeat the previous steps. The proposed MPC controller is constructed as a nonlinear constrained optimization problem and solved by dynamic programming methods to achieve the energy consumption minimizing of PEMFC + SC vehicles, FC durability improvement, and supercapacitor SOC maintenance.

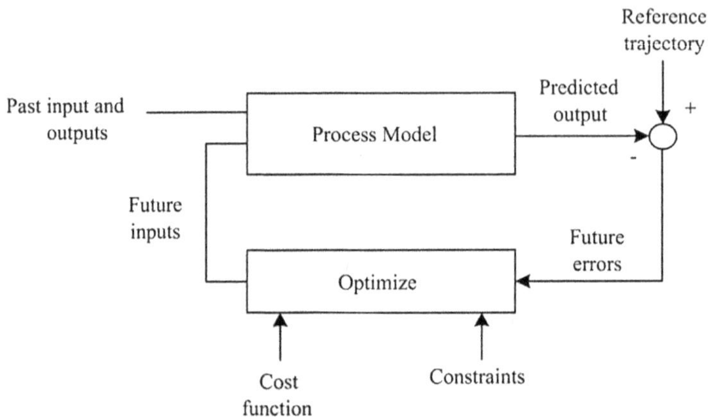

FIGURE 3.15 Block diagram of energy management based on MPC

3.3.3 Learning-Based

The learning-based energy management strategy performs data mining on large datasets containing real-time and historical information to achieve optimal control. Learning-based algorithms are well-learned and adaptive, without absolute model knowledge, but creating an accurate database with a direct impact on control performance and size is difficult and time-consuming. In the application of energy management for fuel cell vehicle energy management, there are algorithms such as reinforcement learning, rule-based learning, and neural network.

1) Reinforcement learning
   With the rise of artificial intelligence, reinforcement learning is also gradually being applied to the energy management of fuel cell vehicles. The subject of reinforcement learning includes the agent and the environment, where the agent and the environment exchange the actions and get the corresponding rewards. The agent uses the exchanged information learning decision rules to maximize the long-term accumulated rewards, and finally produces and identifies the goal of the best action that can bring the maximum return, as shown in Figure 3.16. The reinforcement learning algorithm can independently choose the best operation without any prediction and prior knowledge. Compared with the rule-based EMS, the reinforcement learning algorithm can provide higher precision optimization results. Compared with the optimization algorithm, the reinforcement learning algorithm can run online, greatly reducing the computing time and cost. The Q-learning and deep deterministic policy gradient (DDPG) in reinforcement learning have been widely used in the energy management of fuel cell vehicles.

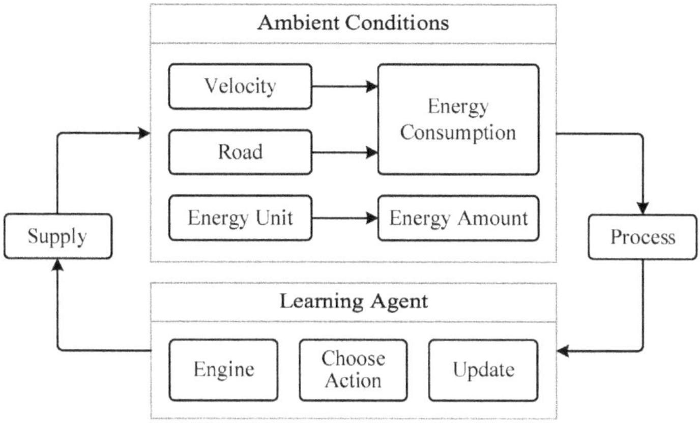

FIGURE 3.16 The basic process of reinforcement learning.

Traditional Q-learning-based reinforcement learning algorithms can successfully perform power allocation under low-state, low-action conditions, but often require more powerful computational performance when the dimension of the state-action space increases, or faces continuous state-action variable problems. To solve these problems, we propose a kind of fast learning algorithm based on Q-learning and ECMS that improves the convergence rate without affecting the optimality of results, a fast Q-learning (SQL) algorithm that adjusts the learning rate, a three-level energy management strategy based on improved Q-learning algorithm to optimize the energy efficiency of electric vehicles by designing low-dimensional state-action table and the system efficiency optimization by designing double reward function.

2) Rule-based learning

Rule-based learning energy management strategies can combine the advantages of rule-based and optimization-based energy management strategies. The purpose of rule learning is to use the rule learning theory based on mathematical algebra logic to achieve the real-time application purpose. In the application of energy management in fuel cell vehicles, the general process of rule learning is: a) under a certain driving cycle, the optimal control sequence was determined by using the offline optimization algorithm; b) to simplify the data set composed of the optimal solution; c) to learn and classify the rules using the rule learning algorithm, such as repeated incremental pruning to produce error reduction algorithm (RIPPER); and d) to apply the regression algorithm to fit the generated rule set and to solve the parameters as shown in Figure 3.17.

3) Neural networks

Neural network is an algorithmic mathematical model of distributed parallel information processing, similar to actual multi-linked neurons, which minimizes the error between the actual output and the

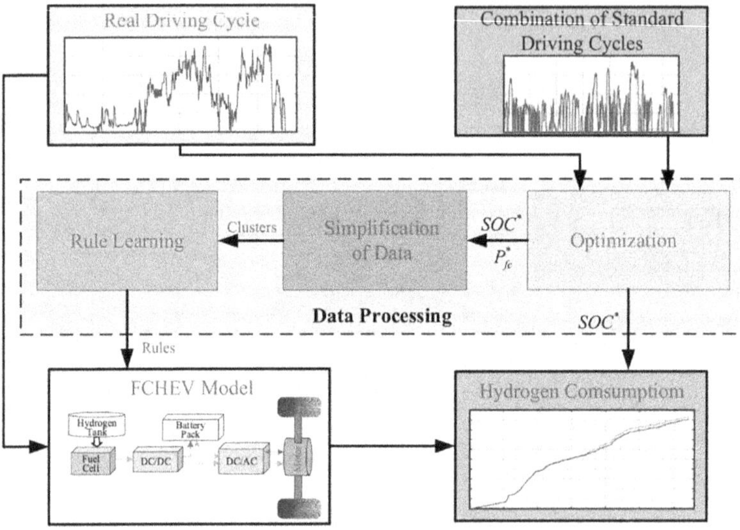

FIGURE 3.17 The application of rule learning in energy management.

predicted output of the training set by adjusting the interconnected relationship between a large number of internal nodes within. The adaptive structure of neural networks makes them suitable for energy management applications in fuel cell vehicles, but requires extensive training data to train the network. Neural network training is used in fuel cell/cell car optimal power control sequence, first based on the gradient method under different conditions to get the corresponding optimal control sequence. Then using neural network training, with the optimal power flow allocation between the fuel cell system and the battery system as the training object, neural network training is used to minimize the total equivalent energy consumption. Neural network is also used in the speed prediction module of energy management, designing a speed prediction method, which uses the back propagation neural network to obtain the initial speed. And the radial basis function neural network predicts the prediction error, adding these two parts to get the final prediction speed.

3.4 ENERGY MANAGEMENT STRATEGY CONSIDERING COMPONENT CHARACTERISTICS

While ensuring fuel economy, in order to effectively extend the service life of power system components and reduce the full life cycle cost of fuel cell vehicles, the characteristics of power system components should be one of the considerations in the design of energy management strategies of fuel cell vehicles, which can be started from the following aspects:

1) Considering component performance benefits
The multi-energy sources of fuel cell vehicles have different performance benefits. Fuel cells can continuously provide high power discharge, but their dynamic response performance is poor, and supercapacitors are just the opposite. Therefore, when designing energy management strategies, the performance differences of the components should be fully considered to achieve vehicle performance and cost optimization. The rule-based energy management strategy and the wavelet-based energy management strategy in the following.
2) Considering component degradation
Considering component degradation in the energy management strategy, on the one hand, the energy management strategy can reduce or avoid the loss cost caused by component degradation if the component degradation is in the target function. On the other hand, it can be updated in real time to make the optimal power control more conducive to fuel economy according to the actual component degradation situation. The following improved ECMS can achieve this goal by real-time optimization equivalent factor.

Some of the energy management strategies that consider the component properties are listed here:

1) Wavelet-transform-based energy management strategy: In the application of fuel cell energy management, wavelet transform can extract vehicle demand power low frequency and high frequency part, fuel cell, or battery bear low frequency part of the power demand, super capacitor bear high frequency part, which can effectively alleviate the dynamic fluctuation of fuel cell power, and greatly improve the service life of fuel cell and battery. The Wavelet transform is incorporated into a rule-based energy management strategy, and it optimizes the rules and control parameters of the low-frequency parts offline using the DP to further improve the fuel economy (Figure 3.18).
2) Improved ECMS: The disadvantage of traditional ECMS is that when the characteristics of the energy unit change, the deviation between the set equivalent factor and the actual optimal equivalent factor will lead to a suboptimal solution. Therefore, the optimal solution can be realized by considering the degradation characteristics of the energy unit and adjusting the equivalent factor to the maximum extent. On the basis of the traditional ECMS, an adaptive ECMS was proposed, through the fuel cell and cell health state model, with the fuel cell and cell SOH value into the equivalent factor equation. Realizing the purpose of

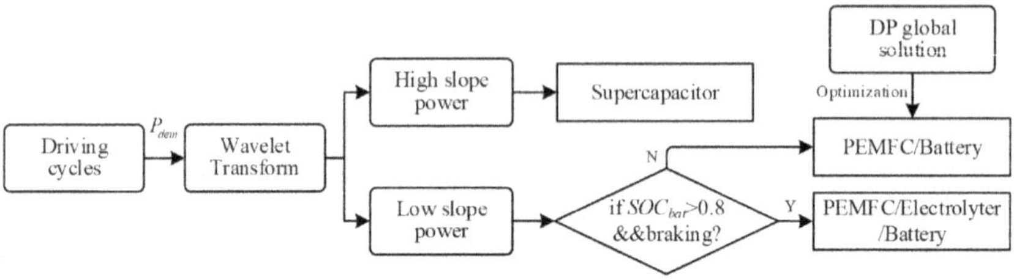

FIGURE 3.18 Application of Wavelet transform in energy management.

adjusting the equivalent factor, with the energy unit degradation, the dynamic power change rate of fuel cell can also be adjusted, in order to reduce the degradation rate of fuel cell, in addition to the AECMS, the SECMS, etc.

In addition, energy management strategies considering component characteristics include stochastic dynamic programming and extremal finding method.

3.5 FLEET MANAGEMENT AND OPERATING COST

In the design process of the full life cycle of fuel cell vehicles, especially for fuel cell buses, the impact of the fleet management mode on the whole life cycle performance and cost needs to be considered. In this process, the one-day driving time and distance management and scheduling management of the fleet will directly affect the vehicle performance and cost. Due to the few studies on fuel cell vehicle fleet management at the present stage, this part also includes the fleet management of pure electric vehicles and plug-in hybrid electric vehicles in the research scope.

3.5.1 Daily Travel Time and Distance

At this stage, for fuel cell vehicles, hydrogen cost is higher than charging cost. With the increase of daily driving time and distance, in order to achieve the expected driving distance, to design the larger auxiliary energy unit capacity, it can effectively improve the fuel economy but also affect the operating cost, and increase the purchase cost of the auxiliary energy unit. Although the purchase cost will be reduced, but the operating cost and use cost caused by the increased hydrogen consumption will also increase. Therefore, determining the optimal daily driving time and distance is of certain significance to the full life cycle performance and design of fuel cell vehicles.

3.5.2 Scheduling Management

Fleet scheduling management is another major factor affecting vehicle performance and operating cost. For fuel cell vehicle fleet, scheduling management mainly includes the formulation of operation schedule, vehicle scheduling management, and energy supply scheduling management, etc. The energy supply scheduling here includes hydrogenation scheduling and charging scheduling. In the process of scheduling and management, it also includes the planning decision of the location and number of charging stations and hydrogenation stations, and this part is not within the scope of this paper.

The scheduling management of fuel cell fleet can be regarded as a multi-layer planning problem. First, determine the team running schedule, considering the constraints of team travel time and limited mileage, with the target of small team running cost for vehicle scheduling. Next, consider the charging and hydrogenation time and limited distance constraints, with the target of smaller team operating cost for energy supply scheduling. In general, fleet scheduling management affects vehicle performance, such as vehicle fuel economy and component degradation life, as well as vehicle use and operating cost, including transportation cost, handling cost, charging or power changing cost, hydrogenation cost, lag penalty cost, etc.

Fleet scheduling management can be achieved through modeling methods. For the scheduling problem of electric bus, the multi-objective optimization integration model of single line electric bus operation is established, including smooth vehicle departure interval and minimizing vehicle number and total charging cost. Constraints include different periods of departure interval, vehicle mileage and charging conditions, compared with the existing schedule and sequence schedule, the integration model cannot only effectively reduce the number of vehicles and total charging cost, but also significantly improve the stability of departure interval, and uniform distribution of vehicle charging cycle in non-peak hours. A multi-objective and two-layer planning model is established to jointly optimize the vehicle scheduling and charging scheduling of traditional vehicles and new energy vehicle hybrid bus teams under the operation conditions of the bicycle field.

4 Vehicle Energy Storage Devices and Their Second-Life Applications

Yuanjian Zhang and Di Zhao

Compared to pure electric vehicles, the manufacturing cost and using cost of fuel cell vehicles are higher at present. In order to reduce the full life cycle cost of fuel cell vehicles, improve market competitiveness, and accelerate the commercialization process, on the premise of ensuring the performance of fuel cell vehicles, the design of fuel cell vehicles considering component performance parameter, component degradation, and second-life applications is particularly important. This chapter will start from the performance parameters of common powertrain components, such as fuel cell, battery, and supercapacitor, analyze the impact of component performance on vehicle performance and manufacturing cost, and explore the prediction method of component remaining service life and the way to improve second-life value from the of degradation mechanism and factors, so as to evaluate and optimize component parameters and energy management strategies.

4.1 FUEL CELL

4.1.1 Performance and Purchase Cost of Fuel Cell

Fuel cell used in vehicles is mainly proton exchange membrane fuel cell (PEMFC). PEMFC reactor is composed of several single fuel cells in series, combined with the auxiliary system to form the fuel cell system. The electromotive force of the fuel cell system is inversely proportional to the current density, and due to the existence of the auxiliary system, when the net power is very small, the system efficiency is relatively low, as shown in Figure 4.1.

The performance of the fuel cell system is mainly characterized by the electric momentum, system net power, system efficiency, and hydrogen storage quantity. They not only directly determine the performance of the vehicle, but also determine the manufacturing cost of the vehicle. Further, the voltage level and one-way DCDC determined by the fuel cell stack series number should match the DC bus. In addition, because the fuel cell system is the main energy source of fuel cell

DOI: 10.1201/9781003302827-5

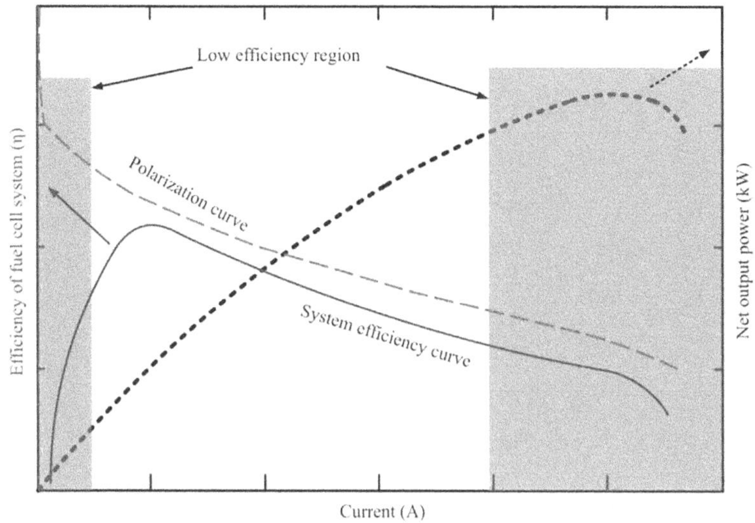

FIGURE 4.1 Fuel cell performance.

vehicles, the maximum net power determines the ultimate performance of the vehicle. However, higher maximum net power also implies increased manufacturing costs, as the fuel cell system constitutes a significant portion of the cost among power system components. Therefore, the reasonable selection of fuel cell system parameters on the vehicle manufacturing cost is huge. The vehicle performance changes caused by different fuel cell system performance parameters have been studied, and the results are shown in Figure 4.2. For power performance, with the increase of the maximum net power of fuel cell, the vehicle maximum speed increases and acceleration time decreases. This is because the vehicle power in the early energy unit maximum power makes it worthwhile to improve the additional manufacturing cost of fuel cell system power output. However, when the maximum net power is further increased, the drive motor will reach its performance limit, increasing net power instead of vehicle performance improvement. This will also result in mass maximum speed reduction and acceleration time. For economics, higher system efficiency and greater hydrogen storage clearly bring in longer driving miles, but it also means higher manufacturing cost. Therefore, the fuel cell performance design should take into account both vehicle performance and component procurement cost.

By 2030, the cost of fuel cell systems is expected to fall from 30 $/kW (500,000 vehicles/year), making it possible to reduce the cost of hydrogen storage bottles to 8 $/kWh by changing fiber materials and manufacturing methods. According to a 2018 report issued by American Strategic Analytics, the cost of fuel cells decreases with higher annual production, fluctuating between 1,000 and 500,000 units, and the cost of fuel cell systems is between 181.07 and 44.58 $/kWnet.

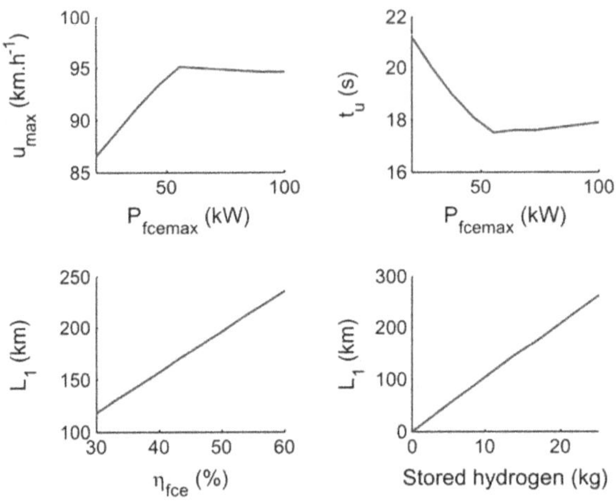

FIGURE 4.2 Influence of fuel cell performance on vehicle performance.

4.1.2 Fuel Cell Degradation Characteristics

4.1.2.1 Degradation Mechanism

The degradation of the fuel cell system is mainly reflected in the degradation of the membrane electrode aggregate, which is manifested by the reduction of voltage or power. Therefore, this part focuses on the degeneration of membrane electrode collections. From the perspective of degradation pathway, the degradation of fuel cell system can be reflected in both chemical degradation and mechanical degradation.

Chemical degradation

Due to the decline of the membrane electrode aggregate performance caused by the harmful substances produced or free radicals from the cathode and anode in the process of electrochemical reaction, the chemical degradation of the membrane electrode aggregate takes place. The typical free radicals include peroxide, carbon monoxide, etc., which will cause anode poisoning. In addition, the common chemical degradation also includes Pt sintering and dissolved carbon carrier corrosion.

Mechanical degradation

The mechanical degradation of the membrane electrode aggregate is caused by performance decline or membrane failure due to pressure, mechanical stress, humidity, thermal stress, reaction gas supply, and working state. Due to manufacturing defects, such as excessive or uneven pressure caused by mechanical degradation, it may cause early failure of the fuel cell system. In addition, if the fuel cell system experiences thermal cycle and humidity cycle, it may cause additional mechanical stress, membrane thickness drop, cracking, tear, and perforation. When there is MEA perforation, hydrogen and oxygen in the electrode catalyst surface cross and react, and fuel cell voltage will quickly drop down, causing battery polarization.

4.1.2.2 Degradation Factors

In addition to the fuel cell system degradation caused by manufacturing defects, water management, thermal management, gas management, and working state management are the main factors affecting the chemical degradation and mechanical degradation of the fuel cell system, which is closely related to the fuel cell system design and the fuel cell vehicle energy management strategy design.

Effective water management is essential to prevent issues such as poor proton conductivity, high membrane resistance, and membrane tearing caused by either membrane drying or flooding. Since almost all humidification and generated water exit through the cathode outlet, liquid water often accumulates in the catalyst layer and the gas diffusion layer, hindering the delivery of oxygen and hydrogen to the reaction surface. This also leads to the coverage of active catalyst sites and the formation of larger water channels in the polymer electrolyte, which accelerates catalyst degradation. Consequently, a voltage drop occurs, leading to a significant decline in fuel cell performance and irreversible material degradation.

As with water management, thermal management can maintain thermal balance during the electrochemical reaction of the fuel cell system. When the membrane is completely hydrated, the increased reactor temperature will cause the average battery voltage to increase; otherwise it will cause a sharp drop in the reactor voltage so the appropriate operating temperature can guarantee the performance of the fuel cell system. Extremely high temperature will accelerate the degradation of the catalyst and the membrane, while extremely low temperature will affect the progress of electrochemical reactions and may cause flooding, intensifying the degradation if the fuel cell is exposed to sub-zero temperature for a long time.

The main purpose of gas management is to provide appropriate hydrogen and oxygen pressure to improve the efficiency of fuel cell systems and avoid gas shortage. Lack of gas is one of the most important reasons of proton exchange membrane fuel cell life decay, while lack of oxygen will lead to carbon carrier corrosion, battery inversion, and unequal current distribution of a series of serious consequences. And excessive oxygen will lead to auxiliary system power increase, lower net output power, while lack of hydrogen will lead to carbon carrier corrosion, platinum particles agglomeration, and irreversible degradation of the anode.

Working state management is to make the fuel cell system work in a relatively stable and efficient state, which is related to the fuel cell load change cycle, start and stop cycle, no-load time, and high load time. Frequent changes in work loads can lead to gas scarcity and accelerate the degradation of fuel cell systems, while long periods of high-load work can also cause it. Frequent starting and stopping of the fuel cell cycle, especially when the catalyst is exposed to reverse current conditions, can lead to catalyst degradation. However, if the fuel cell is stopped and the voltage quickly disappears, the impact of the start-stop cycle on the lifespan of the fuel cell can be ignored. But passing the shutdown of the residual gas in the flow field will make it difficult to eliminate the voltage quickly and completely. Since the reactant consumption is zero under no-load conditions, this will greatly increase the probability of the cathode and anode gas crossing, resulting in higher rates of membrane degradation. Therefore, when designing the energy management strategy of fuel cell vehicles, the fuel cell system degradation caused by frequent and large load changes, frequent start and stop, long idle speed, and high load should be avoided to the maximum extent, so as to reduce the loss cost and the second-life application cost.

4.1.2.3 Remaining Useful Lifetime Prognostics

Lifetime prognostics aims to predict future states using historical and current performance data, but accurate life prediction remains challenging due to the high nonlinearity of the component system. Fuel cell life prediction methods include data-driven approach, model-driven approach, and hybrid approach, in which the hybrid approach is a combination of the first two approaches, with the advantages of both.

1) **Data-driven approach**
   Data-driven approach can study the degradation rules of fuel cell systems from historical operating data obtained by sensors and build empirical or semi-empirical mathematical models through statistical methods or artificial intelligence technologies to predict the fuel cell system degradation. Because data-driven approach does not use physics-based specific models, but rather builds mathematical models or acquires weight coefficients based on training, it is more flexible and applicable, but must have historical data and typical operational data. Data-driven approach can be divided into statistics-based and artificial intelligence technologies.
2) **Model-driven approach**
   The model-driven approach can construct accurate physical and mathematical models from the degradation behavior of the fuel cell system, and the model can target the multi-physical field, multiphase, and multiscale of the fuel cell system. The model driven approach is able to represent the complex and nonlinear relationships between the data, and it is relatively difficult to build accurate mathematical models to predict the decay on the premise that the degradation principle of the fuel cell system is not fully understood. The modeling objects of the model-driven approach can be the catalytic layer, the proton exchange membrane, and the diffusion layer.

4.1.3 Second-Life Applications of Fuel Cell

Secondary recycling after end of life (EoL) is considered to be one of the effective ways to reduce the fuel cell vehicle life cycle cost and improve market competitiveness. EoL of fuel cell can be defined from two aspects. One is that the fuel cell no longer meets the requirements, including power density, hydrogen consumption, and system efficiency. The second is a definitive end of life threshold, for which DoE defines the fuel cell EoL as a 10% loss of initial performance. When fuel cells reach the end of their life cycle, the main second-life application methods are overall recycling and material recycling.

Overall recycling of fuel cell system

The overall recycling of the fuel cell system is the overall transplantation of the fuel cell system from the retired vehicles to the second-life application sites. Before the second-life application, not only to the fuel cell system performance evaluation but alsothe second-life application demand performance parameters to match. These determine the second-life application value of the fuel cell system. Therefore, when designing fuel cell vehicles, the fuel cell performance degradation and second-life application of performance optimization should be considered. This may be done by considering the following points:

1) **Reduce fuel cell degradation**
The degree of fuel cell degradation directly affects the feasibility of the fuel cell with overall recycling. Therefore, in order to meet the requirements of fuel cell performance in the second-life application scenario, it is necessary to reduce the performance degradation of fuel cell from the perspective of vehicle design, which echoes the fuel cell degradation mentioned previously. When taking into account the external factors of the fuel cell, degradation is associated with load changes, start-stop, no-load, high-power loads, and air pollution. Therefore, according to the vehicle performance requirements and driving cycles, component parameters and energy management strategies should be reasonably designed to reduce fuel cell degradation. Looking at the internal factors of the fuel cell, either a continuously high operating temperature or a continuously dry polymer film can cause some damage to the fuel cell, which might cause performance degradation. Therefore, by the rational design of the fuel cell hydrothermal management system, its running time should be strictly reduced under dry conditions, while real-time of the fuel cell system for hydrothermal balance should be monitored. And to reduce the degradation, the value of second-life application should be improved.
2) **Performance optimization considering second-life applications**
The performance parameters of the fuel cell system should be fit with the required parameters of the second-life application scenarios to reduce the transplantation and secondary development cost. Therefore, in the design process of vehicles, especially in the design of component performance parameters, the constraints of the proposed application scenario can be taken into account in the constraints of the optimization of component parameters when the vehicle performance requirements are met.

Material recycling of fuel cell system

End plates and fastening bolts in fuel cell stack are made of ordinary steel and alloy and can be directly recycled. The simple reuse of membrane electrode aggregate MEA is unrealistic because fuel cell failure or degradation is usually caused by MEA, especially membrane degradation caused by dehydration, pinhole, or pollutant accumulation. So physical or chemical recycling can be used with MEA materials, mainly platinum group metals, rare earth elements, and polymer film. The recovery process of membrane electrode aggregate can be summarized into four kinds, namely a) high temperature combustion treatment; b) acid dissolution treatment; c) recovery treatment based on electrochemical process; and d) alcohol treatment. Recycling Pt from MEA can reach a 76% recovery rate, and the polymer film recovery rate is nearly 100%, but whatever process is used, there are still challenges in removing trace elements, which also increases the second-life application cost to some extent. Therefore, in the design process of fuel cell vehicles, the pollution of harmful substances to the fuel cell system can be reduced by purifying the reaction gas and ensuring the consistent pretension force of the fuel cell stack so as to achieve the reduction of the material recovery cost in the process of second-life application.

4.2 BATTERY

4.2.1 Performance and Purchase Cost of Battery

Lithium-ion batteries are widely used in electric vehicles. The main performance indicators of the battery include charge state SOC_{batt}, open-circuit voltage U_{ocv_batt} (V), charge and discharge internal resistance R_{batt} (Ω), capacity C_{batt} (Ah), etc. Because the vehicle battery is connected in series and parallel, the performance of the battery pack is related to the combined form.

Compared with fuel cell performance parameters design, cell performance parameters can also affect the component degradation process. When the battery capacity is larger, the battery can output power, fuel cell system maximum net power can be smaller, and the dynamic response to fuel cell system performance requirements and manufacturing cost is more favorable. With the longer the fuel cell life, large capacity battery means higher manufacturing cost, so more component parameters combination optimization will be needed, in order to seek the lowest total manufacturing cost, which will be introduced in detail later. The vehicle performance changes caused by different battery performance parameters are also studied (Figure 4.3). When the number of battery packs connected in parallel increases, thanks to the increased maximum capacity of the battery, with the higher capacity increasing the maximum speed and reducing the acceleration time, the maximum driving range in the battery mode increases. When the internal resistance of the battery pack increases, the output power decreases, and the energy loss increases, resulting in the reduced maximum speed, the increased

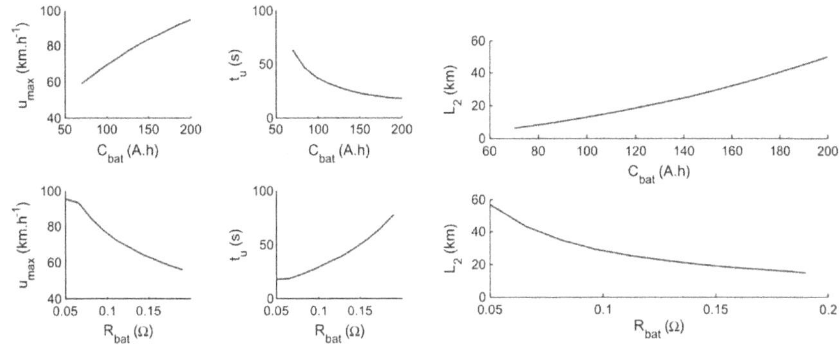

FIGURE 4.3 Influence of battery performance on vehicle performance.

nonlinear acceleration time, and the reduced driving range. When the internal resistance of the battery is large, the acceleration time increases rapidly.

According to the 2019 Bloomberg NEF Market report, lithium-ion battery packs are currently priced at 156 \$/kWh. In 2021, the price of lithium-ion batteries fell from 1,160 \$/kWh to 156 \$/kWh, and it is expected to fall to 62 \$/kWh by 2030. The price of lithium-ion batteries is predicted to fall from 257 \$/kWh in 2018 to 143 \$/kWh in 2028.

4.2.2 Battery Degradation Characteristics

4.2.2.1 Degradation Mechanism

The degradation of lithium-ion batteries is caused by the irreversible changes of the electrolyte, anode, and cathode characteristics, as well as the structure of the battery use components. The specific degradation modes include the lithium-ion loss, the loss of anode and cathode active materials, and the electrolyte loss. The main degradation mechanism is shown in Figure 4.4. In terms of graphite anode lithium-ion batteries, the aging process of the battery is divided into three stages. Stage 1: During the previous charge and discharge cycles, the surface of the graphite electrode forms a passivation protective layer of the solid electrolyte interface membrane SEI, which will lead to a rapid decline in the battery capacity, especially in the first life cycle. Since the SEI layer allows only lithium ions to pass while blocking electrons, it helps slow down the degradation of the electrolyte, allowing the battery to stabilize thereafter. Stage 2: With the increase of battery charge and discharge cycle times, lithium ion insertion and stripping process will lead to electrode volume change, resulting in SEI membrane rupture and graphite/lithium cobalt oxide electrode surface fragmentation, lithium graphite, and electrolyte contact and reaction between SEI membrane generation and thickening. At the same time, due to the fact that SEI membrane thickening diaphragm pore blockage causes uneven current distribution and lithium plating, these may couple side reaction process causing lithium ion and electrolyte loss. Because the battery

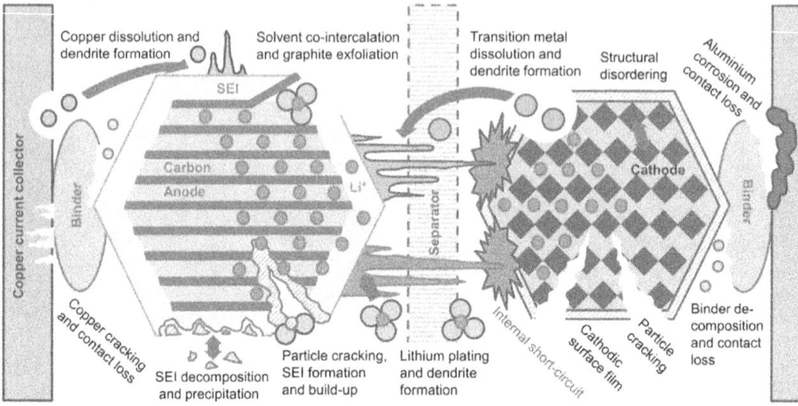

FIGURE 4.4 Main degradation mechanism of lithium-ion battery.

capacity is directly determined by the active substance and lithium ion content, the battery capacity will continuously decrease and the internal resistance will continuously increase, and the battery performance tends to decline linearly. Stage 3: When the cycle number increases to near end of life, SEI film has completely covered the carbon particles on graphite electrode, increased mechanical stress will cause graphite loss (electrode disintegration and contact loss between carbon particles), collecting fluid degradation (contact loss between fluid and carbon particles), adhesive degradation (contact loss between adhesive and carbon particles), resulting in the rapid loss of active material, swift decline in capacity and internal resistance, and battery performance with nonlinear decline.

4.2.2.2 Degradation Factors

The degradation process of lithium-ion battery is determined by both the battery design method and the battery working state.

1) Battery design method

   Design of appropriate battery materials, cell module parameters, battery structure combined with suitable scenarios of fuel cell vehicles can reduce the battery manufacturing and procurement cost, use cost, loss cost, and the cost of second-life application.

   Electrode material: The performance of lithium-ion batteries depends on the various characteristics of internal materials. Reasonable design of battery materials can effectively reduce the side reactions inside the battery to improve the battery life. Studies have shown that the formation of the SEI membrane that determines the battery degradation is highly correlated with the electrode material, and that the graphite anode will lead to the formation of the SEI membrane, while lithium titanium oxide (LTO) will not. In addition, the volume change of the lithium iron phosphate (LFP) cathode is less than that of the lithium manganese oxide (LMO), so its structural deformation is also less.

Battery module parameters: From the review of the battery degradation mechanism, it can be seen that the battery module parameters, such as the thickness and proportion of the anode and cathode active materials, porosity, particle size, and electrolyte composition directly affect the mechanical stress and side reactions inside the battery, thus affecting the battery life. In addition, the battery performance parameters determined by the battery module parameters also affect the life. The C / $LiCoO2$ battery is tested in the range of rated voltage, and the results show that the capacity degradation also increases with the maximum voltage of the battery.

Battery structure: The inhomogeneity of the battery structure may cause lithium plating in small characteristic areas, and the resulting lithium ion loss will rapidly change the electrode balance, resulting in divergent loss of active materials and lithium ions. This can cause a sudden drop in capacity. In addition, the battery structure determines the current distribution and temperature distribution inside the battery, and a large number of uneven temperature rises and uneven current and temperature distribution will affect the battery life.

2) Battery working state

Battery working state also affects the battery degradation, which is often coupled between each state. With these working states including temperature, charge and discharge rate, state of charge, discharge depth and overcharge and overdischarge, through the reasonable design of energy management strategy of fuel cell vehicles, the life brought by the battery working state degradation can be effectively avoided.

Temperature: Capacity degradation has a strong correlation with temperature. With the increase of temperature, the internal side reaction of the battery accelerates, especially the continuous generation and thickening of the SEI film. The more serious the capacity decay is, the increase of internal resistance brought about by battery degradation also further aggravates the temperature rise. If the temperature is too high, it may trigger a thermal runaway. When the battery works at a low temperature, the lithium-ion deposition on the negative electrode will increase the loss of lithium ion, and the nonlinear area of the battery degradation will appear earlier, seriously shortening the life cycle of the lithium-ion battery. Therefore, the battery life decay can be slowed down by appropriately increasing the battery temperature before the nonlinear characteristic occurs.

Charge and discharge rate: Under the low charge and discharge rate, the experimental results show that the capacity loss is affected by time and temperature, and under high charge and discharge rate, the influence of charge and discharge rate, high discharge rate capacity degradation, this may be because the high charge and discharge rate brings fatigue and damage of active material crystal structure and faster internal side reaction.

State of charge, depth of discharge and overcharge and overdischarge: When the battery SOC is high or overcharged, too low anode potential increases the side reaction rate such as SEI membrane thickening, and lithium deposition may occur; when the battery SOC is low or overdischarged, although the anode potential is higher than the cathode potential is conducive to the battery life, too low SOC will cause the corrosion of the anode copper collector fluid (anode copper current collector) and the cathode active material structure will collapse (disordering), which will greatly affect the battery life. However, even more unfortunately, battery degradation will lead to continuous overcharging, which exacerbates battery degradation. Furthermore, monomer cells cycling at DOD greater than 50% were shown to reach a defined end-of-life state faster than at lower DOD. Since these mechanisms rely on battery potential, from the perspective of vehicle design, especially in the design of energy management strategy, it is necessary to consider both battery life and vehicle performance requirements, design a reasonable working range for the battery, delay the nonlinear aging characteristics, and prolong the life of lithium-ion battery.

4.2.2.3 Remaining Useful Lifetime Prognostics

As with the fuel cell life prediction method, the residual service life prediction of the battery can also be divided into data-driven method, model-driven method, and fusion approach/hybrid method.

1) Data-driven approach

The data-driven approach can be divided into data-driven method based on statistics and data-driven method based on artificial intelligence technology.

Data-driven method based on statistics: Data-driven method based on statistics can establish the mapping relationship between the remaining lifetime RUL and the battery properties, such as the exponential and logarithmic types. Here the battery characteristics can be capacity, internal resistance, and stress factors, such as temperature, charge state, charge and discharge ratio, discharge depth, etc. Statistics-based data-driven method is often combined with the filtering algorithm, such as the extended Kalman filter, unscented Kalman filter (Unscented KF), particle filtering, unified particle filter (UPF), Bayesian Monte Carlo method, etc., to update the model parameters according to the latest battery data. To achieve higher prediction accuracy and stability, this method is also known as the adaptive model of update parameters.

In the application of battery degradation model, data-driven method based on artificial intelligence technology can be support vector machine, relevance vector machine (RVM), support vector regression (SVR), Gaussian process regression (GPR), artificial neural network, autoregressive integrated moving average (ARIMA), fuzzy logic learning system, etc.

2) Model-driven approach
The modeling starting point of model-driven approach can be the mechanistic models and equivalent circuit models of battery degradation. The mechanical model can be built based on the battery electrochemical reaction process, combined with the reaction kinetics and the porous electrode theory, and the equivalent circuit model can be established by generating a combination of circuit elements with the same electrical behavior as the battery.

3) Fusion approach/hybrid approach
Fusion approach/hybrid approach can leverage the advantages of both data-driven methods and model-driven methods. To overcome the limitations of an empirical model of life cycle, based only on equivalent circuits, an attempt to couple the conventional empirical model of capacity loss with a Newman porous composite electrode model containing electrochemical reaction dynamics and material/charge balance can be used to estimate battery life cycle for specific applications.

4.2.3 Second-Life Applications of Battery

Most scholars believe that when electric vehicle batteries degenerate to 70–80% of its initial capacity, with power reaching 50% of the initial value (namely 100% impedance increment), they are at the end of its life, no longer suitable for vehicle energy storage units, but some scholars believe that the battery scrapped to the state that it will no longer meet the driver's daily travel needs is more accurate. When the battery reaches its EoL, there is an opportunity to reuse the battery in fixed applications with low performance requirements to produce greater social and economic benefits, while avoiding the environmental pollution and resource waste caused by landfill recycling, which is also known as the second life. The secondary use of fixed applications can be a commercial and residential power grid standby energy storage or regional regulation system, photovoltaic and wind renewable energy power station power balance application, smart grid fast charging stations, etc. The secondary battery has shown that together with the integration of photovoltaic renewable energy, in addition to reduce the electricity cost of the end user because of the battery price and the reduction of investment cost. In the second-life application process of the battery, the performance index of the single battery and the battery pack at the end of the first life cycle will affect the second-life application performance. Therefore, in the process of vehicle design, the following points should be started to maximize the second-life application value after the end of the first life cycle:

1) Reduce battery degradation
After the end of the first life cycle of a fuel cell vehicle, the lower the degradation degree of batteries, the higher the value of second-life

application. The study shows that for the individual cells with different degrees of degradation, the battery capacity of the degradation turning point decreases rapidly, the DC internal resistance increases rapidly, and the aging trend will not be slowed down in the lower demand of second-life applications. Therefore, in the design of the vehicle during the first life cycle period, the degradation of the battery or super-capacitor influence factors should be fully considered to minimize the degree of degradation during the second-life application. The degradation of batteries is related to component performance, energy management strategy, and single battery degradation characteristics. In terms of component performance, fuel cell battery combination and power level, voltage level and power level should be reasonably designed in combination with component degradation characteristics and operating environment differences to reduce degradation.

2) Reduce the degradation difference of single cell
Second-life application batteries must be split and recombined to meet the new operating environment requirements, and the difference of single cell degradation during the first life, will greatly increase the matching cost of single cells during second-life application. If the consistency of the single battery is poor during second-life application, the available capacity of the battery pack mainly depends on the minimum capacity of the battery pack, and the battery with poor consistency is easy to overcharge and discharge, leading to serious safety problems. Therefore, in the design process of fuel cell vehicles, the status detection of single cell can be done well to reduce the degradation difference of single cell, which can reduce the cost of second-life application.

3) Monitor battery degradation data
Battery degradation data monitoring is a prerequisite for reasonable matching of single battery second-life applications. At the same time, accurate periodic and calendar aging characteristic data is the basis for forming and optimizing the effective second-life application system control strategy. The retired battery is applied to the smart grid electric vehicle fast charging station, by analyzing the correlation between SOC interval and the first life cycle aging, using two replacement battery pack, to realize the optimal SOC operating range control, and the battery pack discharge rate, second-life application of battery life and grid stability and the relationship between battery temperature and internal resistance as the basis for improving the second-life application control strategy. Therefore, in the design of fuel cell vehicles, monitoring the degradation data, such as battery pack capacity and internal resistance changes can accurately understand the influencing factors and degree of degradation of batteries, so as to provide guidance for the matching and control strategy optimization of second-life application and improve the value of second-life application.

4.3 SUPERCAPACITOR

4.3.1 Performance and Purchase Cost of Supercapacitor

Supercapacitors is divided into electric double-layer capacitors, Pseudo-capacitors and hybrid supercapacitors. Currently, the supercapacitors commercialized in fuel cell vehicles are double-layer capacitors, so this paper only studies the commonly used double-layer capacitors based on graphite and organic electrolyte. Similar to lithium-ion batteries, the main performance indicators of supercapacitors include energy state SOE_{sc}, open-circuit voltage U_{ocv_sc} (V), charge and discharge internal resistance R_{sc} (Ω), and capacity C_{sc} (F). When the supercapacitors charge state SOC_{sc} drops from 1 to 0.5, the energy state SOE_{sc} drops from 1 to 0.25. At this time, the supercapacitors should not be further discharged, so the working range of the supercapacitors charge state is generally controlled above 0.5, which is of great significance to the design of energy management strategy.

The supercapacitor performance metrics can also affect manufacturing cost, vehicle performance, and component life. Supercapacitors can provide large changes of power demand in a short period of time, so the larger its capacity, the lower the dynamic response performance of fuel cell system and battery requirements, which is almost with the influence of the same rules. In addition, the wide working temperature range of supercapacitors makes the vehicle cold start performance better, but the supercapacitors manufacturing cost is higher than the battery.

In 2019, supercapacitors are currently priced at 8,000-10,000 $/kWh and 8–12$/kW, as compared to 250$/kWh batteries.

4.3.2 Supercapacitor Degradation Characteristics

The degradation of supercapacitors is mainly manifested in the increase of capacitor loss, storage energy loss, and equivalent series resistance. When the equivalent series resistance of supercapacitors usually increases to twice its initial value or when the capacitor is less than 80% of its initial value, the manufacturer defines it as defective, and 80% of the capacitor loss is always reached faster than the internal resistance double. Supercapacitors age more on power density, so aging supercapacitors can again serve applications that do not have high power requirements.

4.3.2.1 Degradation Mechanism

The fabrication process of the supercapacitor electrodes causes the residue of impurities on the electrodes. These impurities may be metal impurities and surface functional groups, etc. During the cycle of supercapacitor charge and discharge, surface impurities will react with the electrolyte, thus, producing solid and gas products and causing electrolyte loss. At the same time, solvent decomposition caused by overvoltage, electrolyte evaporation caused by high temperature, and the electrolysis of trace amounts of water in the electrode can also lead to the gaseous products. The blockage of the electrode pore by the solid product and the adsorption of the gas product by the electrode reduces the contact area between the electrode and the electrolyte. The Cathe surface area loss of porous activated carbon electrode can reduce porosity and electrolyte conductivity. Thus, it causes the loss of the capacitance and energy. These solid and gas products may also cluster in the diaphragm, preventing the migration of the ionic charges. In addition, the increased internal pressure caused by the gas product will cause electrode cracks and structural changes in the supercapacitor, damaging the fluid collector. Meanwhile, the solvent decomposition will produce free radicals and further fluoridation, resulting in a reaction with the fluid collector and causing further damage to the fluid collector. The preceding processes together trigger an increase in the contact resistance of the electrolyte, the contact resistance between the carbon layer and the fluid collector, and the AC resistance, resulting in an increase in the equivalent series internal resistance.

4.3.2.2 Degradation Factors

The factors affecting the degradation of the supercapacitor also include the design method and the working status.

1) Supercapacitor design method
   As mentioned in the degradation mechanism of supercapacitors, the impurities in the manufacturing process of supercapacitors will affect the degradation, reducing the trace impurities, such as water and oxygen in the supercapacitors, which is conducive to extending the service life. In addition, the composition of the electrode material also has a strong impact on the degradation process because the functional groups on the carbon electrode surface cause instability during aging when floating at a high potential, which causes a series of side reactions.
2) Supercapacitor working state
   The working state affecting the degradation of supercapacitors mainly includes the working temperature, working voltage, working current, and storage time. Among them, except that the operating current affects the cycle degradation of supercapacitors, other factors all affect the calendar degradation.

Both increased working temperature and voltage promote the degradation of supercapacitors, with their effects on degradation. From the mechanism of supercapacitor degradation, overvoltage will lead to solvent decomposition, which produces by-products and causes damage to the fluid collection. Too high of a temperature will accelerate electrolyte evaporation and solvent decomposition, thus accelerating the degradation of supercapacitors. In addition, if the temperature is too high or too low, it can cause an equivalent increase in series internal resistance because the adhesive damage caused by high temperature can lead to the formation of a poorly conductive intermediate layer between the fluid collector and the active surface, and the low temperature leads to an increase in the electrolyte viscosity, reducing the electrolyte conductivity.

In addition to voltage and temperature, cycling conditions (mainly current) are also another prominent factor affecting the degradation of supercapacitors. Comparing the relationship between supercapacitor degradation and cycle conditions under the same voltage and temperature, one can conclude that about 30 A will increase the equivalent series internal resistance, and the life of the current cycle is shortened by two times, and the role of the cycle in the acceleration of supercapacitor degradation is verified. In addition, a novel method to quantify the acceleration process of cycle degradation is put forward.

With the increase of storage time, it will bring about the three-dimensional structure collapse of the electrode surface caused by self-oxidation and reduction and the cracking or separation of the interface between the electrode and the fluid collector, resulting in the reduction of the pore and specific surface area of the active substances, resulting in the degradation of the supercapacitors.

4.3.2.3 Remaining Useful Lifetime Prognostics

Since the empirical model development for the life prediction of supercapacitors is still in its infancy, this section is only briefly described in the existing life prediction methods. The first type of life prediction model is based on supercapacitor degradation mechanism developed by a mathematical model and influencing factors. Based on the influencing factors, the prediction model can be categorized into types such as operating temperature, operating voltage, and operating current. These factors are treated as independent variables to establish various exponential relationships with battery life. By combining empirical formulas, the model can account for the influence of multiple factors on degradation. The second type is about life prediction based on artificial intelligence technology.

4.3.3 Second-Life Applications of Supercapacitor

Since the characteristics of second-life applications of supercapacitor are the same as those of battery, this section does not delve into that. The analysis of second-life applications of supercapacitor can be referred to the applications of second-life batteries as detailed in Section 4.2.3.

4.4 COST OPTIMIZATION

The optimization of the performance parameters of fuel cell vehicle components is to find the optimal combination of the component performance parameters in order to meet the vehicle performance requirements, take into account the operation efficiency and life of the components, and minimize the manufacturing cost, use cost, and loss cost. Component performance reduction, although good for manufacturing cost, may limit the vehicle performance, component inefficient operation and component degradation, and cause the whole life cycle cost not to drop but rise, such as in lower fuel cell system power limits vehicle performance, lower auxiliary energy unit power improves the fuel cell inefficient area operation probability and degradation rate, and component performance improvement will increase unnecessary quality and manufacturing cost. Therefore, the component optimization and energy management optimization of fuel cell vehicles are of great significance to reduce energy consumption, extend component life and reduce cost. These two problems are usually combined. This section will focus on component optimization.

These three energy unit parameters were optimized based on the PEMFC+Batt+SC fuel cell vehicle. First, it determines the discrete parameter vector of 3,024 sets of fuel cell, battery and supercapacitors according to the vehicle performance requirements and driving conditions, calculating the manufacturing cost of each power component combination according to the average fuel cell system, battery, and supercapacitors cost. And the PMP strategy is proposed to calculate the use cost to ensure close to the optimal value and reduce the calculation amount. The optimization results show that the increased power of the fuel cell system increases manufacturing cost and causes slightly lower use cost (Figure 4.5a); the increase of battery series and parallel number leads to increased manufacturing cost and reduced use cost (Figure 4.5b); in the case of fuel cell power and battery parameters fixed, the inflection point of supercapacitor parameters corresponds to the optimal supercapacitor parameters (Figure 4.5a). In the process of optimization, the envelope curve is the fuel cell system, battery and supercapacitor parameter optimization problem, namely when the manufacturing cost reduction, use cost may be greatly increased, so in the selection of component performance parameters, which should be the cumulative results, namely the total cost of life cycle. Similar results were obtained in PEMFC+SC fuel cell vehicles (Figure 4.6a), and as the number of supercapacitors increases, the global efficiency of the power system increases and eventually maintains at a high level (Figure 4.6b), and the fuel cell system can always operate in the high efficiency range (Figure 4.6c).

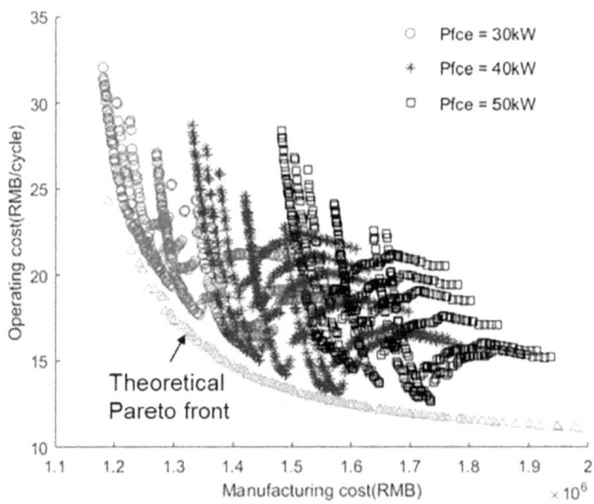

FIGURE 4.5a Impact of component performance on manufacturing cost and use cost.

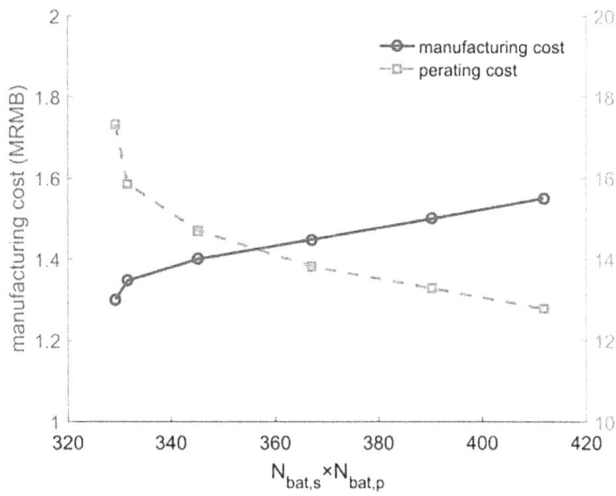

FIGURE 4.5b Impact of component performance on use cost.

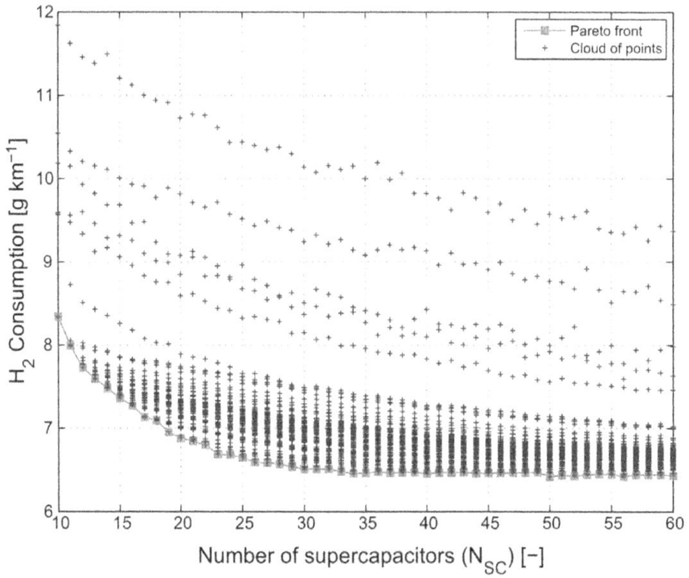

FIGURE 4.6a Effect of supercapacitor performance on fuel consumption.

However, faced with the multi-objective problem of optimization of vehicle cost and component life, the combined optimization of component performance parameters and energy management strategies is often needed, and even the optimized objects include working conditions because they are usually strongly coupled.

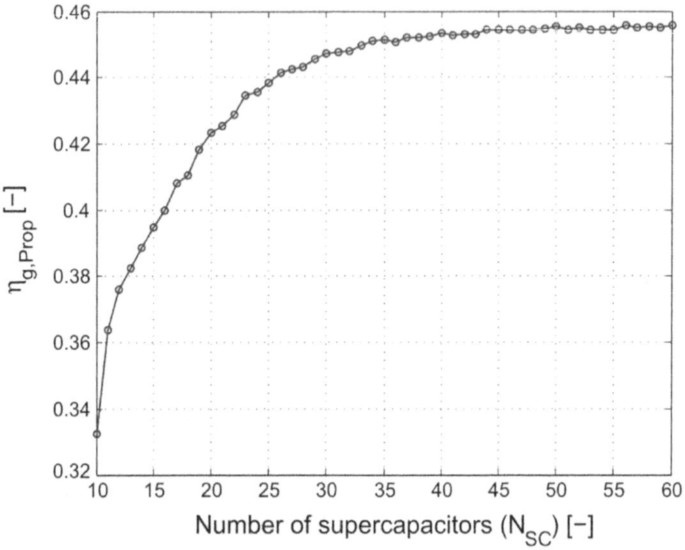

FIGURE 4.6b Propulsion system energy efficiency.

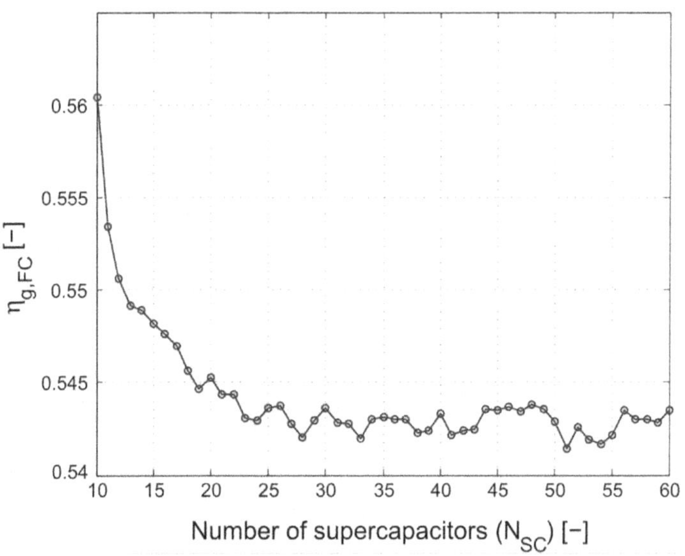

FIGURE 4.6c Fuel cell system energy efficiency.

Based on PEMFC/battery/supercapacitor fuel cell vehicle, a coupling optimization problem of component performance parameters and energy management strategy is established. The principle is shown in Figure 4.7. The objective function of the optimization problem includes two parts: the manufacturing cost and the cost composed of the use cost and the loss cost, as shown in Equation 1.1. First,

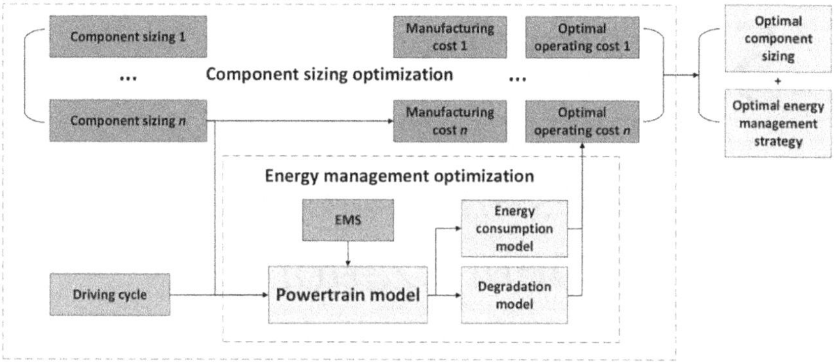

FIGURE 4.7 Multi-objective optimization framework for component performance parameters and energy management strategies.

it determines the 3,024 discrete parameter groups of fuel cells, cells, and supercapacitors according to the vehicle performance requirements and driving cycles, calculates the total manufacturing cost based on their average manufacturing cost, and serves as the first part of the target function. Another part of the target function is composed of fuel cell hydrogen consumption and cell and supercapacitor consumption and fuel cell and battery degradation. Cost, such as Equation 1.2, is hydrogen consumption and power consumption through component power. Equation 1.3 is loss cost according to the actual degradation model of component loss converted into the manufacturing cost, such as Equation 1.4. The global optimal solution is implemented by 2-D DP (two-dimensional DP optimization method), and the quasi-global optimal solution of the energy management strategy and component parameters is implemented by 2-D PMP (two-dimensional PMP) to ensure close to the optimal value and reduce the computation.

$$J = J_m + J_{op} \tag{1.1}$$

$$J_{op} = \{J_{1,ele}, J_{1,H_2}, J_{2,bat}, J_{2,fc}\} \tag{1.2}$$

$$\left.\begin{aligned} J_{1,ele}(k) &= \left(\frac{P_{bat}(k)}{\eta_{bat}} + \frac{P_{sc}(k)}{\eta_{sc} \cdot \eta_{sc,DC}}\right) \cdot C_{ele} \cdot dt \\ J_{1,H_2}(k) &= \Delta m_{H_2} \cdot C_{H_2} \end{aligned}\right\} \tag{1.3}$$

$$\left.\begin{aligned} J_{2,bat}(k) &= \frac{\Delta Q_{loss,bat} \cdot price_{bat} \cdot Q_{bat} \cdot V_{bat}}{1000 Q_{loss,eol}} \\ J_{2,fc}(k) &= \frac{\Delta P_{loss,fc} \cdot price_{fc} \cdot P_{fc,0}}{P_{loss,eol}} \end{aligned}\right\} \tag{1.4}$$

5 Digital Twin

An Effective Big Data Processing Tool for the Optimization of Electric Vehicles

Quan Zhou, Ji Li, Cetengfei Zhang and Hongming Xu

Decarbonization requires global actions from almost all industry sectors. The data from the International Energy Agency (IEA) shows that the transport sector contributed to one-fifth of the total carbon emissions; therefore, decarbonization in the transportation sector is in urgent demand. Electrification of vehicle powertrains is the mainstream method to mitigate carbon emissions in road transport. One of the great changes in the automotive industry is that the new vehicle will be certified based on the evaluation of their real-world performance. The conventional design of experiments (DoE) method makes it difficult to meet the increasing demands for R&D of high-performance and low-cost vehicle products. Digitalization of the R&D of new vehicle products based on digital twin, Internet-of-Things (IoT), and artificial intelligence (AI) is now under rapid development. This chapter will introduce the challenges in the automotive industry and define the technical terms in digital twins for automotive applications. A case study of DT applications on vehicle control will be discussed before giving an outlook on future technology trends and research directions.

5.1 CHALLENGES AND CHANGES IN THE AUTOMOTIVE INDUSTRY

5.1.1 Global Actions in Decarbonization

Conventionally, transportation is largely dependent on energy from fossil fuels. Since the 1970s, global emissions from fossil energy combustion have increased by 90%, which has led to excessive greenhouse gas (GHG) emissions. According to the 2021 EV outlook published by the International Energy Agency (IEA), more than 20 countries around the world have electrification targets or internal

DOI: 10.1201/9781003302827-6

combustion engine (ICE) bans for cars over the next 10–30 years, and 8 countries (Canada, Chile, Fiji, Korea, New Zealand, Norway, United Kingdom, Sweden) plus the European Union have announced net-zero pledges (Figure 5.1). According to the IEA's prediction, there would be over 3 billion electrified vehicles on the road by 2050 [1]. The electrified vehicles include battery electric vehicles, fuel cell vehicles, and plug-in hybrid vehicles, mainly driven by electric motors so that the propulsion system can operate with much higher energy efficiency than conventional internal combustion (IC) engine-driven vehicles.

Different types of vehicles will require different powertrain solutions based on their energy and power demand. An example of the power and energy demands from different types of vehicles is given in Figure 5.2 [2]. Typically, light-duty vehicles with a total weight of less than 3.5 T require low to medium power and energy while buses and coaches need medium to high energy and low to medium power. The design of heavy goods vehicles and off-highway vehicles is more challenging since they have a wide range of energy and power demand. For light-duty urban transport, the main technical solutions will be battery electric vehicles (BEV) and range-extended electric vehicles (ReEV). For light-duty vehicles with long-range mobility requirements and coaches and buses, plug-in hybrids (PHEV) and fuel cell vehicles (FCEV) are the main solutions. For this application scenario, BEV would also be considered if there would be a breakthrough in battery technology. For heavy-duty applications, PHEVs, full hybrids, fuel cells, and dedicated ICE with zero-emission fuels would provide flexible solutions for different applications.

5.1.2 Real-World Driving Economy and Emissions Evaluation

Vehicle economy and emissions are key indicators for vehicle certification. The ways of driving can significantly impact the vehicles' performance in terms of

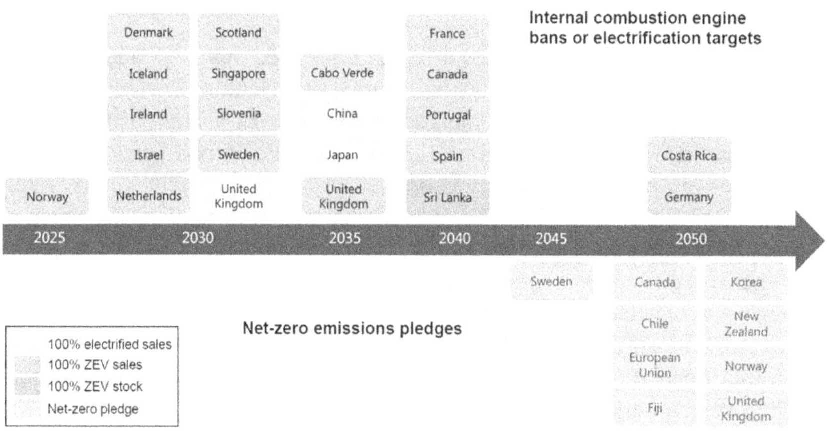

FIGURE 5.1 International combustion engine bans or electrification targets [1].

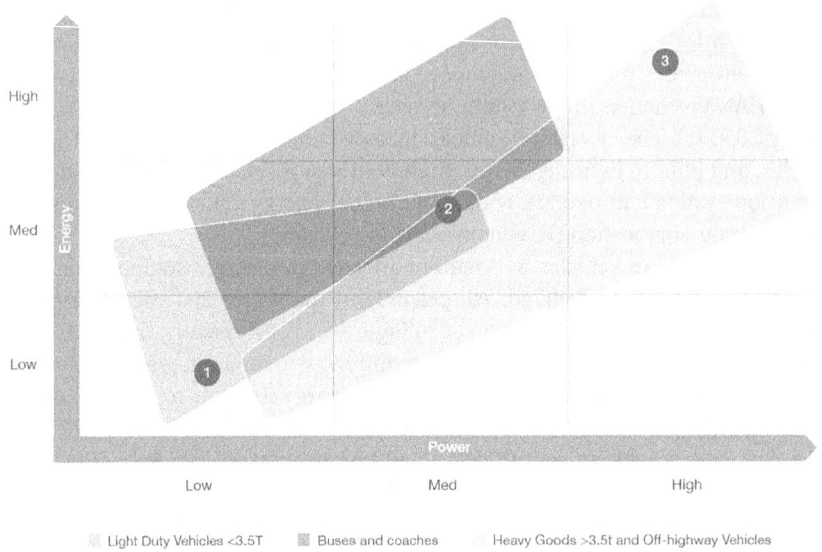

FIGURE 5.2 Energy and power demand of different types of vehicles [2].

energy consumption and emissions. Driving cycles are used to provide a relatively fair evaluation of the vehicle's performance in the R&D stage. Conventionally, there are two types of driving cycles: powertrain cycles (or engine cycles) and chassis cycles. Both are vehicle speed profiles versus the time (normally in km/h or mph). The former transfers the vehicle speed profile into the rotation speed of powertrains to allow the powertrain to be tested on a dynamometer (load provided by hydraulic, eddy current, or an AC motor) and the latter converts the vehicle speed into wheel speed to test the prototype vehicle on a chassis dynamometer.

Table 5.1 listed a summary of the driving cycles used in the US and EU countries. The statistic indicators (maximum speed/acceleration/deceleration, and average speed) are compared. Compared to the New European Driving Cycle (NEDC), which was officially used in Europe for emission certification and fuel economy testing in light vehicles, Worldwide Harmonized Light Vehicles Test Procedure (WLTP) cycles, which are currently used for EU vehicle certification, is more aggressive and involves more trainset operation points [3]. To enable a more comprehensive evaluation, there are three classes of WLTP cycles (Class 1–3) for different types of vehicles based on the vehicle power-to-mass ratio (PMR). WLTP Class 3 is the most radical of the suburban driving cycles. There is a trend that more trainset operations and high-frequency stop-and-go conditions will be considered in vehicle certification.

One of the main changes in the new EU6d emission regulation is the implementation of real driving emissions (RDE) testing as an additional requirement from 2017 onwards. In the future, the vehicles will be evaluated in more dynamic real-world driving conditions, which brings more challenges than conventional testing based on driving cycles. The testing facilities for chassis dynamometer testing (based on driving cycles) and RDE testing are compared in Figure 5.3. RDE legislation adds the road as an environment for emission testing and certification. From Figure 5.3(c), we can see that there is a significant increase in the powertrain (engine) working points when the vehicle testing is transiting from NEDC to US06 and to RDE cycles. The vehicle fuel/energy consumption and emissions will be also impacted by several environmental dynamics (e.g., wind speed, altitudes, temperature), which conventionally can be controllable in the chassis dynamometer testing but unpredictable in RDE testing. This would also increase the working load for the development of vehicle powertrain products.

TABLE 5.1
Summary of Driving Cycles Used in the US and the EU Countries

				WLTP		
	UDDS	**FTP-75**	**NEDC**	**Class1**	**Class2**	**Class3**
Max. speed (km/h)	91.25	91.25	120.00	64.40	85.20	131.30
Average speed (km/h)	31.51	25.86	33.60	28.47	35.72	46.50
Max. acceleration (m/s^2)	1.475	1.475	1.042	0.764	0.958	1.583
Max. deceleration (m/s^2)	1.475	1.475	1.389	1.000	1.111	1.486

FIGURE 5.3 (Continued)

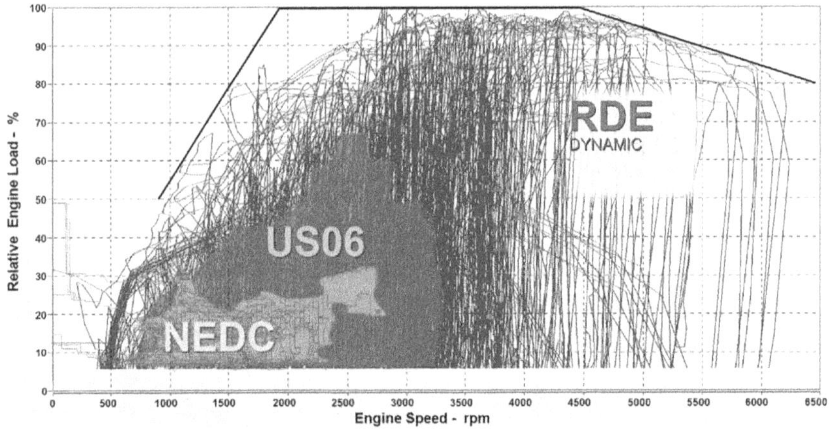

FIGURE 5.3 Vehicle emissions testing: (a) chassis testing [4]; (b) real-world driving emission (RDE) testing (picture from AVL's website); and (c) working points for different testing regulations (picture from AVL's report).

Source: https://www.avl.com/en/testing-solutions/all-testing-products-and-software/emission-analysis-and-measurement/avl-move#-downloads

5.1.3 Model-Based Development for Automotive Products

The development of automotive products follows a system engineering logic, as illustrated in Figure 5.4, which conventionally implements model-based development (MBD) to resolve R&D tasks efficiently at low cost [5]. MBD for automotive is a process of designing, simulating, and testing automotive systems using models that represent the behavior and functionality of the system. This approach uses computer models to simulate the behavior of the system and verify that it meets the intended requirements before it is physically built.

In MBD, the system is first designed as a model using a graphical modeling language such as Simulink, which is widely used in the automotive industry. The model is then tested and refined through simulation to ensure that it behaves correctly and meets the required performance specifications. Once the model is verified, it can be used to generate software code that will be used to control the system. The general process of MBD for automotive engineering includes eight main steps as follows:

1) Requirements gathering: The first step in MBD is to gather and define the requirements for the system being developed. This includes functional and performance requirements as well as any constraints on the system.
2) Model design: Using a graphical modeling language such as Simulink, engineers design a model of the system. The model includes components that represent the various parts of the system, such as sensors, actuators, and controllers.
3) Simulation: The model is then simulated to test its behavior under various conditions. This allows engineers to verify that the model is functioning correctly and to identify any design issues or performance problems.

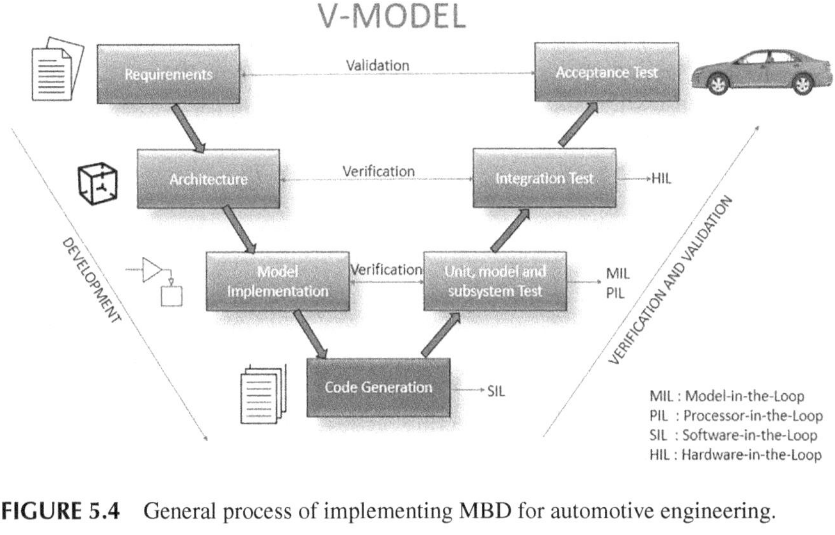

FIGURE 5.4 General process of implementing MBD for automotive engineering.

4) Model refinement: Based on the results of the simulation, the model can be refined and improved to address any issues or optimize performance.
5) Code generation: Once the model is verified and refined, it can be used to generate software code that will be used to control the system.
6) Testing and validation: The software code is then tested and validated to ensure that it behaves correctly and meets the required performance specifications. This includes testing under various conditions and verifying that the software code is compatible with the hardware and other components of the system.
7) Deployment: Once the software code has been validated, it can be deployed in the final product. This includes integration with the hardware and other components of the system as well as testing of the complete system to ensure that it meets all of the requirements and functions as intended.
8) Maintenance and updates: As the system is used in the field, it may require maintenance and updates to address issues or improve performance. MBD allows for easy updates to the software code based on changes to the model or requirements. This helps to ensure that the system continues to function correctly and meet the needs of its users.

MBD is particularly useful in the automotive industry because it allows engineers to design and test complex systems more quickly and accurately than traditional methods. It also enables early detection and resolution of design issues, reducing the need for expensive and time-consuming physical prototypes and testing. Nevertheless, there are also some potential drawbacks that need to be considered.

1) Complexity: Developing and maintaining MBD models can be complex and require specialized knowledge and expertise. The models may also become very large and difficult to manage, which can make it challenging to ensure that they accurately represent the system being developed.
2) Verification and validation: While MBD can help with early detection of design issues, verifying and validating the models can be time-consuming and expensive. It is important to ensure that the models are tested thoroughly and that the results are accurate before they are used in the final product.
3) Tool dependency: MBD relies heavily on software tools such as Simulink and other modeling and simulation software. This can lead to tool dependency, where the models and software are tightly coupled and changes to one may require changes to the other.
4) Cost: MBD can require significant investment in terms of time, resources, and software tools. This can make it more expensive than traditional development methods, especially for small projects or companies with limited budgets.
5) Human error: Like any engineering process, MBD is subject to human error. Mistakes in the modeling process can lead to incorrect results or faulty designs, which can be costly and time-consuming to correct.

5.2 DEFINITION OF DIGITAL TWIN FOR AUTOMOTIVE APPLICATIONS

The digital twin (DT) is built as a virtual counterpart that can represent the dynamics and performance of physical assets. This concept was initially conceived in 2003, then it was rapidly developed in the 2020s [6]. Specifically, the DT is applied to the automotive industry in recent years [7]. In this section, the definition of DT will be first demonstrated. Then relative applications of DT in automotive engineering will be introduced according to each component. Figure 5.5 gives an architecture of the DT system for automotive applications, which consists of a physical entity (PE), a virtual representation (VR), a database with interfaces, and connected intelligence.

5.2.1 Physical Entitites

Physical entities in digital twins refer to tangible objects, equipment, or systems that are replicated in a virtual environment to represent their real-world counterparts. These entities are modeled with a high degree of accuracy and detail, using various data sources such as sensors, IoT devices, and other monitoring systems. By creating a digital twin of physical entities, it becomes possible to simulate their behavior, performance, and interactions with other entities in a controlled environment. This enables various applications such as predictive maintenance, virtual testing, and optimization of processes, which can help improve efficiency, reduce costs, and minimize risks associated with the physical entities in question.

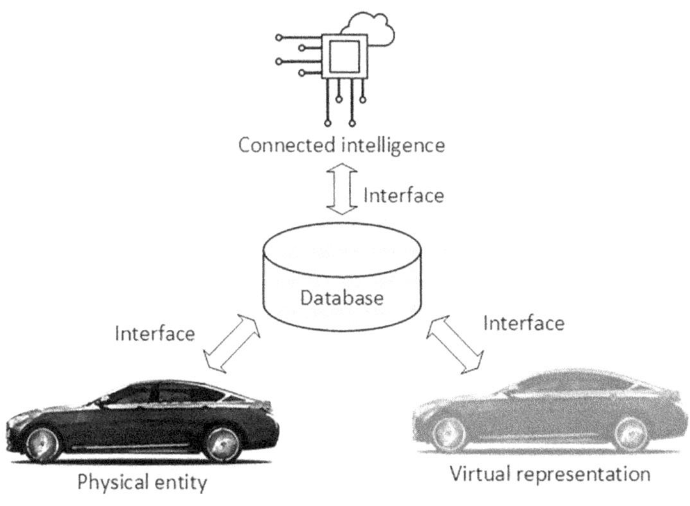

FIGURE 5.5 The frame of the digital twin applied to the vehicle system.

The physical entity referred to in this chapter is the vehicle system, including the whole vehicle system and its subsystems, such as the powertrain system. The development of physical entities can be based on some existing techniques, like hardware-in-the-loop testing. There are some on-shelf solutions from industry companies, for example, AVL and Siemens. In academic research, Rassolkin et al. proposed a test platform for the electric propulsion drive systems for the autonomous vehicle [8]. In the research of Ruba et al., a test bench is built for the permanent magnet synchronous machine based on field programmable gate arrays in establishing a motor digital twin system [9].

5.2.2 Virtual Systems

A virtual system in digital twin refers to the digital replica of a physical system or process. It is a digital model that mimics the behavior, characteristics, and interactions of a physical system, such as a manufacturing plant, a power grid, or a transportation network. A virtual system is typically created by combining various data sources, such as CAD files, simulation models, and sensor data, to develop a complete and accurate representation of the physical system. By creating a virtual system, it is possible to simulate the behavior of the physical system under various conditions and scenarios, enabling predictive analytics, optimization, and decision-making. This can help improve efficiency, reduce downtime, and optimize operations, making it an essential tool for various industries such as manufacturing, energy, and transportation.

The virtual system in the automotive digital twin is developed based on their digital models, which are similar to the models that were used for model-based development. The main difference between the virtual system and the models in MBD is that the virtual system is not constant. The virtual system is capable of the perception of physical entities and can be self-adapted through statistical learning. The recent update of some industry software like MATLAB/Simulink and AVL CRUISE provides the functions to allow the model to be self-optimized with experimental data. In academia, there are also some studies on the development of virtual systems. Zhang et al. present a digital framework to represent the EV fleet behavior to optimize the charging scheduling and positioning [10]. In the research of Magargle et al., virtual vehicle braking is modeled for simulation with various dynamic factors. By this approach, the predictive maintenance information can be obtained in the field [11]. While Li et al. proposed a virtual system of EV battery packages, the battery parameters can be estimated with links to the physical entity [12].

5.2.3 Data Interface

The data interface in a digital twin refers to the method used to transfer and exchange data between the physical system and the digital twin. It enables the integration of data from various sources such as sensors, IoT devices, and other monitoring systems into the digital twin, allowing for real-time monitoring and analysis of the physical system. The data interface is responsible for collecting, processing, and transforming data into a format that is compatible with the digital twin. It also enables the digital twin to send commands or instructions back to the physical system to control or adjust its behavior. The data interface plays a critical role in ensuring the accuracy and reliability of the digital twin by ensuring that the data used in the virtual model reflects the behavior of the physical system as closely as possible.

For automotive applications, there are some existing communication protocols and data interfaces that can be used to bridge physical entities, virtual systems, and connected intelligence. One of the most used interfaces is the control area network (CAN) or CAN bus, which was developed by Bosch in the 1980s. The CAN bus enables high-speed communication between devices on a network, allowing them to exchange messages in real time. It uses a bus topology, where all devices are connected to a shared communication line, and each device can send and receive messages. However, the CAN bus usually has bandwidth limits of 1 Mbps, which limits its application in the era of connected and autonomous driving. More effective Ethernet-based methods (e.g., EtherCAT and time-sensitive networking (TSN)) are in rapid development and can support data rates of up to 10 Gbps. Ethernet-based networking also enables more advanced features such as over-the-air updates, remote diagnostics, and advanced driver assistance systems (ADAS). These functions are enabling technologies for better connectivity of the key components in an automotive digital twin system.

5.2.4 Connected Intelligence

Connected intelligence for digital twins refers to the ability to gather, analyze, and share data from various sources to create a more comprehensive and accurate digital representation of a physical system or process. It involves the integration of data from sensors, IoT devices, and other monitoring systems into the digital twin, enabling real-time monitoring and analysis of the physical system. Connected intelligence enables the digital twin to simulate the behavior of the physical system more accurately, providing insights into its performance, potential failures, and opportunities for optimization. It can also enable predictive analytics and real-time decision-making, allowing for the more efficient and effective operation of the physical system.

Connected intelligence also facilitates collaboration and knowledge-sharing across different teams and departments, enabling more effective problem-solving and decision-making. It can help break down silos and foster a more integrated approach to system design, development, and management. Connected intelligence for digital twins is a key enabler of the digital transformation of industries such as manufacturing, energy, and transportation, allowing them to operate more efficiently, reduce costs, and improve sustainability.

Based on the key milestones of AI applications in automotive systems, we can categorize the development of AI into four levels as shown in Figure 5.6. For Level I, AI models are used to model the nonlinearity of the vehicle systems, such as the engine performance [13]. The AI models are expected to assist in some R&D tasks

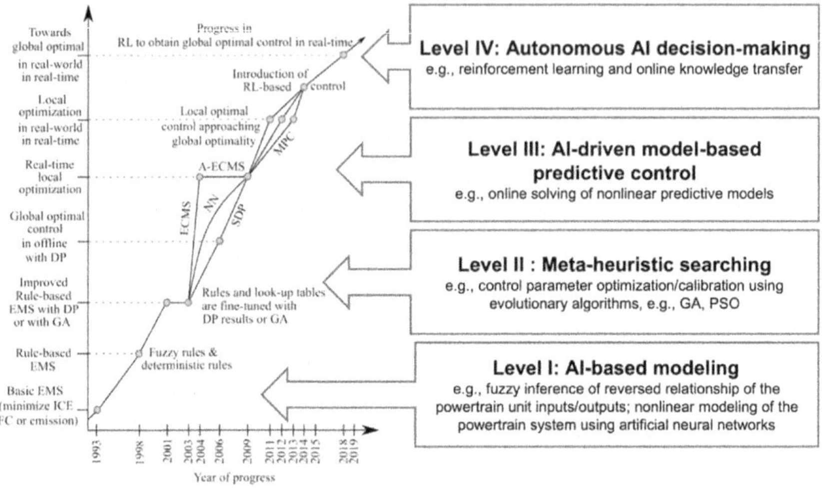

FIGURE 5.6 Stages of AI development for automotive applications.

(e.g., component sizing [14] and control calibration [15]) if they have the capability of offline optima searching (Level II). By incorporating AI-based modeling and AI-based optimization, the Level III AI models can deal with model predictive control tasks that allow the vehicle system to be optimized online [16, 17]. In Level IV, AI models would have the capability of self-learning and adaptive learning based on the recent development in the internet of vehicles and advanced algorithms (e.g., the reinforcement learning [18] and transfer learning [19]).

5.3 DT-BASED AUTOMOTIVE POWERTRAIN CONTROL OPTIMIZATION

Digital twin technology also can contribute to the development of automotive powertrains. There are several successful applications of DTs in the design, control, and maintenance of large and expensive industry products: Wang et al. developed a DT for real-time monitoring and fault diagnosis of offshore wind turbines [20]. Zaccaria et al. proposed a DT with multilayer intelligent approaches for the health management of aircraft engines [21]. Jiang et al. developed a DT with a five-dimension structure for prognostic and health management of a smart grid [22].

Currently, most DT research for vehicle applications focuses on improving model accuracy [12] and data connection [23]. Li et al. incorporate an H-infinity filter with the particle swarm optimization (PSO) algorithm to improve the model accuracy of the digital counterpart of batteries for EVs [12]. Zhou et al. developed a DT of an energy-harvesting shock absorber to optimize its design parameters using the PSO algorithm [24]. Zhou et al. proposed a human-knowledge–augmented Gaussian process regression method to build a battery DT for the state-of-health estimation [25]. It is essential for OEMs to attain robust and reliable design in the R&D stage harnessing DT because it is impossible for vehicle manufacturers to monitor and control all vehicles centrally through centralized cloud computing [26]. According to the outlook from two recent review papers for DTs [27, 28], research into DT-based robust design with self-adaption is in high demand.

5.3.1 The Digital Twin System for Optimization

The DP system for powertrain control optimization is illustrated in Figure 5.7, which consists of a physical entity (the PHEV), a digital model of the PHEV, a service model for EMS optimization, a database, and interfaces between each component. This sector provides a workflow of transferring testing data to vehicle digital twin and digital twin empowered control optimization, and the details of the digital modeling and the energy management strategy optimization will be introduced as follows.

5.3.1.1 The Physical Entity of the Vehicle

The physical entity in this study is a plug-in hybrid electric vehicle powered by a 125 kW traction motor. The primary power source of the powertrain is a 60 Ah battery pack. A 26.6 kW range-extender is the alternative power unit, and it supplies the power for battery charging. The vehicle specifications are summarized in Table 5.2.

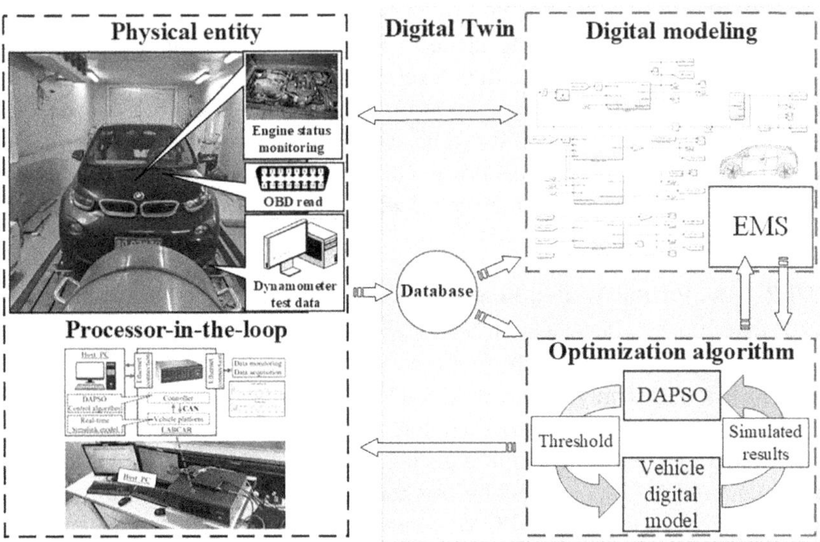

FIGURE 5.7 Digital twin–based vehicle optimization.

TABLE 5.2
The Specifications of the Vehicle

Vehicle Specifications	Parameters	Values
	Weights	1400 kg
	Wheelbase	2570 mm
Vehicle body	Length/Width/Height	3398/2040/1577 mm
	Front area	2.38 m^2
	Air drag coefficient	0.3
Tires	Size F/R	175/70 R19
Engine	Displacement	0.647 L
(DOHC- 8 valve I2)	Maximum power	25 kW
Traction motor	Maximum power	125 kW
(PM AC Synchronous)	Maximum torque	250 Nm
	Number of cells	96
Battery	Nominal system voltage	355 V
(Lithium-Ion)	Pack capacity	60 Ah
	Pack energy	22 kWh
Drivetrain (Rear wheel drive)	Final drive ratio	9.7

The power flow of the vehicle is controlled through a rule base determined by three SoC thresholds. The base control logic is that the vehicle runs as a pure EV when the battery SoC is high. When the SoC level drops below 16%, the range-extender starts and gradually increases its power output to maintain the battery SoC level. The working mode of the range-extender is determined by three different thresholds of the SoC level, namely SoC low limit, SoC medium limit, and SoC high limit, which are also defined as the SoC_1, SoC_2, and SoC_3 in the rest of this paper. When the vehicle's speed is lower than 16.1 km/h, the range-extender is completely shut down. Otherwise, the range-extender may have low speed, medium speed, and high speed with low torque or high torque, six pairs of working conditions.

5.3.1.2 The Virtual Model, Real-Time Environment, and Data Interfaces

The virtual vehicle is developed in MATLAB/Simulink, which will be running on a real-time computer to allow the real-time simulation as illustrated in Figure 5.8. The vehicle's digital model was downloaded from a host PC to the LABCAR through an Ethernet connection. Then, the algorithm was transferred into a code compiler and then implemented into the controller unit of the LABCAR for real-time validation. A CAN connection was made between the control unit and the vehicle platform so that the LABCAR could represent the CAN connection in the vehicle. In the processor-in-the-loop (PiL) test, the data was transferred between the controller and the vehicle platform in real time

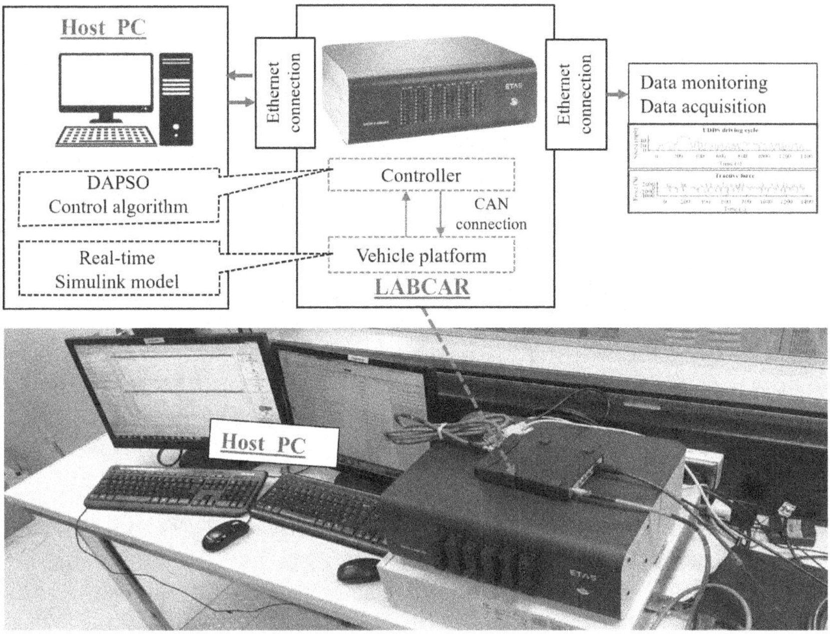

FIGURE 5.8 Running vehicle model in real-time computer.

5.3.2 Dedicated Particle Swarm Optimization for DT-Based Optimization

This section introduces the optimization problem from the mathematical modeling aspect. A dedicated adaptive particle swarm optimization (DAPSO) algorithm is proposed to deal with the DT-based optimization in this section. Distinguished from conventional particle swarm optimization, the proposed method is capable of self-adaption through interaction with the digital twin.

5.3.2.1 The Optimization Problem

Three battery SoC thresholds, SoC_1, SoC_2, and SoC_3, will be varied to achieve the minimum equivalent consumption of fuel and electricity while maintaining battery SoC. The mathematical representation of the optimization problem can be formulated as

$$
\begin{gathered}
\left[SoC_1, SoC_2, SoC_3\right] = \arg\min\left(J_{\text{com}}\right) \\
s.t.\left\{\begin{array}{c}
J_{\text{com}} = \varphi \cdot \left| \dfrac{soc_{\text{final}} - soc_{\text{std}}}{soc_{\text{ref}}} \right| + (1-\varphi) \cdot \dfrac{EC_{\text{final}}}{EC_{\text{ref}}} \\
SoC_1 < SoC_2 < SoC_3 \\
SoC_1, SoC_2, SoC_3 \in \left[SOC^-, SOC^+\right]
\end{array}\right.
\end{gathered}
\tag{5.1}
$$

where J_{com} is the objective function; SoC_{final} and EC_{final} are the final SoC and equivalent consumption at the end of a given driving cycle, respectively; φ is the weight factor to balance the preference of cycle-end SoC control against the cycle's equivalent consumptions. The SoC_{std} is a desired final SoC level taken from the rule-based control. The SoC_{ref} and EC_{ref} are normalization factors, where $SoC_{\text{ref}} = SoC^{+} - SoC^{-}$, the EC_{ref} is the equivalent fuel consumption under the conventional rule-based control. The SoC = 13.5% and SoC^{+} = 60% are battery SoC's lower and higher boundaries, respectively. The equivalent consumption of fuel and electricity can be calculated by

$$\left.\begin{aligned} EC_{\text{final}} &= \int_0^{T_t} (\dot{m}_{\text{ice}}\left(P_{\text{ice}}(t)\right) + \dot{m}_{\text{em}}\left(P_{\text{em}}(t)\right))dt \\ \left[P_{\text{ice}}(t), P_{\text{em}}(t)\right] &= \mathbf{DT}\left(SoC_1, SoC_2, SoC_3\right) \end{aligned}\right\} \tag{5.2}$$

Where $\dot{m}_{ice}$ is the fuel consumption rate in g/s; P_{ice} and P_{em} are the power demand of the engine and the traction motor in kW, respectively. $P_{\text{ice}}(t)$ and $P_{\text{em}}(t)$ can be obtained with the DT with different settings of the SoC thresholds. $\dot{m}_{em}$ is the equivalent fuel consumption transferred from electricity in g/s, it can be calculated by

$$\left.\begin{aligned} \dot{m}_{\text{em}}\left(P_{\text{em}}(t)\right) &= \gamma \cdot s_{\text{dis}} \frac{1}{\eta_{\text{batt}}\eta_{\text{em}}} \frac{P_{\text{em}}(t)}{H_{\text{LVH}}} \\ &\quad + (1-\gamma) \cdot s_{\text{chg}} \cdot \eta_{\text{batt}}\eta_{\text{em}} \frac{P_{\text{em}}(t)}{H_{\text{LVH}}} \\ \gamma &= \frac{1 + \text{sign}\left(P_{\text{em}}(t)\right)}{2} \end{aligned}\right\} \tag{5.3}$$

Where H_{LVH} is the fuel's lower heating value; S_{dis} and S_{chg} are the equivalent factors for battery discharging and charging, respectively; and $S_{\text{dis}} = S_{\text{chg}} = 2.5$ are predefined; η_{batt} and η_{em} are the energy efficiency for battery and motor respectively.

5.3.2.2 The Interactive Optimization Processes

Like the PSO algorithm, the DAPSO uses computer agents (particles) to retrieve the optimal objective function values and return the best particle position. Initially, the DAPSO randomly allocates p within a search space defined in Eq. (5.1). The position of each particle is $x_i^t = \left[SoC_1^{i,t}, SoC_2^{i,t}, SoC_3^{i,t}\right]^T$, where i = 1,2,3, . . .;p is the index number for individual particles; k= 1,2,3, . . .; t is the number of iterations; and τ is the maximum number of iterations. Based on the position information, each particle will compute its objective function using the DT, and the best positions will be returned to update the particles' positions following the mechanism of the particle swarm optimization

$$\left.\begin{aligned} v_i^{k+1} &= \omega \cdot v_i^k + \alpha \cdot \in_1 \left[g^* - x_i^k \right] + \beta \cdot \in_2 \left[x_i^{*(k)} - x_i^k \right] \\ x_i^{k+1} &= x_i^k + v_i^{k+1} \end{aligned}\right\} \tag{5.4}$$

Where v_i^t and x_i^t represent the velocity and position of the ith particle at the tth status; g^* represents the global best of the particles among all the results; $x_i^{*(t)}$ is the local best of the current swarm at the tth iteration; $\in_1$ and $\in_2$ are two random numbers between 0 and 1; α and β are two attraction factors, and $\alpha = \beta = 2$ satisfy the general cases; ω is the inertia factor which is used to balance the exploration and exploitation, it is normally drawn from 0.6~0.9. We proposed an initial set of 0.65 for this case.

5.3.2.3 Adaptive Control Through Mode-Switching

The state machine regulates settings of the DAPSO based on three modes: a) exploration, b) exploitation, and c) termination. The main difference between DAPSO and conventional PSO is that DAPSO can explore the best swarm moving strategy during the heuristic optima search process through a new exploration and exploitation mechanism. In the exploration mode, DAPSO will attempt new search areas while the individuals of the DAPSO will follow the inertia to move in the exploitation model. The detailed settings for the three modes are illustrated as follows.

5.3.2.3.1 Exploitation Mode

The algorithm would switch to the exploitation model if the criteria defined in threshold 1 are met. Threshold 1 evaluates the battery state of charge and equivalent consumption by:

$$\frac{\left| soc_{\text{opt}}^k - soc_{\text{ref}} \right|}{soc_{\text{ref}}} \leq 5\% \text{ and } EC_{\text{opt}}^k \leq EC_{\text{ref}} \tag{5.5}$$

Where soc_{opt}^k and EC_{opt}^k are battery state of charge and equivalent energy consumption obtained by the DT with the best particle position at iteration t; SoC_{ref} is the reference battery SoC level, which is 16% in this example; and EC_{ref} is the reference equivalent consumption for each cycle.

The fulfillment of both criteria indicates the present process is heading in the desired direction, so the exploitation continues with the ω remaining the same. The particles will update their velocity and position.

5.3.2.3.2 Exploration Mode

To prevent the algorithm's entrapment in local optima, the DAPSO implements an exploration mechanism based on the differences of the cost function value in five neighboring iterations at an early stage (before the first 20 iterations),

$$\begin{aligned} \in = &\left| J_{com}^{\ k} - J_{com}^{\ k-1} \right| + \left| J_{com}^{\ k-1} - J_{com}^{\ k-2} \right| + \ldots \\ &+ \left| J_{com}^{\ k-4} - J_{com}^{\ k-5} \right| \end{aligned} \tag{5.6}$$

If $\in = 0$, the exploration will be activated, and the inertia factor (at the first iteration after reset) will be reset to its initial value as

$$\omega^1 = 0.3 \tag{5.7}$$

otherwise, the DAPSO will use a logistic chaos map to generate the inertia factor as follows

$$\omega^k - 2.6 \cdot \omega^{k-1} \cdot \left(1 - \omega^{k-1}\right) \tag{5.8}$$

5.3.2.3.3 Termination Mode

Conventionally, the termination conditions were mainly categorized by four types of criteria in Table 5.3 following the work of Zielinski et al. [29]. The algorithm may stop at a certain number of iterations (θ_{iter}) regardless of the optimized results, also if the improvement of the objective function does not reach a given threshold (θ_{imp}) within a certain range of iterations or if the movement (dimensionless quantity in search spaces) of single particles is less than a certain threshold θ_{mov}, the algorithm will be terminated. The distances of each particle to the instant best particle (which represents the position of the best solution) are evaluated, and this indicates how close the particles gather towards the optimal position; if the average distance is below a certain threshold θ_{dis}, the algorithm will be terminated.

In this example, all of four criteria are considered. The improvement-based termination condition is applied to decide the exploration and exploitation actions, where the $\theta_{\text{imp}} = \in$. The rest of the three criteria are the termination threshold (threshold 3), the algorithm will terminate when the termination threshold (threshold 3) is fulfilled, as indicated by the achievement of any of the following three conditions

$$\left.\begin{array}{l} k \geq \theta_{\text{iter}} \\ \bar{P}_{\text{mov}} \leq \theta_{\text{mov}} \\ \bar{P}_{\text{dis}} \leq \theta_{\text{dis}} \end{array}\right\} \tag{5.9}$$

TABLE 5.3
Different Termination Conditions of PSO

Termination conditions	Objects of consideration	Thresholds
Maximum-iterations	Number of iterations	θ_{iter}
Improvement-based	The improvement of objective function between iterations	θ_{imp}
Movement-based	The movement of single particles	θ_{mov}
Distribution-based	The average distance of the swarm	θ_{dis}

where the k is the iteration number, the $\overline{P}_{mov}$ is the average particles' movement, the is the $\overline{P}_{dis}$ average distance between the position of each individual particle and the position of the best particle. θ_{iter}, θ_{mov}, and θ_{dis} are their corresponding thresholds. Mathematically, $\overline{P}_{mov}$ and $\overline{P}_{ids}$ are calculated as:

$$\left.\begin{aligned} P_{mov} &= \sum\nolimits_{i=1}^{N} (v_i^{k+1} - v_i^k) \\ P_{dis} &= \sum\nolimits_{i=1}^{N} (x_i^k - x_i^{*(k)}) \\ \overline{P}_{mov} &= \frac{P_{mov}}{N} \\ \overline{P}_{dis} &= \frac{P_{dis}}{N} \end{aligned}\right\} \tag{5.10}$$

where the v_i^k, x_i^k and $x_i^{*(k)}$ are from Eq. (5.4), the N is the number of particles in the swarm. the P_{mov} is the velocity increment of each particle from the k-th iteration to the $(k+1)$-th iteration. Its average value $\overline{P}_{mov}$ indicates the magnitude of particle movement. The P_{dis} is the distance between positions of each particle (x_i^k) with the position of the best particle $(x_i^{*(k)})$ at the k-th iteration. Similarly, the $\overline{P}_{dis}$ indicates the magnitude of distribution for particles near the best solution.

5.3.3 Optimization Performance

Optimality, computational efficiency, and trustworthiness are the three most important performance indicators for evaluating AI algorithms [30]. A comprehensive scoring system is established in this paper by calculating the weighted sum value of the functions affecting these three factors. To measure the performance of the DT system, the overall score $S$is defined based on the weighted sum of three components, which evaluate optimality, computation efficiency, and success rate, respectively, as follows

$$S = \mathrm{w}_1 \underbrace{(1 - J_{\mathrm{com}})}_{optimality} + \mathrm{w}_2 \underbrace{\left(1 - \frac{T_{\mathrm{opt}} - T_{\mathrm{ave}}}{T_{\mathrm{ave}}}\right)}_{Computation\ efficiency} + \mathrm{w}_3 \underbrace{\left(\frac{\aleph_{opt}}{\aleph_{total}}\right)}_{Success\ rate} \tag{5.11}$$

where, J_{com} is the measure of vehicle performance (i.e., the objective function defined in Eq. 5.1), which is the core of the component representing the optimality of the DT-based system. In the second component (representing the computation efficiency), T_{opt} is the time consumption of a single optimization, and T_{ave} is the average time consumption of all studied PSO algorithms. The smaller T_{opt} is, the higher computational efficiency will be. In the third component (representing the success rate), $\aleph_{opt}$ is the number accounted for the success and $\aleph_{total}$ is the total number of the trials ($\aleph_{total}$ = 20 in this example). The ratio of $\aleph_{opt}$ to $\aleph_{total}$ measures the robustness of the DT-based system; and w_1, w_2, and w_3 are weighting factors for optimality, computation efficiency, and success rate, respectively. This example demonstrates how DAPSO can find the maximum value of the overall score function, and the setting of weighting factors, $w_1 = 0.5$, $w_2 = 0.25$, and $w_3 = 0.25$, are selected as an example. The setting of the weighting factors can be adjusted for different optimization preferences.

5.3.4 Local Optimization Performance in Learning Cycles

The optimization results obtained under the local learning cycle are shown in Table 5.4, with the best optimization results underlined. In Table 5.4, the system optimality is evaluated based on the objective function (defined in Eq. 5.1) value that measures the vehicle performance by considering: 1) the difference between the final SoC and target SoC and 2) the equivalent fuel consumption (EC) improvement rate. The better the vehicle performance is, the higher the optimality value

TABLE 5.4
Local Learning Results

		Single scoring targets					
Driving cycle	**PSO type**	**Optimality**	**Iteration of convergence**	**Time spends (s)**	**Computational efficiency**	**Success rate (%)**	**Overall Score**
UDDS	PSO1	0.606	50	989.5	0.77	<u>100</u>	0.745
	PSO2	0.586	36	724.6	1.10	65	0.730
	PSO3	0.610	<u>27</u>	<u>548.5</u>	<u>1.32</u>	65	0.797
	PSO4	<u>0.617</u>	47	947.7	0.82	80	0.713
	DAPSO	0.610	30	611.0	1.24	<u>100</u>	<u>0.864</u>
Highway	PSO1	<u>0.506</u>	50	650.9	1.43	40	0.709
	PSO2	−0.020	103	1343.8	0.82	70	0.369
	PSO3	−0.074	117	1533.7	0.65	60	0.275
	PSO4	0.349	77	1010.8	1.11	85	0.664
	DAPSO	0.419	<u>34</u>	<u>445.4</u>	<u>1.61</u>	<u>90</u>	<u>0.836</u>
US06	PSO1	0.497	50	662.8	1.58	65	0.806
	PSO2	0.485	117	1554.9	1.01	70	0.671
	PSO3	0.500	253	3364.6	−0.13	60	0.366
	PSO4	0.162	54	724.6	1.54	50	0.591
	DAPSO	<u>0.543</u>	<u>31</u>	<u>412.7</u>	<u>1.74</u>	<u>80</u>	<u>0.906</u>
NEDC	PSO1	0.019	50	866.7	1.37	40	0.453
	PSO2	0.392	59	1037.3	1.25	80	0.708
	PSO3	0.487	153	2653.8	0.08	65	0.426
	PSO4	0.111	55	969.5	1.30	80	0.580
	DAPSO	<u>0.575</u>	<u>39</u>	<u>676.9</u>	<u>1.51</u>	<u>90</u>	<u>0.890</u>

will be. The computation efficiency is evaluated based on the second component of Eq. 5.11; the faster the calculation is, the higher the computation efficiency will be. The number of iterations for convergence are compared in Table 5.4. The overall score, S, gives the overall performance of the DT system by considering its optimality, calculation efficiency, and success rate, simultaneously.

For the UDDS cycle, PSO4 generates the highest optimality of 0.617. PSO3 is the least time-consuming at 548.5 s, and the DAPSO and the PSO1 have the highest success rate of 100%. For the Highway driving cycle, the PSO1 has the best optimality but only has a success rate of 40%. The DAPSO has the highest success rate of 90%, with the smallest average time consumption of 445.4 s. For the US06 cycle learning process, the DAPSO attains the highest optimality with 0.543, the least time consumption of 412.7s, and the highest success rate of 80%. The local optimization results for the NEDC cycle are shown in the fourth part of Table 5.4. Again, the DAPSO has the best optimality of 0.575, the smallest average time consumption of 676.9 s, and the highest success rate of 90%. The overall scores of the DAPSO and PSO1–4 are shown in Table 5.4, and it indicates that the DAPSO has the highest overall scores, which are 0.864, 0.836, 0.906, and 0.890 for UDDS, Highway, US06, and NEDC cycles, respectively.

5.3.5 Cross-Validation in Standard Cycles

After the local learning process, the optimal local strategy obtained under each driving cycle is validated by the other three driving cycles to study the global performance. Through the cross-validation, the best thresholds settings of SoC_1, SoC_2 and SoC_3 are 9.65%, 14.80%, and 16.02%, respectively, under the Highway cycle. The overall scores obtained in both optimization and validation cycles are compared in Figure 5.9. The strategies optimized by the DAPSO and PSO1 have similar performance in optimization and validation cycles. The average optimality is calculated (with Eq. 5.11) to indicate the robustness of the heatmaps and presented in Table 5.5, where an SoC-corrected fuel consumption control is evaluated. It can be seen that the DAPSO is distinguished from its counterparts by achieving the highest value of optimality (0.6180). To demonstrate the effectiveness of the proposed method, the performances (fuel consumption and battery SoC trajectory) of

TABLE 5.5
The Average Optimality of Different EMS Control Settings

PSO type	Optimality values
PSO1	0.6004
PSO2	0.5580
PSO3	0.4983
PSO4	0.5948
DAPSO	**0.6180**

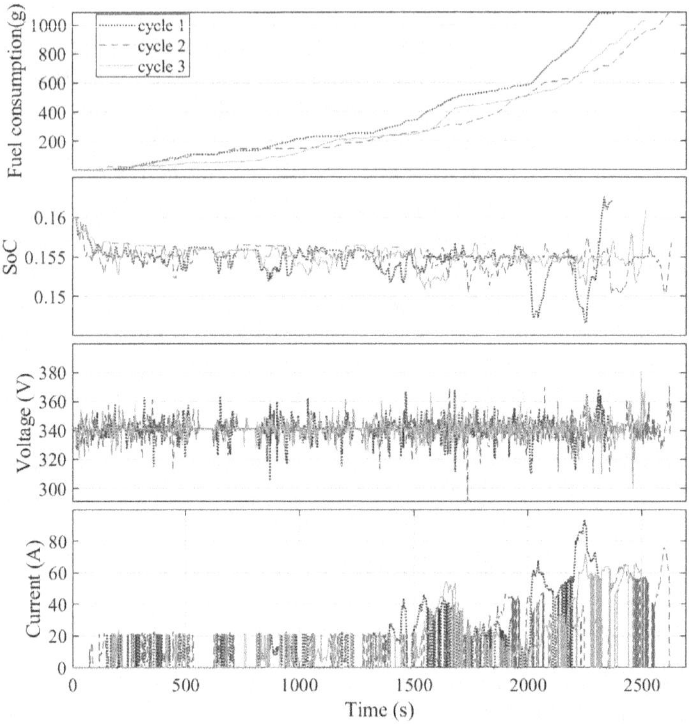

FIGURE 5.9 The PiL results for the three validation cycles: (a) the fuel consumption of each cycle, (b) the SoC level, (c) the battery voltage variation, (d) the current generation of the range-extender.

the PHEV optimized by DAPSO under the Highway cycle are compared with the theoretical best performance (obtained by dynamic programming, DP) and the baseline rule-based method. According to the results, the proposed DAPSO algorithm is shown effective in the optimization of PHEV control performances: More than 11.4% of fuel consumption is saved compared to the rule-based method. With the capability of maintaining similar SoC levels with DP results (error within 0.63%), the PHEV optimized by DAPSO only consumes 6% more fuel than the theatrical best performance.

5.3.6 Global Performance in Simulated Real-World Driving

The EMSs with thresholds optimized by DAPSO based on the cross-validation under the standard cycles are implemented on the vehicle running under the random driving cycles and the results are compared in Table 5.6. The effectiveness of the proposed DAPSO is shown via the optimality calculations of the PHEV. The optimality scores are calculated based on the final SoC control and the fuel

TABLE 5.6
Global Optimization Comparison

	DAPSO optimized thresholds			
Optimality	**UDDS**	**Highway**	**US06**	**NEDC**
simulated real-world driving cycle 1	0.5723	0.5935	0.5556	0.5735
simulated real-world driving cycle 2	0.5812	0.6001	0.5492	0.5794
simulated real-world driving cycle 3	0.5875	0.5925	0.5562	0.5726
Average optimality score	**0.5803**	**0.5954**	**0.5537**	**0.5752**

economy as mentioned in Eqs. (5.1) and (5.11), which indicate the performances of the EMS optimized by the DAPSO. The EMS optimized based on the Highway cycle has the highest average optimality value of 0.5954, which is 2.6% higher than the average value of the UDDS (0.5803), 7.5% higher than the value of the US06 cycle (0.5537), and 3.5% higher than the average value of the NEDC cycle (0.5752). It can be obtained that the EMS threshold settings generated by DAPSO based on the Highway cycle hold the highest robustness in general driving conditions.

Processor-in-the-loop (PiL) tests were conducted to validate real-time control functionalities, and the vehicle's real-time performances under the three selected driving cycles are illustrated in Figure 5.8. The initial battery SoC was set to 16% for a charging sustaining scenario for the vehicle with high electricity demand. The random driving cycles 2 and 3 are gentler than random cycle 1; thus, less arduous working conditions of the range-extender are generated in these two cycles, which caused more minor variations for their SoC. It can also be obtained that when SoC drops significantly in the period of 2000s to 2600s, the EMS is shown to maintain the battery voltage in a reasonable range (4.5%). According to Figure 5.9 (d) and (c), it can be obtained that the variations of SoC and voltage are mild in most urban and rural driving conditions, indicating that the battery works primarily in gentle scenarios, which helps mitigate the degradation of battery [31].

5.4 SUMMARY AND OUTLOOK

The advancement of transportation electrification has led to the emergence of advanced control and virtual engineering. One notable outcome is the ability to update vehicle control software through over-the-air (OTA) service. Automotive original equipment manufacturers (OEMs) now have access to users' extensive data, necessitating the use of digital twin technology as an effective tool for processing this data and optimizing vehicle design. This chapter provides an overview of the significant changes and challenges in the automotive industry, defines the concept of the automotive digital twin, and showcases the advantages of using digital twin technology in developing vehicle control software by introducing a dedicated optimization algorithm for digital twin–based control optimization.

Based on the authors' observations, it is anticipated that in the future, digital twin technology will deeply integrate with e-mobility systems and artificial intelligence. Stages I and II AI methods have already demonstrated their strong capabilities in handling offline optimization tasks in automotive product development, such as system integration, component sizing, and control parameter calibration. These AI methods are on the verge of commercialization, with DT-based optimization using particle swarm optimization (PSO) being a notable example. It is conceivable that higher-level AI methods, such as reinforcement learning and transfer learning, will further enhance the big data processing capability of digital twins, consequently driving product evolution, reducing research and development costs and time, and significantly influencing the future of the automotive industry.

REFERENCES

1. International Energy Agency, "Global EV outlook 2021—Accelerating ambitions despite the pandemic," 2021 [Online]. Available: https://iea.blob.core.windows.net/assets/ed5f4484-f556-4110-8c5c-4ede8bcba637/GlobalEVOutlook2021.pdf
2. Advanced Propulsion Centre, "Light duty vehicle below 3.5t roadmap 2020," February 2021 [Online]. Available: www.apcuk.co.uk/app/uploads/2021/02/Exec-summary-Product-Roadmap-LDV-final.pdf
3. J. Pavlovic, A. Marotta, and B. Ciuffo, "CO2 emissions and energy demands of vehicles tested under the NEDC and the new WLTP type approval test procedures," *Appl. Energy*, vol. 177, no. 2016, pp. 661–670, 2016, doi: 10.1016/j.apenergy.2016.05.110.
4. E. G. Giakoumis, "Driving and engine cycles," *Driv. Engine Cycles*, pp. 1–408, 2016, doi: 10.1007/978-3-319-49034-2.
5. M. H. N. Tayarani, X. Yao, and H. Xu, "Meta-heuristic algorithms in car engine design: A literature survey," *IEEE Trans. Evol. Comput.*, vol. 19, no. 5, pp. 609–629, 2015, doi: 10.1109/TEVC.2014.2355174.
6. F. Tao, "Digital twin and its potential application exploration," *Comput. Integr. Manuf. Syst.*, vol. 24, 2018.
7. A. Parrott, and L. Warshaw, *Industry 4.0 and the digital twin*. Deloitte University Press, pp. 1–17, 2017.
8. A. Rassolkin, T. Vaimann, A. Kallaste, and V. Kuts, "Digital twin for propulsion drive of autonomous electric vehicle," *2019 IEEE 60th Annu. Int. Sci. Conf. Power Electr. Eng. Riga Tech. Univ. RTUCON 2019—Proc.*, pp. 31–34, 2019, doi: 10.1109/RTUCON48111.2019.8982326.
9. M. Ruba *et al.*, "Digital twin real-time FPGA implementation for light electric vehicle propulsion system using EMR organization," *2019 IEEE Veh. Power Propuls. Conf. VPPC 2019—Proc.*, 2019, doi: 10.1109/VPPC46532.2019.8952428.
10. T. Zhang, X. Liu, Z. Luo, F. Dong, and Y. Jiang, "Time series behavior modeling with digital twin for internet of vehicles," *Eurasip J. Wirel. Commun. Netw.*, vol. 2019, no. 1, pp. 1–11, 2019, doi: 10.1186/s13638-019-1589-8.
11. R. Magargle *et al.*, "A simulation-based digital twin for model-driven health monitoring and predictive maintenance of an automotive braking system," *Proc. 12th Int. Model. Conf. Prague, Czech Republic, May 15–17, 2017*, vol. 132, pp. 35–46, 2017, doi: 10.3384/ecp1713235.

12. W. Li *et al.*, "Digital twin for battery systems: Cloud battery management system with online state-of-charge and state-of-health estimation," *J. Energy Storage*, vol. 30, p. 101557, June 2020, doi: 10.1016/j.est.2020.101557.
13. J. Li *et al.*, "Fuzzy-tree-constructed data-efficient modelling methodology for volumetric efficiency of dedicated hybrid engines," *Appl. Energy*, vol. 310, December 2021–2022, doi: 10.1016/j.apenergy.2022.118534.
14. Q. Zhou *et al.*, "Intelligent sizing of a series hybrid electric power-train system based on chaos-enhanced accelerated particle swarm optimization," *Appl. Energy*, vol. 189, pp. 588–601, 2017, doi: 10.1016/j.apenergy.2016.12.074.
15. Z. Li, Q. Zhou, Y. Zhang, J. Li, and H. Xu, "Enhanced intelligent proportional-integral-like fuzzy knowledge-based controller using chaos-enhanced accelerated particle swarm optimization algorithm for transient calibration of air-fuel ratio control system," *Proc. Inst. Mech. Eng. Part D J. Automob. Eng.*, 2019, doi: 10.1177/0954407019862079.
16. Q. Zhou *et al.*, "Cyber-physical energy-saving control for hybrid aircraft-towing tractor based on online swarm intelligent programming," *IEEE Trans. Ind. Informatics*, vol. 14, no. 9, pp. 4149–4158, 2018, doi: 10.1109/TII.2017.2781230.
17. J. Li *et al.*, "Dual-loop online intelligent programming for driver-oriented predict energy management of plug-in hybrid electric vehicles," *Appl. Energy*, vol. 253, 2019.
18. T. Liu, X. Hu, W. Hu, and Y. Zou, "A heuristic planning reinforcement learning-based energy management for power-split plug-in hybrid electric vehicles," *IEEE Trans. Ind. Informatics*, vol. 15, no. 12, pp. 6436–6445, 2019, doi: 10.1109/TII.2019.2903098.
19. Q. Zhou *et al.*, "Transferable representation modelling for real-time energy management of the plug-in hybrid vehicle based on k-fold fuzzy learning and Gaussian process regression," *Appl. Energy*, vol. 305, p. 117853, 2022, doi: 10.1016/j.apenergy.2021.117853.
20. M. Wang *et al.*, "Recent progress on reliability analysis of offshore wind turbine support structures considering digital twin solutions," *Ocean Eng.*, vol. 232, p. 109168, January 2021, doi: 10.1016/j.oceaneng.2021.109168.
21. V. Zaccaria, M. Stenfelt, I. Aslanidou, and K. G. Kyprianidis, "Fleet monitoring and diagnostics framework based on digital twin of aero-engines," *Proc. ASME Turbo Expo*, vol. 6, pp. 1–10, 2018, doi: 10.1115/GT2018-76414.
22. Z. Jiang, H. Lv, Y. Li, and Y. Guo, "A novel application architecture of digital twin in smart grid," *J. Ambient Intell. Humaniz. Comput.*, 2021, doi: 10.1007/s12652-021-03329-z.
23. G. Bhatti, H. Mohan, and R. Raja Singh, "Towards the future of ssmart electric vehicles: Digital twin technology," *Renew. Sustain. Energy Rev.*, vol. 141, p. 110801, January 2021, doi: 10.1016/j.rser.2021.110801.
24. Q. Zhou *et al.*, "Global optimization of the hydraulic-electromagnetic energy-harvesting shock absorber for road vehicles with human-knowledge-integrated particle swarm optimization scheme," *IEEE/ASME Trans. Mechatronics*, vol. 4435, no. c, p. 1, 2021, doi: 10.1109/tmech.2021.3055815.
25. Q. Zhou *et al.*, "Human-knowledge-augmented Gaussian process regression for state-of-health prediction of lithium-ion batteries with charging curves," *J. Electrochem. Energy Convers. Storage*, vol. 18, pp. 1–10, August 2021, doi: 10.1115/1.4050798.
26. M. Liu, S. Fang, H. Dong, and C. Xu, "Review of digital twin about concepts, technologies, and industrial applications," *J. Manuf. Syst.*, vol. 58, no. PB, pp. 346–361, 2021, doi: 10.1016/j.jmsy.2020.06.017.

27. A. Rasheed, O. San, and T. Kvamsdal, "Digital twin: Values, challenges and enablers from a modeling perspective," *IEEE Access*, vol. 8, pp. 21980–22012, 2020, doi: 10.1109/ACCESS.2020.2970143.
28. A. Fuller, Z. Fan, C. Day, and C. Barlow, "Digital twin: Enabling technologies, challenges and open research," *IEEE Access*, vol. 8, pp. 108952–108971, 2020, doi: 10.1109/ACCESS.2020.2998358.
29. K. Zielinski, and R. Laur, "Distribution," *Inform.*, vol. 31, no. 1, pp. 51–59, 2007, doi: 10.31449/inf.v31i1.127.
30. W. Pedrycz, A. Sillitti, and G. Succi, "Computational intelligence: An introduction," *Stud. Comput. Intell.*, vol. 617. pp. 13–31, 2016, doi: 10.1007/978-3-319-25964-2_2.
31. G. Suri, and S. Onori, "A control-oriented cycle-life model for hybrid electric vehicle lithium-ion batteries," *Energy*, vol. 96, pp. 644–653, 2016, doi: 10.1016/j.energy.2015.11.075.

Part II

Design, Management, and Control of the Powertrain of Electric Vehicles

6 Topologies Design and Component Sizing of Electrified Powertrains

Weichao Zhuang, Haoxuan Dong and Fei Ju

6.1 INTRODUCTION

The attention towards environmental sustainability has undergone a great increase in recent years. The governments worldwide legislate stricter and stricter fuel economy and CO_2 emission regulations since the transportation sector is one of the largest contributors to the greenhouse gas emissions. To meet the regulations, automakers and researchers around the world developed eco-friendly vehicle techniques to reduce emissions over the past decades, including downsized engines, electrified powertrain, lower rolling resistance tires and the use of lightweight materials. Among them, the electrified powertrain is recognized as the most promising and effective approach.

Powertrain electrification refers to the process of replacing vehicle components that operate on a conventional energy source with components that operate on electricity, which has been one of the biggest trends in the automotive industry today. Despite the benefits of electric powertrains, electrification has been slower than expected due to limited driving range, long charging time, and inadequate charging infrastructure. Recent developments in power electronics and especially in lithium-based battery technology have overcome most of the technical challenges, enabling the large-scale promotion of pure electric vehicles (PEV).

In addition to the pure electric-drive powertrain, hybridization is considered a viable step towards powertrain electrification. The idea of utilizing a hybrid powertrain dates back to 1898, when Ferdinand Porsche built his first car, the Lohner Electric Chaise, which was powered by both an internal combustion engine (ICE) and an electric motor. The main purpose of the hybrid powertrain in the early stage was to improve the launching performance by using the electric machine to assist the ICE. Owing to the cost and performance constraints of battery packs, the hybrid design had not been accepted by the retail market until decades later, in the late 1990s. In 1997, the Toyota Prius was introduced with two power sources: a gasoline engine and a battery pack. This model rapidly became successful owing to its significant fuel-saving benefits. Since then, numerous hybrid electric vehicles (HEV) have been launched such as the Honda Insight in 1999 and the Ford

DOI: 10.1201/9781003302827-8

Escape Hybrid in 2000. The additional power source allows for greater flexibility in engine use while meeting the requirement of the driver's demand. Moreover, supernumerary techniques such as regenerative braking and engine shut-down provide alternative methods for achieving better fuel economy and emissions reduction.

In this context, electrified powertrain refers to the use of electric power to operate a vehicle with or without electrical energy storage. Therefore, it covers full electric and hybrid electric powertrains.

6.1.1 Powertrain Electrification Level

Powertrain electrification level that is normally indicated by battery voltage, stored energy, and power determines the capacity of the electric path. In general, the electrified powertrain can be divided into six types according to the electrification level: micro HEV, mild HEV, full HEV, plug-in HEV (PHEV), extended range EV (ER-EV) as well as pure EV (PEV). Table 6.1 shows their main distinctions.

Micro HEV is the simplest and most economic hybrid solution, which usually contains an electric motor often in the form of a small integrated or belted alternator/starter. This electric motor is not involved to drive the vehicle when running but is used to shut down the engine at a complete vehicle stop, and then restart when the brake pedal is released. Mild HEV uses a larger electric motor, which enables torque assistance and regenerative braking to achieve better fuel savings. In addition, mild HEV is typically featured in high-voltage electrical systems (e.g., 48 V or 90 V). Full HEV has larger batteries and stronger electric motors than those in micro and mild hybrids, which are often high in cost. They enable the usage of the purely electric mode for a short duration in city driving.

TABLE 6.1
Comparison of Micro, Mild, Full, and PHEV

Function or component parameters	Types					
	Micro	Mild	Full	PHEV	ER-EV	PEV
Idle stop/start	◆	◆	◆	◆	◆	-
Electric torque assistance		◆	◆	◆	-	-
Energy recuperation		◆	◆	◆	◆	◆
Electric drive			◆	◆	◆	◆
Battery charging (during driving)			◆	◆	◆	-
Battery charging (from the grid)				◆	◆	◆
Battery voltage (V)	12	48 ~ 160	200 ~ 300	300 ~ 400	>300	>300
Electric machine power (kW)	2 ~ 3	10 ~ 15	30 ~ 50	>60	>60	>60
Pure electric driving range (km)	0	0	5 ~ 10	>10	>200	>200
CO_2 estimated benefit	5 ~ 6 %	7 ~ 12 %	15 ~ 20 %	>20 %	>20 %	>20 %

Compared to these three HEV types without the charger, another three types (PHEV, ER-EV, and PEV) have larger battery packages with normally over 300 V voltage to store electricity from the grid. This allows ER-EV and PHEV to be operated in PEV mode for longer periods than full HEV. The former generally uses a series configuration, in which the gasoline engine only generates electricity and the electric machine drives the vehicle; an example is the BMW i3 with a range-extender. The engine is not fired until the battery is depleted. In the latter, the engine is usually engaged to directly power the vehicle. An electric machine acts as a motor/generator based on the driving demand and battery state of charge level, such as that installed in the Chevrolet Volt. PEV is exclusively driven by the electric motor that is powered by electricity stored in the battery. Generally, higher electrification levels can lead to bigger fuel economy benefits.

6.1.2 Chapter Organization

This chapter provides a comprehensive review of topologies design and component sizing of electrified powertrains, which is organized as follows. We start by discussing the characteristic of the electrified powertrain, then show the powertrain architectures of PEV and HEV, respectively. Next we explain the hybrid powertrain system component sizing method followed by an exploration of HEV modeling and configuration. Finally, we discuss the challenges and future works and conclude.

6.2 PEV POWERTRAIN ARCHITECTURES

The main components of an PEV powertrain are the motor, gearbox, clutch, and differential. These components may be combined in many ways to provide various PEV powertrain topologies. The vehicle dynamic, energy efficiency, and price of PEV are significantly influenced by the design of their powertrain architectures. Depending on how the motors are installed on the vehicle chassis, there are two categories of powertrain designs for PEV: centralized drive and dispersed drive.

6.2.1 PEV with Integrated Drive Powertrain

The first electric vehicle was created in 1880, but advancement in electric vehicle technology was modest until 1973, and ICE vehicles dominated. People were influenced by internal combustion engine vehicles and only explored replacing the ICE with an electric motor without increasing the study on the drive structure, when PEV were once again extensively concerned with the energy crisis and environmental pollution. The electric vehicle with a centralized drive powertrain was developed as a result of this concept. The centralized drive powertrain design is similar to that of an internal combustion engine vehicle, except instead of the ICE, an electric motor is used. Through the gearbox and differential, the power output must be delivered to the wheel end. The rich technical basis of internal combustion engine vehicles has accelerated the development of the centralized drive in PEV, and automakers have invested much in safety and driving performance research.

The architecture of a centralized drive powertrain can be divided into three categories, as shown in Fig. 6.1. The first is the clutch-equipped gearbox (Fig. 6.1(a)), which normally has two to three gears and interrupts power while reducing shift stress during gear changes. The clutch is eliminated in the second kind (Fig. 6.1(b)), and the gearbox with a fixed gear ratio connects the electric motor directly. This architecture allows for additional vehicle interior space by lowering the mass of the gearbox as well as the size of the transmission device. In the third kind (Fig. 6.1(c)), the motor, gearbox, and differential are all united. The left and right half-shafts, respectively, drive the matching side wheels.

Although the centralized drive powertrain offers features such as a simple structure and technological maturity, making it convenient for PEV deployment, there are certain drawbacks, including a huge powerplant, limited vehicle interior space, low transmission efficiency, and complex management. Furthermore, its complicated chassis construction and long mechanical connections significantly impair power responsiveness and energy-efficiency. The inherent difficulties of high size and low efficiency of the transmission system in internal combustion engine vehicles have not been overcome by the centralized drive powertrain design; therefore, the challenge of optimizing and updating the powertrain architecture of PEV remains unsolved.

6.2.2 PEV with Distributed Drive Powertrain

As more people became aware of PEV, research into this new form of vehicle grew, expanding from the original engine substitution to the updating and optimization of the powertrain structure. PEV drive structures have begun to shift to a distributed drive with various drive actuation systems. Researchers are attempting to perfect distributed drive architectures in order to fully profit from the advantages of electric drive systems.

As early as 1900, the German automotive expert Ferdinand Porsche had already developed the concept of a distributed drive for PEV. The traditional form of a single power source driving a single or dual axle is transformed into a power configuration where each wheel is powered by an independent power source by distributed drive powertrain, which breaks the chassis structure of the centralized drive. PEV have two types of distributed drive powertrains: in-wheel motor drive and wheel-side motor drive, as shown in Fig. 6.2. To provide distributed drive, the wheel-side motor drive places the motor on the side of the wheel and links it to the wheel through the transmission assembly (Fig. 6.2(a)). Although the drive shaft and differential are all reduced, the multi-stage reducer assembly is kept, and the hysteresis and torsional vibration issues produced by mechanical transmission are still present. As a result, the wheel-side motor type is an "incomplete" electric vehicle chassis streamlining option.

The in-wheel motor drive uses a hub motor as the drive unit and removes the transmission system. Its most important feature is that the in-wheel motor,

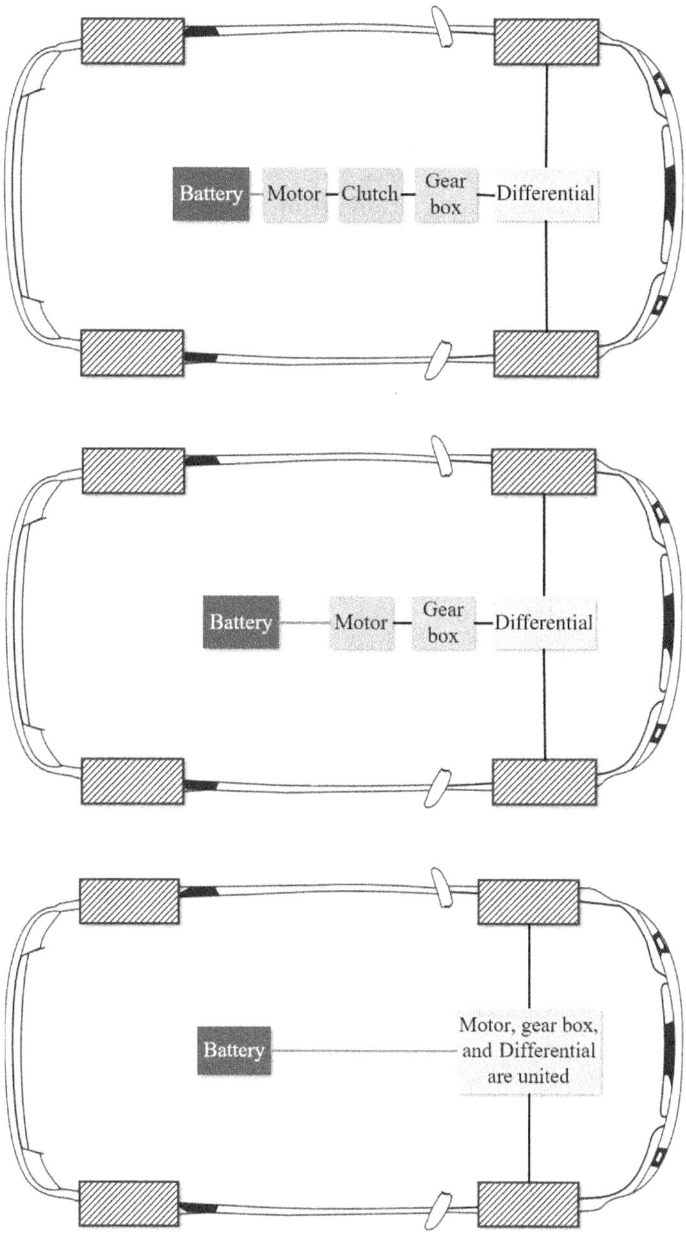

FIGURE 6.1 PEV with integrated drive powertrain. (a) Clutch-equipped gearbox (b) The clutch is eliminated (c) The motor, gearbox, and differential are all united.

hydraulic braking, and suspension systems are integrated into the hub, so the mechanical part of the electric vehicle is fully simplified (Fig. 6.2(b)). The in-wheel motor drive removes the complicated transmission system, increases body space release, decreases powertrain bulk, and enhances power transmission efficiency. The vehicle steering mechanism is straightforward to build because the chassis arrangement approach was concentrated on the wheel, and the coexistence of electrical and friction brakes allows for full use of regenerative braking. One of the most promising electric vehicle architectures has been identified as the in-wheel motor drive with independent driving, braking, and steering capabilities.

6.2.3 Remarks and Summary

For PEV, the centralized drive is still the most common powertrain configuration. Distributed drives, particularly in-wheel motor drives, have been widely

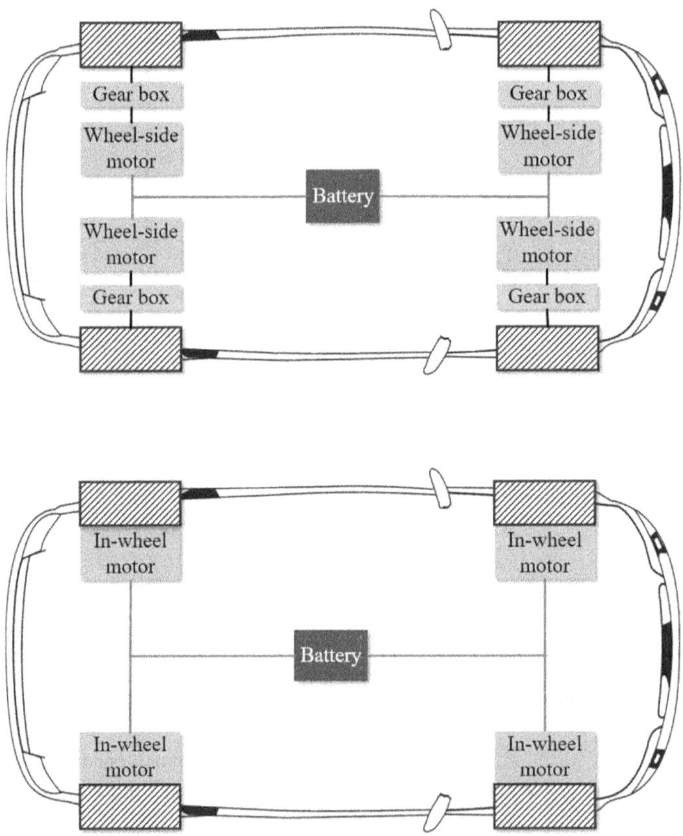

FIGURE 6.2 PEV with distributed drive powertrain. (a) In-wheel motor drive (b) Wheel-side motor drive.

developed due to their numerous advantages. The advantages of distributed drive with in-wheel motor are mainly reflected in the following aspects:

1) High efficiency of transmission
   Because of the centralized drive's long mechanical construction, friction loss in the torque transmission decreases over time, resulting in overall transmission efficiency of only 85% ~ 90%. When compared to the centralized drive, the in-wheel motor drive can avoid transmission path losses to the greatest extent possible and increase powertrain efficiency by 8% ~ 15%, achieving the effect of energy savings from the structural level.
2) Accurate torque control and quick reaction
   The response time of an in-wheel motor can reach 10 ms, which is ten times faster than that of an ICE, and the in-wheel motor drive eliminates the hysteresis caused by mechanical transmission, resulting in a faster power generation speed for an electric vehicle with distributed drive. At the same time, because of the powertrain's high integration, it's simpler for the in-wheel motor to provide accurate control and real-time feedback of wheel-end torque, laying the groundwork for the development and execution of chassis control algorithms.
3) More flexible chassis control
   The chassis' freedom has been substantially increased because of the independent control of the in-wheel motor. Differential steering allows for a more flexible design of steering characteristics, as well as giving the vehicle the ability to steer in place. A torque distribution method that aims for both power and efficiency maximizes tire adhesion margins while lowering total vehicle energy consumption and assuring stable vehicle operation. The decoupled regulation of motor and friction braking allows for faster energy recovery throughout the vehicle braking process while minimizing the detrimental influence of friction braking hydraulic variations on driver comfort.

With the innovation of motor integration technology, the vibration, water, and heat recession problems of hub motors have been effectively solved, and the increase of under spring-mass has been proven to be under control and can be improved by optimizing the suspension structure. At the same time, the closed-spoke shape of the in-wheel motor allows air to pass through the body more smoothly, thus minimizing the effect of wind resistance. Thus, the in-wheel motor drive represents the future form of electric vehicle configuration.

6.3 HEV POWERTRAIN CONFIGURATIONS

Due to the presence of two power sources, namely an ICE and electric motor, as well as a sophisticated power coupling system, the powertrain components of HEV are more complex than those of EV. The powertrain architecture of HEV is influenced by the price, economy, and dynamics of the vehicle as well as the complexity of the power management system. The powertrain configurations of HEV may be broken down into four categories based on power coupling and transfer type: parallel, series, power-split, and multi-mode.

6.3.1 Parallel Hybrid Powertrain

6.3.1.1 Operation Mechanism

In parallel HEV, both the ICE and motor/generator (MG) are connected mechanically with the output shaft (Fig. 6.3) and can simultaneously provide power to operate the vehicle. The available MG is used to shift the engine operating points to a higher-efficiency area. It acts as a generator at low power demand and as a motor at high power demand. In this way, the engine can work at higher efficiency than that in a conventional vehicle. In addition, parallel hybrids must include a transmission to match high engine speed and low vehicle speed.

6.3.1.2 Sub-Types and Typical Models

The parallel configuration can be considered as an incremental add-on to a traditional powertrain, and its design requires relatively little investment and engineering effort. Parallel hybrid powertrains can be further classified into five subtypes according to the location and size of the MG: P0, P1, P2, P3, and P4 (Fig. 6.4).

Subtype P0 refers to the configuration in which a motor is installed before the ICE and is connected to ICE by a belt. Therefore, it is also known as a belt-driven starter/generator HEV. Owing to the torque limitation of the belt, the starter/generator is always small and can fulfill only the start-stop function.

The P1 subtype refers to the configuration in which the motor is mounted on the crankshaft of the engine. Here, the motor is always referred to as an integrated starter generator. The installation position of the integrated starter generator

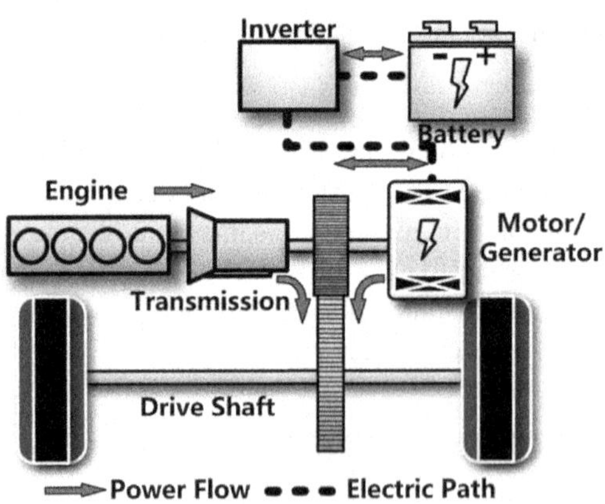

FIGURE 6.3 Configuration of parallel HEV.

always restricts its size; this limitation does not allow the integrated starter generator to provide high torque to operate the vehicle. Only some functions can be fulfilled such as start-stop, regenerative braking, and acceleration assistance.

Subtypes P2 and P3 are the two most popular variations of parallel HEV. In these configurations, the motor is mounted on the input and output of the transmission, respectively. The motor in P2 and P3 is much larger than that in P0 and P1 and has the ability to operate the vehicle at relatively high speeds. Recapturing more regenerative braking and eliminating engine drag result in better energy-efficiency than that in other subtypes. Many European and Korean automakers have released P2-type HEV such as the Volkswagen Passat hybrid and the Hyundai Sonata hybrid. In China, BYD used the P3 subtype in the BYD Qin.

The P4 subtype refers to a parallel hybrid in which the motor is mounted directly on the drive shaft or is incorporated into the hub of a wheel using in-wheel motor technology (Fig. 6.4(e)). P4 is generally not used independently but is combined with other parallel subtypes, P2 and P3, particularly in four-wheel drive vehicles.

Previous studies have compared the performance of different parallel HEV. A comparative study through dynamic programming was also conducted in which P2 is shown to have better fuel economy than P1 owing to its larger motor, and P2 and P3 have similar fuel economy benefits.

6.3.1.3 Limitations and Challenges

Parallel HEV is efficient during city stop-and-go conditions. However, this might not be the most efficient configuration because a mechanical connection still exists between the ICE and the output shaft. In addition, because MG cannot be used to simultaneously charge the battery and assist in powering the engine, the power assist and PEV operations must be controlled carefully to avoid battery depletion. This problem is exacerbated during city driving, in which frequent start–stops can consume a significant amount of battery energy and force the engine to generate power in its low-efficiency area. Because of these drawbacks, parallel HEV has a smaller market share percentage even though a variety of models have become available.

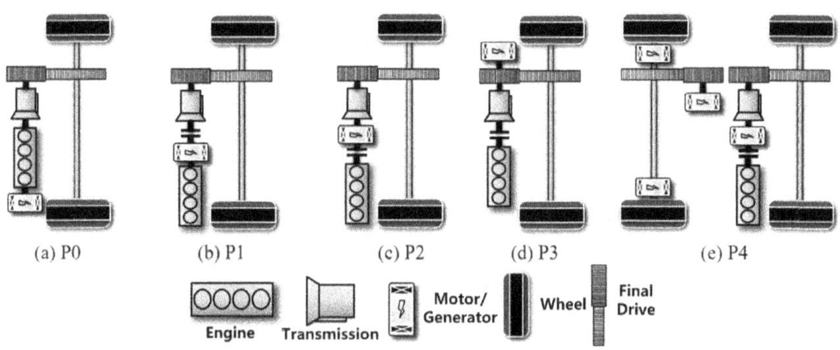

FIGURE 6.4 Configurations of subtypes P_0 to P_4 in parallel HEV.

6.3.2 Series Hybrid Powertrain

6.3.2.1 Operation Mechanism

Series HEV generally use traction motors to operate the vehicle alone, whereas the ICE is connected to a generator (Fig. 6.5). The motors are powered by the battery and the generator and can be placed on both front and rear axles to realize electric all-wheel-drive functionality. Since no mechanical coupling exists between the ICE and vehicle drive axle, the ICE could operate in its best efficiency area regardless of the vehicle speed and power required by the driver. Moreover, the traction motor has a wider operating range and higher efficiency than the ICE. Therefore, a transmission, which is a necessary component in a conventional vehicle, might not be necessary for a series HEV. Thus, the series hybrid powertrain is simpler compared with other types, including in configuration and energy management.

6.3.2.2 Typical Models

Only a few HEV in the market use the series configuration except range-extended HEV. The most successful model of this type in the market is the BMW i3, which provides an optional gasoline-powered range-extender auxiliary power unit. Recently, some original equipment manufacturers (OEM) have developed electric cars with range-extended techniques. A typical example is the Nissan e-Power, which has a 1.2 L gasoline engine that acts solely as a generator for battery charging.

6.3.2.3 Limitations and Challenges

The fuel economy of series HEV can be better than that of conventional vehicles. However, high energy conversion losses can occur because 100% of the engine power must first be converted into electricity. Part of the electricity is stored in the battery, and the remainder powers the motors to propel the vehicle. Even though the MG has relatively high efficiency and the ICE operates at high efficiency, the

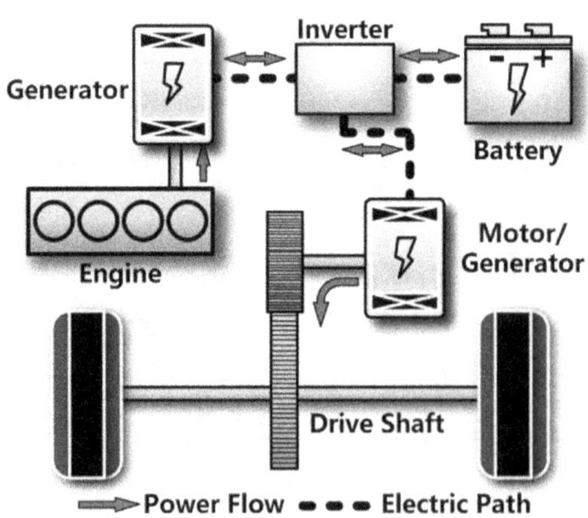

FIGURE 6.5 Configuration of parallel HEV.

multiple energy conversions still result in low overall efficiency. Additionally, the series configuration requires a large traction motor to meet the torque requirement because the motor is the only traction device.

6.3.3 Power-Split Hybrid Powertrain

6.3.3.1 Operation Mechanism

Power-split HEV usually employs one or multiple planetary gears (PG) sets to couple the ICE, two MGs and the driveshaft together (Fig. 6.6). The PG sets are the heart of the power-split hybrid powertrain, which is usually referred to as a power-split device. The power split device decouples the ICE from the vehicle speed and acts as a continuously variable transmission (CVT), which results in efficient engine operation regardless of the vehicle speed. Therefore, the power-split device in power-split HEV is also referred to as an electronic-CVT (E-CVT). Because of this decoupling function, power-split HEV generally shows better fuel economy than both series and parallel HEV, particularly in city driving conditions.

The power split device allows for power flow from the engine to the driveshaft: either through the mechanical path or the electrical path (Fig. 6.7). In the electrical path, the power-split device operates as a series HEV. Part of the ICE power is converted into electricity first by a generator, which drives the motor or charges the battery. In the mechanical path, the power-split device also enables the system to operate as a parallel HEV, in which the ICE can generate power flow to the driveshaft directly. Therefore, the power-split HEV combines the advantages of both series and parallel hybrids.

6.3.3.2 Sub-Types and Typical Models

The power-split hybrid powertrain can be further classified into three subtypes according to the point of the power split execution: input-split, output-split, and compound-split. In an input-split HEV, the ICE power is split at the input to the

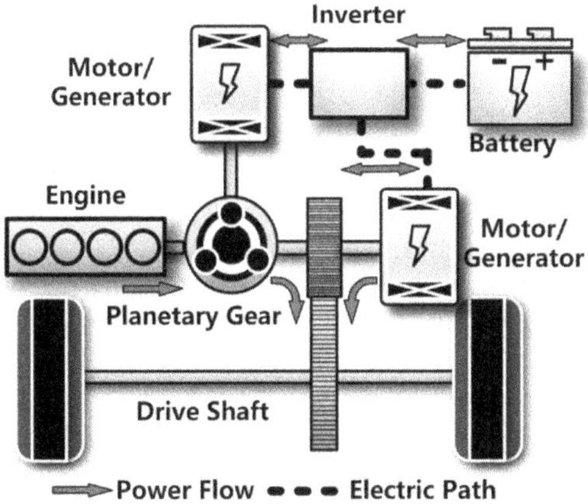

FIGURE 6.6 Configuration of power-split HEV.

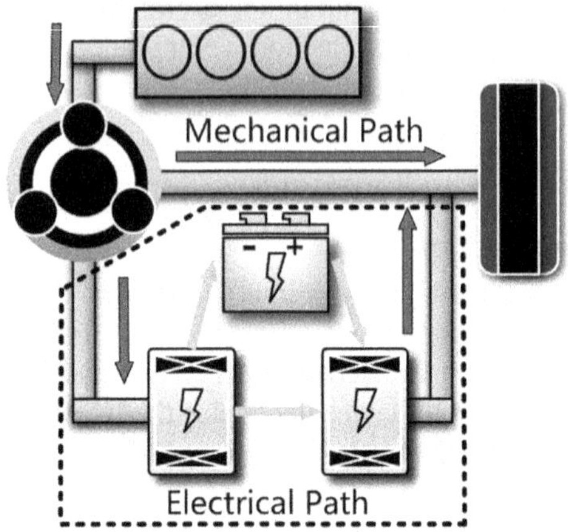

FIGURE 6.7 Power flow of power-split hybrid powertrain.

transmission by collocating an MG with the output shaft and sometimes with an additional set of gears in between. In an output-split vehicle, the ICE power is split at the output to the transmission by collocating an MG with the ICE, also sometimes with an additional set of gears in between. In a compound-split vehicle, no MG collocation occurs with the output shaft or the engine.

The most successful power-split hybrid on the market is the input-split subtype, such as the Toyota Hybrid System or Hybrid Synergy Drive, introduced by the Toyota Motor Corporation. This subtype was first implemented in the Toyota Prius in Japan in 1997 and was then extended to the company's Camry and Lexus hybrid vehicles in the following years. The second-generation Toyota Hybrid System was announced in 2004, offering increased system efficiency, enhanced power, and improved scalability. Scalability enables the Toyota Hybrid System to adapt to different vehicle sizes by changing the reduction paths of ICE/MG1 and MG2. A similar input-split concept was adopted by the Ford Motor Company in its Fusion Hybrid and C-max models. However, General Motor developed two classes of power-split powertrains by using the other two subtypes in a power-split configuration, the Voltec Hybrid Powertrain with an output-split mode and the Allison Hybrid System with both an input-split and a compound-split mode.

Power-split HEV has a variety of design variations by changing the locations of the employed components. To explore all possible designs, Liu and Peng proposed an automated modeling method to efficiently build the dynamics of a power-split HEV and identified a design with optimal fuel economy. Bayrak *et al.* [1] enumerated all feasible powertrains by using a bond graph and generated complete sets of feasible designs based on an exhaustive search. Kim *et al.* [2] reorganized all the possible single PG configurations into compound-lever design space and screened the optimal design by balancing the fuel benefits and the drivability.

6.3.3.3 Limitations and Challenges

The electrical path incurs higher energy loss than the mechanical path because of the extra energy conversion. More ICE power delivered through the electrical path indicates more energy loss caused by the power-split device. When the speed of either MG is equal to zero, the engine-generator-motor path has zero power transmission, and the energy transition is the most efficient. This condition is known as the mechanical point. The energy dissipation in the electrical path might cause power-split HEV to have greater energy losses than those in parallel HEV in some situations, particularly in high-speed cruising.

6.3.4 Multi-Mode Hybrid Powertrain

6.3.4.1 Operation Mechanism

The multi-mode hybrid powertrain system can be developed by adding clutches to parallel or power-split powertrain systems, which can become any of the three hybrid configurations (series, parallel, and power-split) in the same powertrain. Its subtype is also referred to as the operating mode. The freedom to choose from different modes makes it possible to achieve higher energy efficiency and performance than that realized by using the other HEV configuration types introduced before.

6.3.4.2 Sub-Types and Typical Models

Multi-mode hybrids can be further classified into two subtypes, series-parallel and PG coupling, according to whether PG sets are used to couple the powertrain components.

1) Series-parallel multi-mode configuration
The series-parallel subtype (Fig. 6.8) was first introduced by Honda in 2014 in its i-MMD system, which is installed in the Accord plug-in hybrid. Two MGs are used in this configuration: One is fully coupled with the ICE, and the other connects directly to the drive shaft. A clutch is employed to disengage the connection between the ICE and output shaft, which enables three operating modes: PEV, series, and parallel. The mode shift strategy avoids inefficient engine operation: The PEV mode is used when the battery state of charge is high, and the series and parallel modes operate only at low and high vehicle speeds. A regular transmission is no longer required to reduce the powertrain cost. Since a mechanical connection still exists between the ICE and the output shaft in the parallel mode, ICE cannot operate in its most efficient area at all vehicle speeds.
2) PG coupling multi-mode configuration
The PG coupling subtype is developed by adding clutches between the PG nodes in a power-split configuration. A typical example is the Advanced Hybrid System patented by General Motors in 2002 (Fig.

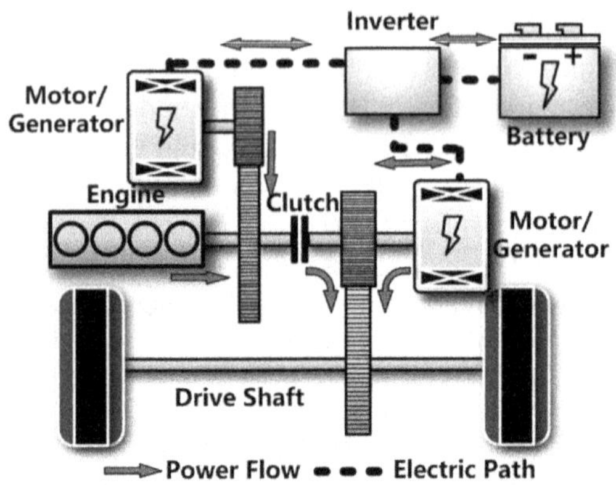

FIGURE 6.8 Configuration of series-parallel hybrid powertrain.

6.9(a)). By switching the two clutches, two operating modes are achieved: the input-split mode and the compound-split mode. The former mode can provide large output torque, which is more suitable for low-speed cruising, whereas the latter tend to have higher efficiency at high speeds by preventing the speed of MG2 from increasing continuously with the vehicle speed. As a result, better fuel economy and launching performance are achieved by proper mode selection, particularly in buses that require higher torque at low speeds.

In addition to Allison Hybrid System, General Motors in 2011 introduced another multi-mode powertrain known as Voltec, which originally had a single PG and three clutches (Gen1) and was later changed to a 2-PG design (Gen2) in 2015 (Fig. 6.9(b)). The three clutches in the Voltec Gen2 enable five operating modes. Of these modes, Voltec Gen2 has an input-split mode and a compound-split similar to that in the Allison Hybrid System, in addition to two PEV modes for plug-in functionality.

Toyota developed a multi-mode hybrid powertrain by adding a Ravigneaux-type PG and two clutches to Toyota Hybrid System. The second MG's gear is switched between high and low ratios for low- and high-speed driving, respectively. To develop a more powerful hybrid system, Toyota combined the Toyota Hybrid System with a four-speed automatic transmission in 2017 to multiply the output torque (Fig. 6.9(c)). This powertrain is the multi-stage hybrid featured in the company's Lexus LC 500h and LS 500h models.

Although many multi-mode configurations have been proposed and patented, many more remain unexplored. A multi-mode hybrid can be generated in two ways. The first involves changing the locations of the powertrain components, including the engine, two MGs, and the output node to the vehicle drive axle. Each device can connect with any node of the PG sets. In the second method, the number and locations of permanent connections and clutches also result in different hybrid powertrains. Fig. 6.10 shows all possible clutches and permanent connections of both double and triple PG sets. The total number of possible clutches that connect two nodes, or a node with the ground, is

$$N_{clutch=}C_{3NP}^{2}+3N_{P}-2N_{P}-1 \tag{6.1}$$

where the first term is the number of clutches that can be added between any two nodes, and the second term represents the number of total possible grounding clutches. Since locking any two of the three nodes in a PG produce identical dynamics, the third term eliminates redundant clutches. Finally, the output node should not be grounded. By changing the locations of the powertrain components and selecting different clutch positions, billions of configurations are available.

6.3.4.3 Limitations and Challenges

The deployment of multiple modes can introduce severe mode shift problems. Improper mode shifting will increase the noise, vibration, and harshness; increase the energy consumption; and reduce the ride comfort and vehicle drivability. To reduce the negative impacts caused by mode shifts, researchers have begun to investigate mode shifts of series-parallel HEV, particularly for transitions from PEV modes to hybrid-drive modes.

In comparison to series-parallel HEV, PG-based multi-mode HEV have a worse performance of noise, vibration, and harshness because they rely on PG sets to couple the engine and the driveline mechanically; no torque converter or clutch is used, which are usually available in series-parallel HEV to passively damp the vibrations and oscillations. Researchers have investigated mode shifts of PG-based multi-mode HEV from two perspectives: mode shift map design and mode transition control. The mode shift duration should be minimized to reduce the torque hole and energy loss during the transition. However, driveline torsional vibrations and torque variation caused by the engine torque pulsations should be suppressed to mitigate the level of noise, vibration, and harshness and improve the vehicle's drivability. Fortunately, the rapid torque response of electricity can compensate for the torque disturbance.

6.4 HYBRID POWERTRAIN SYSTEM COMPONENT SIZING

The mechanical and electrical connection network between the components of the hybrid powertrain determines HEV performance and cost, which is also impacted by the component sizing. HEV has more than two power sources, and the energy management control system is responsible for coordinating the output power of the two sources; if the matching of power parameters is not optimal, the energy of one power source may not be properly used. For example, if the motor matching power of an HEV is too high, the motor does not reach the peak power point during vehicle driving, which is the motor power surplus, and the hybrid powertrain cost

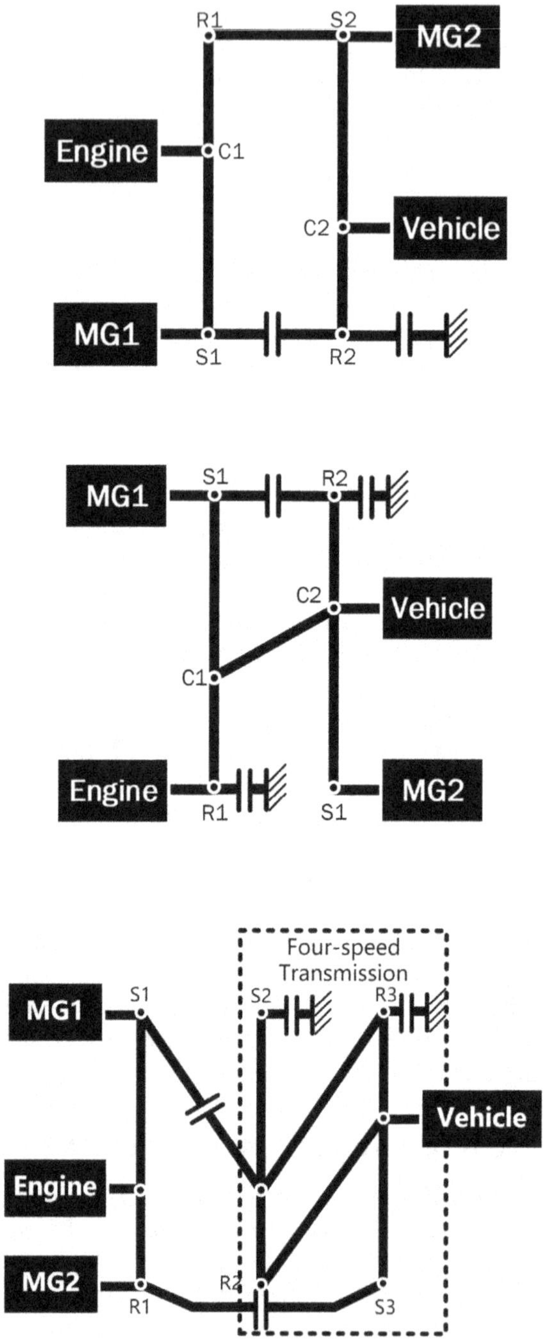

FIGURE 6.9 Lever diagram of different multi-mode HEV. (a) Advanced hybrid system (b) GM Voltec Gen2 powertrain (c) Toyota multi-stage hybrid powertrain.

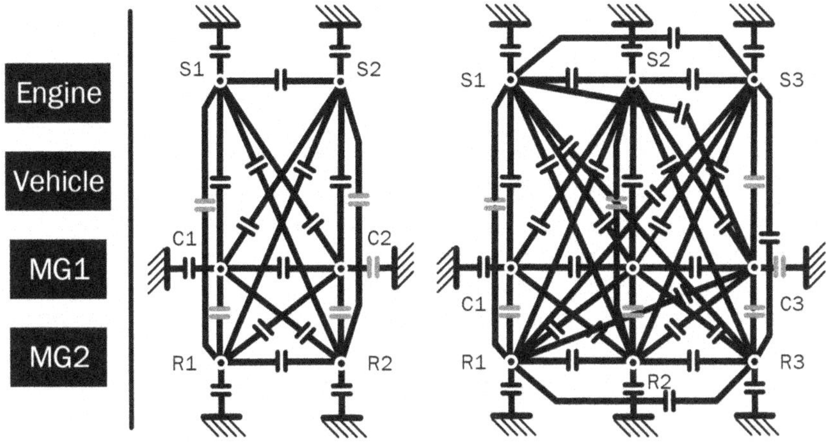

FIGURE 6.10 All possible clutches and permanent connections of both double and triple PG sets.

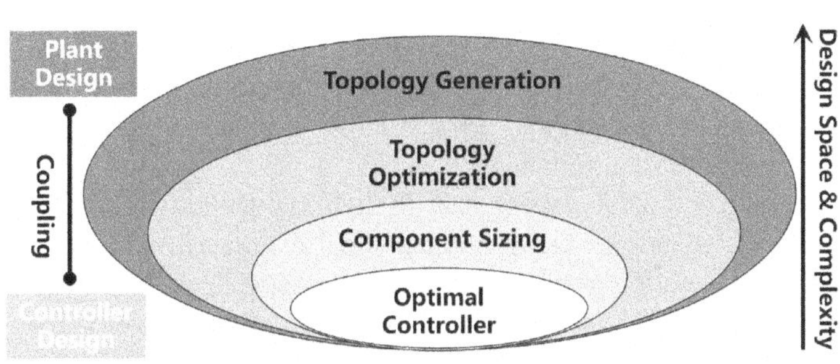

FIGURE 6.11 Component sizing of HEV.

rises; conversely, the hybrid powertrain cannot fully use the vehicle's fuel-saving potential. The component sizing of a hybrid powertrain system is the process of calculating the power source parameters, and it has a substantial impact on the economy and the cost of HEV. The HEV configuration design and energy management system optimal control are strongly connected to hybrid powertrain system component sizing, as shown in Fig. 6.11. In component sizing, the optimization variables are engine power, motor power, battery capacity, planetary gear ring and sun wheel gear ratio, and differential ratio. When hybrid powertrain components parameters are matched, engineers select the final system parameters through many matching tests, which commonly results in engine power deficit or engine power overload during the calibration process. Finally, the effective sizing of the hybrid powertrain component is the focus and challenge of current research in the field of HEV design. Then, a thorough review of component sizing for a hybrid powertrain system is conducted in the next section.

6.4.1 Traditional Component Sizing Methods

In general, there are two traditional component sizing methods: experience-based and equivalent calculation-based, both of which are aimed to satisfy the needs of improving vehicle performance and reduce cost based on experiences, databases, or basic computations. The traditional component sizing methods are plain and easy to understand, considering the planned vehicle performance parameters as well as various constraints such as packaging, component availability, and so on. However, it is not optimal or even sub-optimal.

For the experience-based sizing method, the battery is chosen based on a recommendation from a major automotive manufacturer that the battery pack should not exceed 20% of the vehicle's mass. The ICE and motor are selected with consideration of the maximum power requirement of the vehicle at maximum driving speed while maintaining a specified degree of power excess. In addition, some research looks at the oil-electric mixing ratios to formulate the hybrid powertrain sizing method. For example, S. Lukic *et al.* [3] discovered that both high and low oil-electric mixture ratios are harmful to the hybrid powertrain system energy economy, a high mixture ratio will have a significant impact on the vehicle economy. The lowest energy consumption can be achieved when the mixture ratio is kept between 0.3 and 0.5. These experience-based sizing methods are mainly based on the estimation of previous vehicle performance databases, design experience, or the analysis of trial-and-error simulation or prototyping experimental findings of the accessible components.

In some research, the component sizes for hybrid powertrain systems are determined by vehicle dynamics using mathematical estimates of maximum output power to meet acceleration performance, maximum speed, grade ability, and other requirements. This method is known as equivalent calculation-based sizing. For example, the motor power requirements were determined by the maximum speed and maximum gradient at that speed; in series mode, the ICE power should meet the average power needs; the energy storage system needs are available energy and maximum power, which are computed using the required range. Furthermore, Sinoquet *et al.* [4] quantitatively examined the influence of battery sizes and motor maximum power on fuel economy, and the optimal sizes were chosen based on the results.

6.4.2 Optimization-Based Component Sizing Methods

Because numerous disciplines, such as electrical, chemical, and mechanical, may be involved in such a sophisticated hybrid powertrain system, it is exceedingly difficult to size its components manually or analytically because they rely on sizing experience or simple calculations. The component sizing problem, unlike experience-based sizing and equivalent calculation-based sizing methods, is mostly solved by using an optimization-based sizing method. The goal of the component sizing problem is to reduce cost, electric energy consumption, and fuel in general,

$$\text{minimize } J = \varpi_1 \int_{t_s}^{t_d} P_b dt + \varpi_2 \int_{t_s}^{t_d} \text{m}_f \text{dt} + C_p \tag{6.2}$$

Where ϖ_1 and ϖ_1 are the electricity and fuel price, respectively. p_b is the instantaneous battery power, m_f is the instantaneous fuel consumption, C_P is the price of the hybrid powertrain system. t_s and t_d are the initial and final time of the driving cycle for the component sizing test, respectively. The hybrid powertrain sizing optimization problem is frequently linked to the energy management control of HEV; thus, the research is divided into three categories: alternative optimization, nested optimization, and simultaneous optimization methods, as shown in Fig. 6.12.

The alternative optimization method is rarely employed since it does not ensure the optimality of the optimization results; instead, most studies prefer the nested optimization method, which is frequently also referred to as a two-layer optimization method. The simultaneous optimization necessitates the integration of both vehicle energy management and component sizing simultaneously. A more complicated optimization problem is formulated, which is difficult to solve. Convex optimization algorithm has been used in some research to solve the simultaneous optimization problem of series HEV, but it cannot be used for parallel HEV.

A lot of literature on hybrid powertrain sizing using a two-layer optimization method exists, and the objective of component sizing is vehicle economy in general. In the upper layer, various optimization methods are proposed for hybrid powertrain component sizing, such as sequential quadratic programming, genetic algorithms, and heuristics algorithms. In the lower layer, the energy management strategy is formulated to manage the power distribution between the engine and the motor. Other optimization objectives have been included in some research, for example, A. Malikopoulos's [5] discussion of the influence of hybrid powertrain sizing on CO_2 emissions. A growing number of studies have begun to construct a multi-objective optimization problem that includes the total cost of the hybrid powertrain system as well as the vehicle life cycle cost, to minimize the vehicle's manufacturing or life cycle use costs while maintaining system performance (e.g., energy economy, dynamic performance.). In addition, some research has focused on the impact of battery capacity on hybrid powertrain system economics and cost, proposing optimization for power sources that combine batteries and supercapacitors to lower the total cost of the system (both manufacturing and usage costs) while satisfying the user daily needs (e.g., sufficient range).

6.4.3 Remarks and Summary

In conclusion, relevant researchers have conducted a series of great studies and produced a range of hybrid powertrain components sizing methods, such as experience-based, equivalent calculation-based, and optimization-based. However, because the hybrid powertrain system is complicated and has a variety of control objectives, integrating the optimization problems of topology configuration,

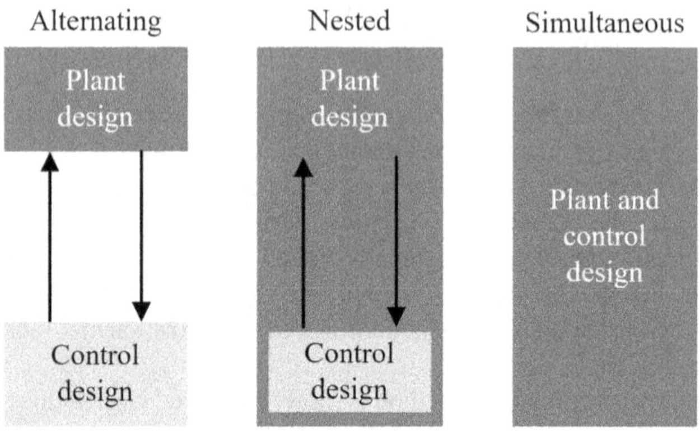

FIGURE 6.12 Optimization-based component sizing methods.

energy management, and component sizing with advanced optimization algorithms to achieve optimal HEV design has become the future research of hybrid powertrain component sizing.

6.5 HEV MODELING AND CONFIGURATION EXPLORATION

Multi-mode HEV, as was previously discussed, offers the highest chance of establishing a balance between vehicle energy efficiency and driving performance. The bulk of the design space has not yet been taken into account, and there are just a few of these designs available on the market. An approach for creating combinations and simulating dynamics is needed to investigate such a large design area. The bond graph approach, automated modeling, and the graph-theoretic technique are typically the three methodologies that are mentioned.

6.5.1 HEV Modeling Techniques

6.5.1.1 Graph-Theoretic Method

Graphs have been used to represent system topologies since the 1700s and are currently adopted in the design of HEV powertrain systems. Silvas *et al.* [6] proposed an undirected connected finite graph to represent the HEV configuration composed of nodes, or components such as power sources and wheels, and edges, or connections between components such as transmissions and PGs. By defining the functionality and cost constraints, a constraint logic programming problem is formulated. The feasible configurations for all four configuration types can be derived. However, the performance of each generated configuration on factors such as fuel economy and acceleration performance cannot be evaluated because the proposed undirected graph cannot be used to model the dynamics. Adam H. Ing [7] used a directed linear graph to represent the powertrain structure. By using the graph theory, a quasi-static model of the configuration can be generated. The fuel economy and acceleration performance of the designs generated are evaluated based on the equations made. However, this method cannot be used to model a hybrid powertrain with multiple modes.

6.5.1.2 Bond Graph Method

Bond graphs are used to model multi-energy domain systems such as mechanical, electrical, and hydraulic systems and have recently been applied to the modeling of HEV. Other than the graph theory, the bond graphs have a notion of causality and allow the modeling of system dynamics.

In bond graphs, power flow is represented by a bond between two nodes and is denoted by a pair of variables known as power variables, that is, flow and effort, the product of which is the instantaneous power of the bond. For example, in a mechanical system, force is the effort variable, and velocity is the flow variable.

On the basis of the bond graphs, Bayrak *et al.* [1] proposed a framework to develop single and multi-mode hybrid configurations. By enumerating all undirected graphs for external junctions and internal junctions, assigning 0 and 1-junctions and the bond weights, the design space of the HEV configurations are generated in the form of bond graph representation. A quasi-static model is generated based on the bond graph in the form of state-space representation as shown in Eqs. (6.3) and (6.4):

$$\begin{bmatrix} \dot{\omega}_{\mathrm{MG1}} \\ \dot{\omega}_{\mathrm{MG2}} \end{bmatrix} = -C_{mode} \begin{bmatrix} \dot{\omega}_{\mathrm{eng}} \\ \dot{\omega}_{\mathrm{out}} \end{bmatrix} \tag{6.3}$$

$$C_{mode}^{-T} \begin{bmatrix} T_{\mathrm{eng}} \\ -T_{\mathrm{out}} \end{bmatrix} = -\begin{bmatrix} T_{\mathrm{MG1}} \\ T_{\mathrm{MG2}} \end{bmatrix} \tag{6.4}$$

where $\dot{\omega}_*$ and T_* are angular acceleration and the corresponding torques of the powertrain devices, respectively, C_{mode} is the kinematic relationship matrix derived from the bond graphs.

In Bayrak's framework [1], the inertia of the engine and MG is ignored, which may have a considerable influence on the performance evaluation of HEV, especially for the mode shift of multi-mode HEV.

6.5.1.3 Automated Modeling Method

Automated modeling refers to the methodology of modeling the dynamics of the hybrid powertrain automatically by following predefined rules. This method was first proposed by Liu and Peng [8] in 2009 for power-split HEV and was extended to multi-mode HEV modeling by Zhang [9] in 2014. The core concept of automated modeling is to model the dynamics of the configuration in the state-space representation as follows:

$$\begin{bmatrix} J & D \\ D^T & O \end{bmatrix} \begin{bmatrix} \dot{\Omega} \\ F \end{bmatrix} = \begin{bmatrix} T \\ O \end{bmatrix} \tag{6.5}$$

where J is a diagonal matrix with dimensions of $3n \times 3n$ reflecting the inertia on each node, where n is the number of PGs. In addition, D is aconstraint matrix with entries determined by the connections of PG nodes with the four powertrain components.

For a multi-mode hybrid powertrain consisting of multiple operating modes, each mode has a dynamic model in the form of $A\dot{\Omega} = T$. To accelerate the modeling, Zhang *et al.* [9] proposed a torque transition matrix M and angular acceleration transition matrix P based on the clutch states. By using the transition matrices, the dynamics of each mode in a multi-mode hybrid powertrain are represented by the characteristic matrix A^* as shown in Eq. (6.6), which governs the relationship between the angular acceleration of powertrain devices $\dot{\omega}_*$ and their corresponding torques T^*.

$$\begin{bmatrix} \dot{\omega}_{out} \\ \dot{\omega}_{eng} \\ \dot{\omega}_{MG1} \\ \dot{\omega}_{MG2} \end{bmatrix} = \begin{bmatrix} a_{11} & a_{12} & a_{13} & a_{14} \\ a_{21} & a_{22} & a_{23} & a_{24} \\ a_{31} & a_{32} & a_{33} & a_{34} \\ a_{41} & a_{42} & a_{43} & a_{44} \end{bmatrix} \begin{bmatrix} T_{load} \\ T_{eng} \\ T_{MG1} \\ T_{MG2} \end{bmatrix} A^* \begin{bmatrix} T_{load} \\ T_{eng} \\ T_{MG1} \\ T_{MG2} \end{bmatrix} \tag{6.6}$$

Based on the derived characteristic matrix, the feasibility, functionality, and characteristics of each configuration can be determined. Thus, in the following section, we discuss automated modeling to explore all possible configurations of the hybrid powertrain.

6.5.2 Configuration Exploration

6.5.2.1 Brute-Force Search and Dynamics Modeling

As discussed in the literature, different powertrain component locations may result in different hybrid powertrains. Thus, for simplicity, we fixed these locations to be the same as that in the GM Voltec, as shown in Fig. 6.13. In this configuration, the ICE connects with the ring gear of the first PG; the vehicle output shaft connects with the carrier gear of the last PG; and the two MGs connect with the sun gears of the first PG and last PG, respectively. In addition, the PG nodes, including the sun, carrier, and ring gears, are numbered in a series in the following description.

Before exploring all possible sub-configurations, the effective number of links that connect PG nodes should be first determined. As the degree of freedom (DoF) of a single PG is 2, the DoF of 2- and 3-PG hybrid powertrains with no connection begins from 4 and 6, respectively. In this study, the system DoF refers to the number of components with independent speed. Since each effective link reduces the DoF by one, the meaningful DoF of the multi-mode hybrids are 1, 2, and 3 representing the parallel mode, the power-split mode, and the engine speed, respectively, where the speed of one of the MGs is free. Therefore, the effective numbers of links for 2- and 3-PG hybrids are between 1 and 3 and between 3 and 5, respectively. In addition, we assigned to each link

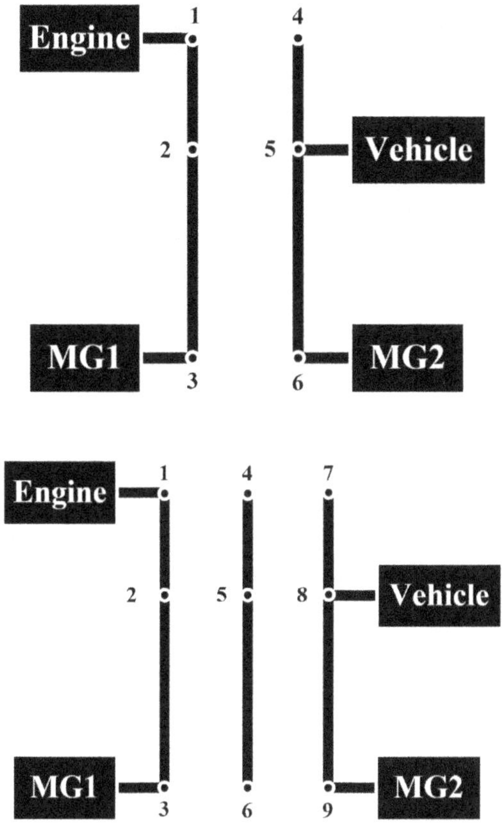

FIGURE 6.13 Locations of powertrain components for (a) 2-PG and (b) 3-PG hybrid powertrains.

a vector composed of the serial numbers of the connected nodes. For example, the link between Nodes 1 and 4 is represented. Node 1 is grounded, with a 0 assigned to indicate the grounded node.

In previous research, all possible combinations of links inside the 2- and 3-PG hybrids are explored exhaustively. It should be noted that any nodes connected by two or multiple links should be first merged. For example, three links in 2-PG hybrids should be combined and represented by the vector. In this study, the depth-first search was used to traverse all possible connected nodes. Finally, the dynamics of each configuration are represented in the state-space form, as shown in Eq. (6.6), by the automated modeling methodology introduced in Section 6.1.3.

6.5.2.2 Mode Classification

For organizing all sub-configurations, several vectors and coefficients were constructed and extracted from the characteristic matrix A*. The four rows of the A* matrix are referred to as H_{veh}, H_{eng}, H_{MG1}, and H_{MG2}, respectively. The elements of H_{veh} are V_{veh}, V_{eng}, V_{MG1}, and V_{MG2}, which represent the torque contributions of each component to the output shaft. V_{veng}, V_{MG1}, and V_{MG2}, can be zero if the powertrain components are not connected with the vehicle output.

In addition, six coupling vectors reflecting the coupling relation among powertrain components are defined as $H_{VE} = [H_{veh}; H_{eng}]$, $H_{VMG1} = [H_{veh}; H_{MG1}]$, $H_{VMG2} = [H_{veh}; H_{MG2}]$, $H_{EMG1} = [H_{eng}; H_{MG2}]$, $H_{EMG2} = [H_{eng}; H_{MG2}]$, and $H_{MG1MG2} = [H_{MG1}; H_{MG2}]$. According to the defined vectors, several parameters are listed as follows.

1) Rank of the characteristic matrix A^*
Since each row of A^* in Eq. (6.6) represents the relationship between the torque input and a component's acceleration, a rank reduction means that the acceleration of one component can be represented as a linear combination of those of the other components. Herein, the number of accelerations that can be represented by other components is determined by the number of the independent accelerations, which represents the DoF defined at the beginning of this section. However, the rank of the characteristic matrix, $rank(A^{*RA})$, is the dimension of the torque input or the component's acceleration on the basis of linear algebra, which also refers to the number of linearly independent accelerations. Therefore, the DoF of the system equals the rank of its characteristic matrix, $rank(A^*)$

$$DoF = rank(A^*) \tag{6.7}$$

2) Rank of the coupling matrix H_{VE}, H_{VMG1}, H_{VMG2}, H_{EMG1} H_{EMG2} and H_{MG1MG2}
Similar to the rank of the characteristic matrix A^*, the rank of the coupling matrix can represent the correlation between the two components in that matrix. For example, if the rank of H_{VE} is equal to 1, the acceleration of the vehicle is proportional to the acceleration of the engine. This means that the vehicle output shaft is coupled with the engine directly. On the contrary, a rank of H_{VE} equal to 2 means that the acceleration of the vehicle and the engine are independent of each other. In the following section, the ranks of the matrices H_{VE}, H_{VMG1}, H_{VMG2}, H_{EMG1}, H_{EMG2}, and H_{MG1MG2} refer to, R_{VE}, R_{VMG1}, R_{VMG2}, R_{EMG1}, R_{EMG2} and R_{MG1MG2}, respectively.

In this section, all configurations explored in Section 6.3 are classified by layers based on the parameters defined by the binary tree, as shown in Fig. 6.14. To summarize, all 14 valid configuration types and their classification criteria are listed in Table 6.2. In addition, the numbers of all configuration types are summarized. The unique configuration in Table 6.2 refers to sub-configurations that share the same characteristic matrix even with different topologies. Fig. 6.15 shows an example of

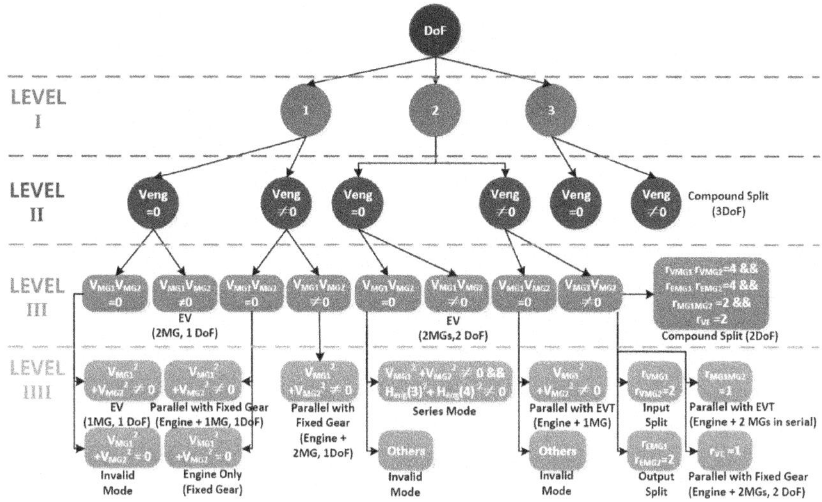

FIGURE 6.14 Configuration classification by binary tree.

multiple topologies having different links but the same dynamics because Nodes C1, R2, C2, S2, and C3 all have the same rotational speed. In this study, configurations sharing the same dynamics are considered to be equivalent; thus, only one is extracted as a unique configuration.

Table 6.2 lists the number of feasible configurations and unique configurations for both 2- and 3-PG hybrid powertrains determined after screening the infeasible and redundant configurations. The 3-PG powertrains have almost 100 times the number of feasible configurations than those of 2-PG powertrains originally. After the equivalent configuration screening, only 102 and 4,041 unique configurations were retained. Performance analysis of all unique configurations has been conducted in previous research.

In this section, all 14 sub-configuration types derived are discussed separately, and examples of a 2-PG system are used to demonstrate their functionality and characteristics.

1) Series configuration
   According to the series configuration introduced in Section 6.3.2, the DoF of the series mode is 2. Fig. 6.16 shows two examples, in which one MG is coupled with the engine mechanically through the first PG set, and the other MG drives the vehicle by grounding the R2 node. Since the engine is not mechanically connected to the wheels, the engine's corresponding coefficient V_{eng} in the H_{veh} is zero. In addition, two MGs are mechanically connected with the engine and wheels separately, leading to $V_{MG1}V_{MG2} = 0, V_{MG1}{}^2 + V_{MG2}{}^2 \neq 0$ and $H_{eng}(3)^2 + H_{eng}(4)^2 \neq 0$.

TABLE 6.2
Classification Criteria and Number of Feasible Configurations for 14 Configuration Types

Configuration type	Classification criteria	Feasible configurations Double PG	Feasible configurations Triple PG	Unique configurations Double PG	Unique configurations Triple PG
Series configuration	DoF = 2, $V_{eng} = 0$, $V_{MG1}V_{MG2} = 0$, $H_{eng}(3)H_{eng}(4) = 0$, $V_{MG1}^2 + V_{MG2}^2 \neq 0$, $H_{eng}(3)^2 + H_{eng}(4)^2 \neq 0$	9	70,978	5	85
Compound split (3 DoF)	DoF = 3	2	88,30	2	650
Compound split (2 DoF)	DoF = 2, $V_{eng} \neq 0$, $V_{MG1}V_{MG2} \neq 0$, $R_{VE} = 2$, $R_{VMG1}R_{VMG2} = 4$, $R_{EMG1}R_{EMG2} = 4$, $R_{MG1MG2} = 2$	4	2,175	4	269
Input split	DoF = 2, $V_{eng} \neq 0$, $V_{MG1}V_{MG2} \neq 0$, $R_{VMG1}R_{VMG2} = 2$	13	12,390	6	172
Output split	DoF = 2, $V_{eng} \neq 0$, $V_{MG1}V_{MG2} \neq 0$, $R_{EMG1}R_{EMG2} = 2$	13	13,227	6	210
ECVT with one MG	DoF = 2, $V_{eng} \neq 0$, $V_{MG1}V_{MG2} = 0$, $V_{MG1}^2 + V_{MG2}^2 \neq 0$	4	2,394	4	106
ECVT with two MGs in series	DoF = 2, $V_{eng} \neq 0$, $V_{MG1}V_{MG2} \neq 0$, $R_{MG1MG2} = 1$	3	2,388	3	82
Engine only	DoF = 1, $V_{eng} \neq 0$, $V_{MG1}V_{MG2} = 0$, $V_{MG1}^2 + V_{MG2}^2 = 0$	17	10,594	4	47
Parallel with fixed gear (2 MGs, 2 DoF)	DoF = 2, $V_{eng} \neq 0$, $R_{VE} = 1$, $V_{MG1}V_{MG2} \neq 0$	3	1,833	3	82
Parallel with fixed gear (2 MGs, 1 DoF)	DoF = 1, $V_{eng} \neq 0$, $V_{MG1}V_{MG2} \neq 0$	330	240,530	21	1,218
Parallel with fixed gear (1 MG, 1 DoF)	DoF = 1, $V_{eng} \neq 0$, $V_{MG1}V_{MG2} = 0$, $V_{MG1}^2 + V_{MG2}^2 \neq 0$	80	68,376	24	666
PEV (2 MGs, 2 DoF)	DoF = 2, $V_{eng} = 0$, $V_{MG1}V_{MG2} \neq 0$	2	1,279	2	55
PEV (2 MGs, 1 DoF)	DoF = 1, $V_{eng} = 0$, $V_{MG1}V_{MG2} \neq 0$	40	35,049	12	331
PEV (1 MG, 1 DoF)	DoF = 1, $V_{eng} = 0$, $V_{MG1}V_{MG2} = 0$, $V_{MG1}^2 + V_{MG2}^2 \neq 0$	76	83,527	6	68
	Sum	596	553,570	102	4,041

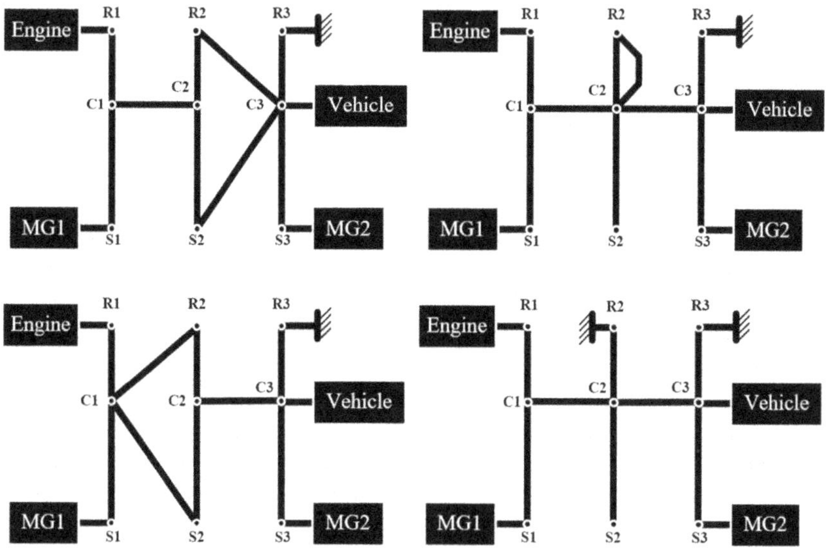

FIGURE 6.15 Example of topologies sharing the same dynamics.

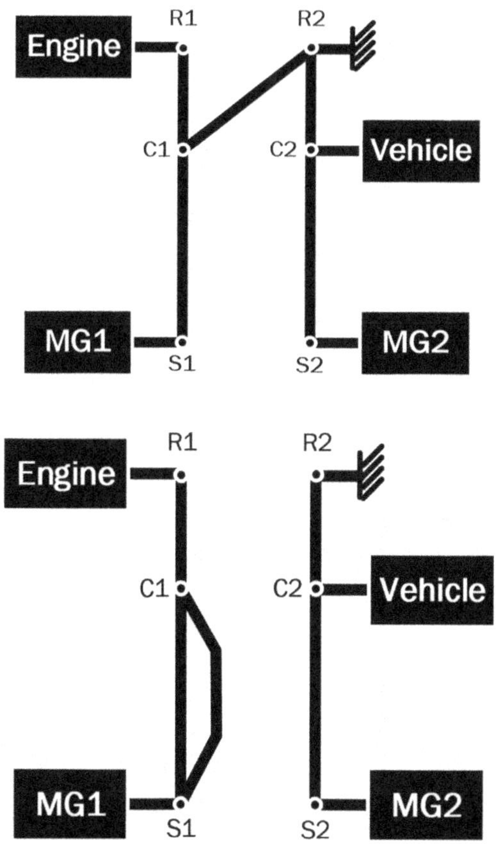

FIGURE 6.16 Examples of series configuration. (a) Topology I (b) Topology II.

2) 3 DoF configuration
The 3 DoF configuration, as shown in Fig. 6.17(a), means that three controllable independent speeds are available in the hybrid powertrain. Therefore, the speeds of the vehicle and engine as well as the speed of one of the two MGs can be controlled independently. That is, three linearly independent equations are required to describe the speed relationship among the powertrain components. Since the number of controllable powertrain components and the DoF are both three, there is no flexibility in the component torque selection when their accelerations are determined. This can also be observed from Eq. (6.6): assuming the first three rows of the A^* matrix are used to calculate the torque commands, the torques from three powertrain components are determined based on the desired component acceleration because the resistance torque T_{load} is not a control variable. This phenomenon may lead to inefficient system operation. Moreover, the component speed will become uncontrollable when the engine is off. Therefore, the 3 DoF configuration is not suitable for topologies with a single output shaft.

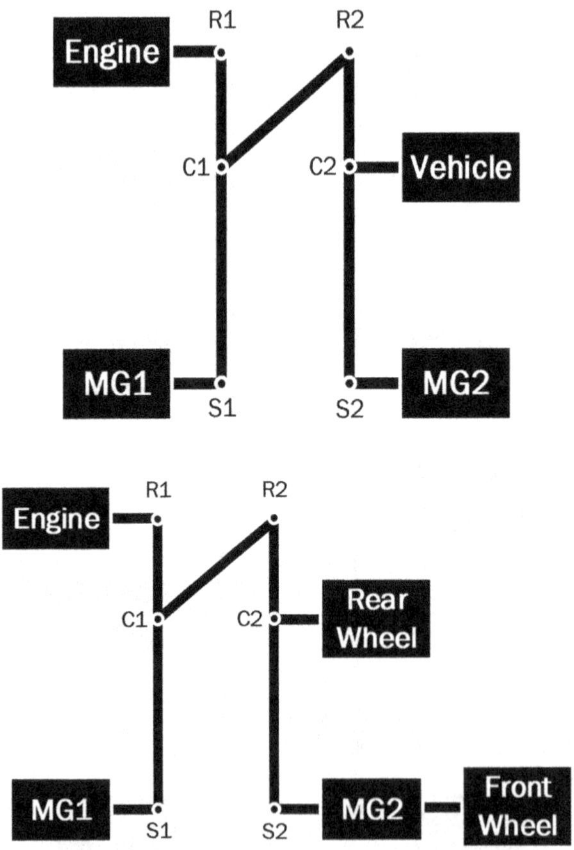

FIGURE 6.17 Examples of 3 DoF configuration. (a) Topology I (b) Topology II.

By adding an extra output shaft, as shown in Fig. 6.17(b), the 3 DoF system will become effective. By connecting the two output shafts with the front-rear or left-right wheels, four-wheel drive or the differential steering function can be achieved.

3) Compound-, input-, and output-split configuration
As introduced in Section 6.3.3, the input-split, output-split, and compound-split configurations all belong to the power-split configuration and are collectively known as electrical continuously variable transmission (ECVT) modes. All of these ECVT modes have 2 DoF, which enables the decoupling of the engine speed from the vehicle speed.

The differences between them are the coupling relationship among the vehicle, engine, and two MGs. In the input-split configuration, the speed of one MG is coupled with the vehicle speed, whereas the speeds of the engine and the other MGs are uncoupled with the vehicle speed, as shown in Fig. 6.18 (i.e., $R_{VMG1}R_{VMG2} = 2$). On the contrary, the engine

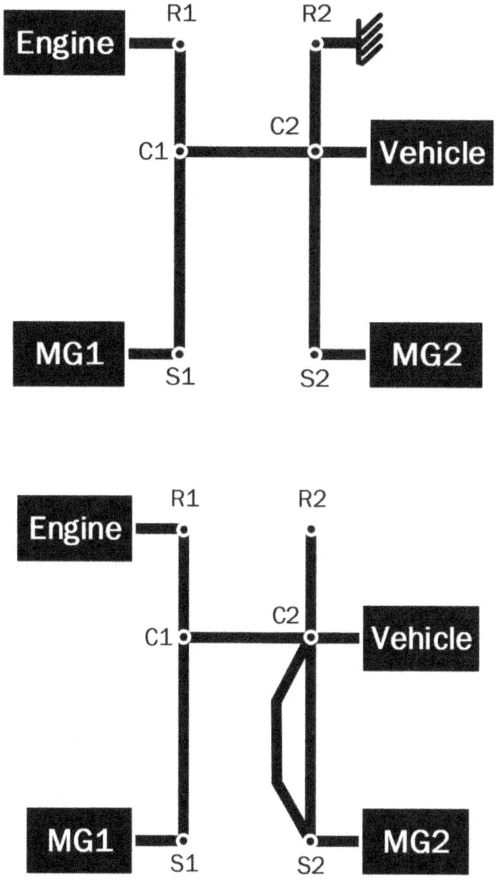

FIGURE 6.18 Examples of input-split configuration. (a) Topology I (b) Topology II.

speed of the output-split configuration is always coupled with one MG and is uncoupled with the speeds of the other MG and the vehicle, as shown in Fig. 6.19 (i.e., $R_{EMG1}R_{EMG2} = 2$). If the speeds of the vehicle, engine, and the two MGs are not coupled with each other, as depicted in Fig. 6.20 (i.e., $R_{EMG1}R_{EMG2} = 4$, $R_{VMG1}R_{VMG2} = 4$, $R_{VE} = 2$, $R_{MG1MG2} = 2$), the compound-split configuration is implied. The structural feature described previously results in different positions of the power split for different ECVT types, as mentioned in Section 6.3.3.

In addition to these structural differences, one of the MGs in an input-split configuration that has a fixed gear ratio with the vehicle output shaft can provide significant torque assist when launching the vehicle. This attribute makes this configuration type more effective than the other two types under low speeds and can still be feasible at high speeds if the motor's maximum speed allows. This configuration is widely

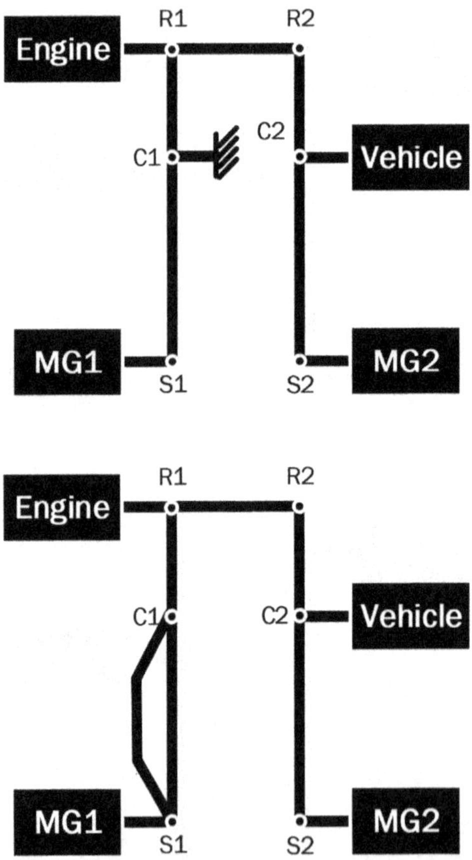

FIGURE 6.19 Examples of output-split configuration. (a) Topology I (b) Topology II.

applied in Toyota hybrid vehicle fleets. In comparison with the input-split configuration, the compound-split configuration can provide flatter output torque and has a wider speed range. In addition, the existence of two mechanical points makes the compound split more efficient under some working conditions, such as that at high speeds. Therefore, it is employed in some multi-mode HEV such as the Chevrolet Volt Gen2 as a high-speed mode to improve the overall vehicle fuel economy. Nevertheless, the output-split configuration has no extraordinary features and is therefore rarely used in vehicle production.

4) E-CVT with one motor/two MGs in serial configuration
An E-CVT with one motor can be viewed as a one-motor case of input-split configuration without MG coupling with the vehicle output shaft, as shown in Fig. 6.21 (a) (i.e., DoF=2; V_{eng} 0, V_{MG1} V_{MG2}= 0, and $V_{MG1}^2 + V_{MG2}^2 \neq 0$).

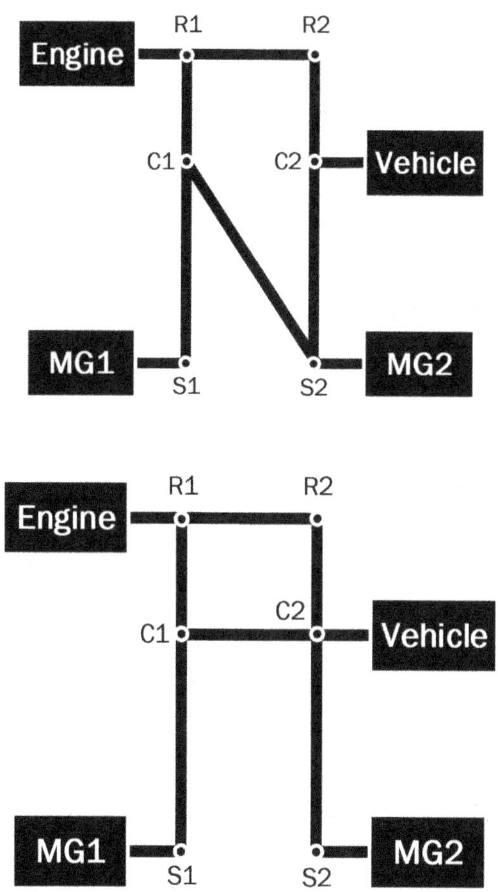

FIGURE 6.20 Examples of compound-split configuration. (a) Topology I (b) Topology II.

However, an ECVT with two MGs in a series is similar to that shown in Fig. 6.21 (b), although the two MGs are connected in a series and can be considered as one larger MG (i.e., $R_{MG_1MG_2} = 1$).

In this configuration type, the vehicle is still propelled simultaneously by the engine and one MG through the PG set. Such a powertrain arrangement provides an ECVT function with the help of the MG so that the engine speed can be controlled regardless of the vehicle speed. However, it does not offer the same flexibility in controlling the engine torque. The DoF of this powertrain is 2, but only two controllable powertrain components are retained, i.e., the engine and one MG. Thus, similar to that in the 3 DoF configuration, the DoF of the torque input is only one, so the engine torque cannot be arbitrarily assigned when the engine is operating at the desired speed.

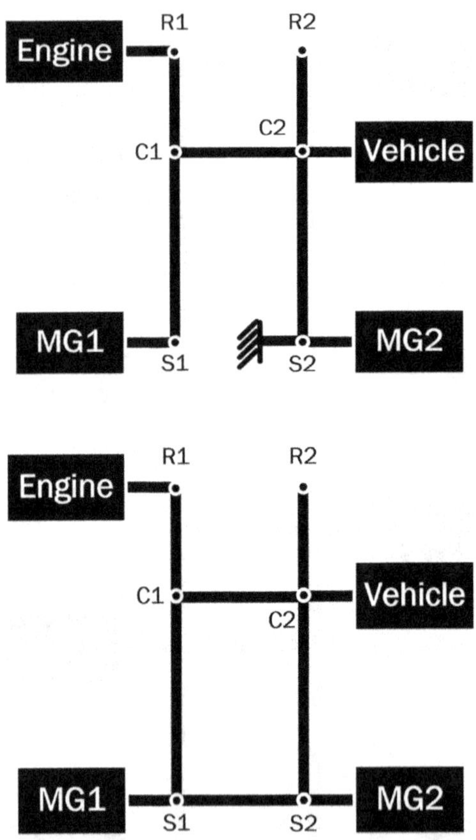

FIGURE 6.21 Examples of ECVTs with (a) one-motor configuration (b) two MGs in serial configuration.

Even considering these limitations, some researchers proposed powertrains including this configuration type because of its ECVT function. Yang *et al.* [10] and Zhu *et al.* [11] both developed power-split hybrid powertrains with a single MG. The uncontrollable engine torque makes such a configuration type limited while the vehicle is operated. As a result, although it is rarely applied in the vehicles produced, it can be used as an intermediate mode in some multi-mode hybrid powertrains when mode shift occurs and the component speeds need to be changed for clutch engagement conditions.

5) Engine-only configuration

In the engine-only configuration, both MGs cannot provide power to operate the vehicle, as shown in Fig. 6.22. In this circumstance, the output shaft is driven only by the engine with a fixed-gear ratio, which is the same as that in a conventional vehicle without MGs. Therefore, the advantages of powertrain hybridization disappear, which makes this configuration less desirable in multi-mode hybrid powertrains.

6) Parallel with fixed-gear configuration (2 MGs, 2 DoFs)

The engine in this configuration type is connected to the drive shaft mechanically with a fixed gear ratio (i.e., $V_{eng} \neq 0, R_{VE} = 1$), whereas the speeds of the two MGs are both decoupled from the vehicle speed through the PG set (i.e., DoF = 2, $V_{MG1} V_{MG2} \neq 0$), as shown in Fig. 6.23.

This topology is recognized as a parallel configuration because the engine speed is coupled to the vehicle speed, which is the same as that in parallel HEV. With the help of MGs, the torque of the engine can be regulated to achieve higher engine efficiency.

In this configuration type, since the speed of the MGs can be manipulated, higher efficiency could be achieved compared with the

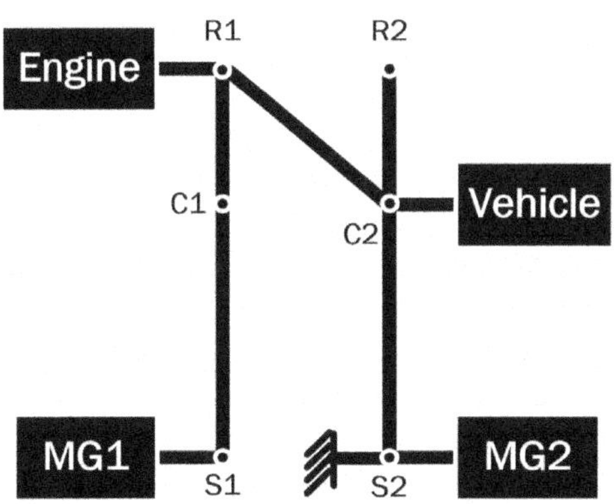

FIGURE 6.22 Example of engine-only mode.

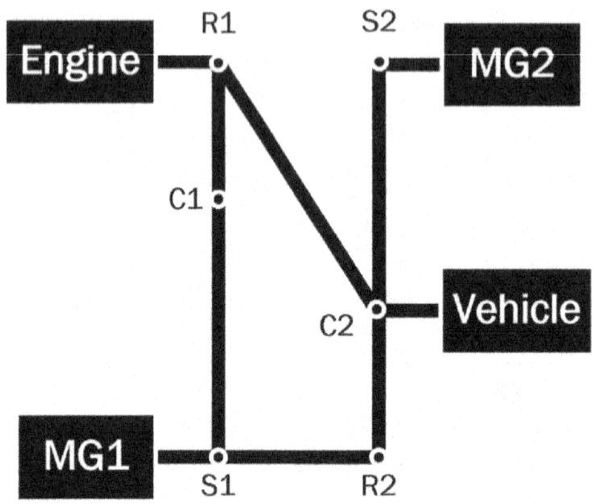

FIGURE 6.23 Example of parallel with fixed-gear mode (2 MGs, 2 DoFs).

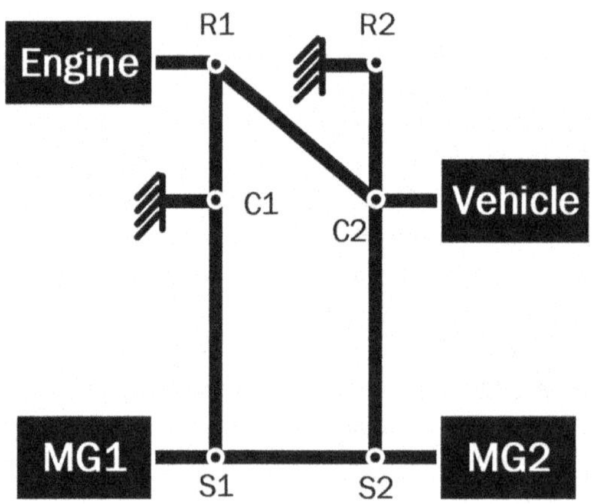

FIGURE 6.24 Example of parallel with fixed-gear configuration (2 MGs, 1 DoF).

configuration type in which the MG speed is proportional to the vehicle speed. However, this configuration was found to be topologically feasible only through an exhaustive search; to the best of our knowledge, it has not been adopted in any commercialized vehicles.

7) Parallel with fixed-gear configuration (1 MG/2 MGs, 1 DoF)
The parallel with fixed-gear configuration is the exact parallel configuration introduced in Section 6.3.1, in which the engine and MGs speeds are all proportional to the vehicle speed. If one of the MGs is grounded, the topology is parallel with the fixed-gear configuration (1 MG, 1 DoF), as shown in Fig. 6.24. Otherwise, the topology is

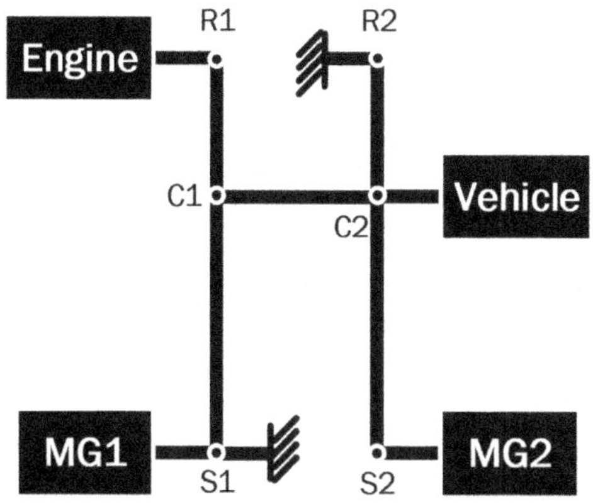

FIGURE 6.25 Example of parallel with fixed-gear configuration (1MG, 1 DoF).

marked as parallel with fixed-gear configuration (2 MGs, 1 DoF), and both MGs can either assist or recuperate energy from the vehicle, as shown in Fig. 6.25.

Different gear ratios between the engine and the output shaft will result in different parallel modes. The 3-PG powertrain can achieve a higher gear ratio (up to 14) if the ring/sun gear ratios are determined. The higher gear ratios are beneficial for buses, sport utility vehicles, and trucks, which require high-traction torque for acceleration, climbing, and towing. In addition, if the gear ratio is appropriate, the parallel configuration may have better efficiency than the ECVT modes at high vehicle speeds because less energy loss occurs in the electrical path, as discussed in Section 6.3.3.

In addition, the parallel with fixed-gear configuration has the maximum configuration numbers among all 14 configuration types, as shown in Table 6.2. Naturally, this configuration is easily combined with other configuration types to form a multi-mode hybrid powertrain.

8) PEV (2 MGs, 2 DoFs)
For one of the PEV modes, the speeds of both MGs are decoupled from the vehicle speed, which is referred to as PEV (2 MGs, 2 DoFs). In addition, the engine is always disabled or grounded, as shown in Fig. 6.26. Compared with conventional PEV mode, in which the MGs are connected with the driveshaft directly, this 2 DoF PEV can tune the speeds of both MGs to potentially achieve higher operation efficiency. Zhang *et al.* [9] proposed a dual-motor-driven electric bus adopting this configuration type that achieved excellent energy efficiency.

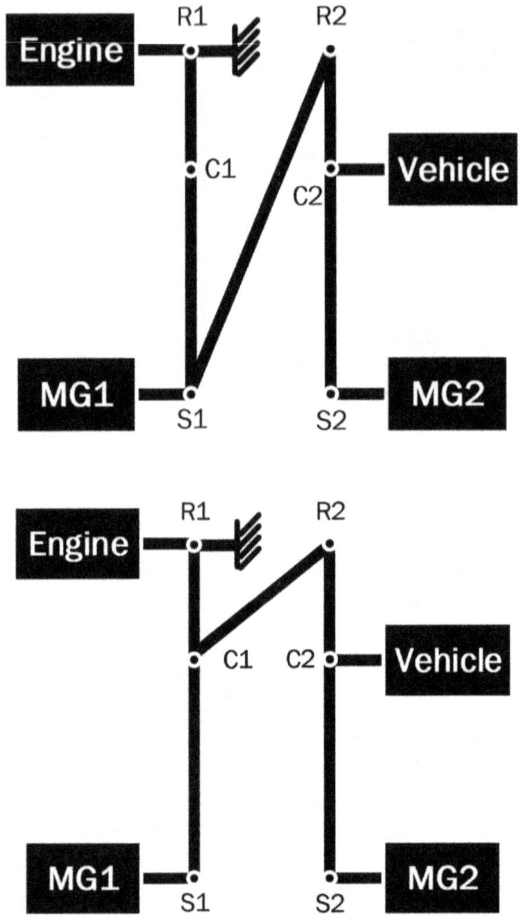

FIGURE 6.26 Example of PEV (2 MGs, 2 DoFs). (a) Topology I (b) Topology II.

9) PEV (1 MG/2 MGs,1 DoF)
   For PEV modes with 1 DoF, the engine is disabled or grounded by a grounding clutch, and the MGs are connected with the output shaft mechanically with fixed gear ratios as shown in Fig. 6.27 in the 2 MG case and Fig. 6.28 in the 1 MG case. Unlike that in the PEV mode with 2 DoFs, the MGs' speeds are coupled with the vehicle speed.

   For the PEV mode with two MGs, the torques of the MGs can be superimposed to achieve improved launching performance without running the engine. Moreover, instead of tuning the MG speed in PEV mode with 2 DoFs, the torques of both MGs can be manipulated to achieve better efficiency while satisfying the driver's demand.

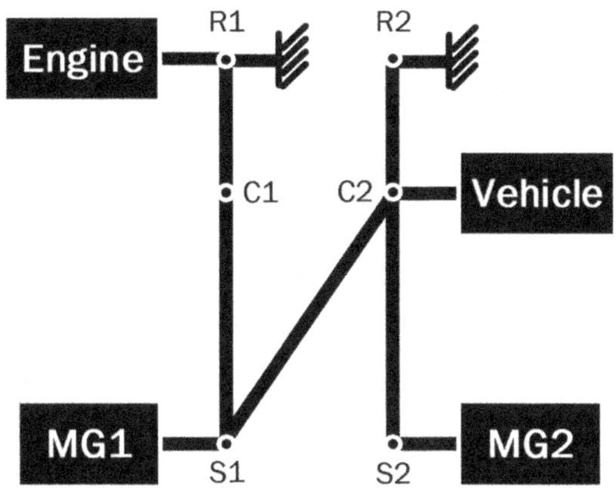

FIGURE 6.27 Example of PEV (2 MGs,1 DoF).

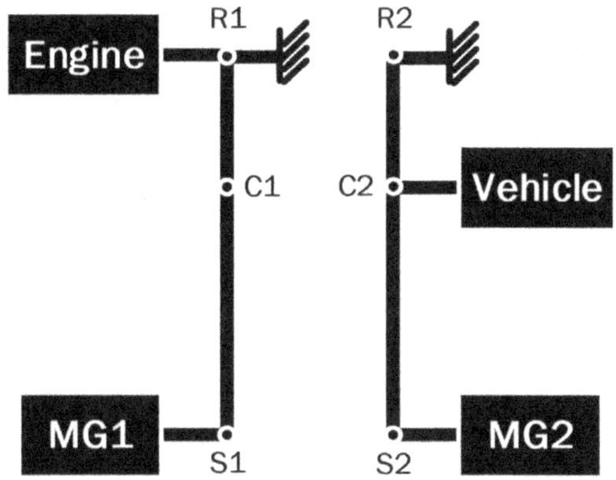

FIGURE 6.28 Example of PEV (1 MG,1 DoF).

6.6 SUMMARY AND CONCLUSION

The growing transportation activities have been not only substantially enhancing the mobility of people and goods but also producing more greenhouse gas emissions and consuming a large amount of energy. Tighter vehicle emission requirements are being implemented to lessen the effects of the vehicles on the environment, which encourages the development of alternative vehicle propulsion and energy storage solutions. One of the potential technologies to lower transportation sector emissions is the electrified vehicle, which may take several forms, such PEV and HEV. Powertrain architecture design, component sizing, and modeling are some of the fundamental technologies used in electrified vehicles. The main technologies of electric vehicle design are reviewed in detail in this chapter, together with their present state of research, difficulties, benefits, and potential future development trends. We anticipate that this chapter will have repercussions on the advancement of technology for electrified vehicles.

As reviewed so far, great efforts have been made in the field of topologies design and component sizing of electrified powertrains. However, developing an electrified powertrain with superior performance remains a challenge. In addition, the emergence of vehicle automation, connectivity, and artificial intelligence technology has provided a great opportunity to further improve vehicle economy, dynamic performance, and mobility. The research gaps and future trends are discussed in the following.

1) Vehicle design with an in-wheel motor
   An in-wheel motor, which integrates driving, braking, steering, and suspension modules into one unit and is independently controlled, has the advantages of high transmission efficiency, quick response time, reasonable axle load distribution, simplified chassis design, and diversified powertrain configuration. The in-wheel motor drive also has a lot of promise in terms of usage and maintenance, active and passive vehicle safety driving control, and intelligent and networked vehicle design. Therefore, designing a reasonable powertrain system with in-wheel motors to formulate an in-wheel motor drive PEV while integrating vehicle safety, energy-saving, and efficient control needs has become difficult a challenge for electrified vehicles design.
2) Multi-objective hybrid powertrain configuration optimization
   The optimal design of a hybrid powertrain is transformed into a nonlinear multi-objective restricted optimization problem by issues like energy efficiency, capital cost, produced emissions, drivability, and ride comfort. Vehicle drivability and ride quality throughout the mode transition are crucial for bringing a multi-mode hybrid powertrain to market. Additionally, a more complex powertrain configuration's effect on transmission efficiency hasn't been rigorously researched yet. In addition, hybrid technology will soon need to be added to trucks and buses due to crucial fuel economy criteria. To satisfy the demands of strong acceleration, hauling, and climbing, four-wheel drive operation is a crucial element for both trucks and buses. Therefore, more research into four-wheel drive solutions that keep the ECVT function is needed.
3) Expansion of optimization to connected and automated electrified vehicles
   Studies on electric powertrain topologies design and component sizing often concentrate on a few common driving cycles, such as inter-urban, in-city, and highway, to assess vehicle performance. However, regulatory testing that uses a regular driving cycle does not effectively account for real-world (or off-cycle) driving behavior. By utilizing look-ahead or preview traffic and vehicle information, advancements in vehicle connection and automation offer more potential for co-optimization of

vehicle dynamics and electrified powertrain control to maximize the vehicle performance in real-world driving. Therefore, it is fascinating to integrate the co-optimization of vehicle dynamics and powertrain control on a single vehicle basis, among cooperating vehicles, or even across the entire vehicle fleet, which can be a promising feature in electrified vehicle design for a certain application. This will improve total cost, energy economy, and vehicle dynamics.

REFERENCES

1. Bayrak, A. E., Ren, Y., & Papalambros, P. Y. (2016). Topology generation for hybrid electric vehicle architecture design. *Journal of Mechanical Design*, 138(8).
2. Kim, M. J., & Peng, H. (2007). Power management and design optimization of fuel cell/battery hybrid vehicles. *Journal of Power Sources*, 165(2), 819–832.
3. Lukic, S. M., & Emadi, A. (2004). Effects of drivetrain hybridization on fuel economy and dynamic performance of parallel hybrid electric vehicles. *IEEE Transactions on Vehicular Technology*, 53(2), 385–389.
4. Sinoquet, D., Rousseau, G., & Milhau, Y. (2011). Design optimization and optimal control for hybrid vehicles. *Optimization and Engineering*, 12(1), 199–213.
5. Malikopoulos, A. A. (2013). Impact of component sizing in plug-in hybrid electric vehicles for energy resource and greenhouse emissions reduction. *Journal of Energy Resources Technology*, 135(4).
6. Silvas, E., Hofman, T., Murgovski, N., Etman, L. P., & Steinbuch, M. (2016). Review of optimization strategies for system-level design in hybrid electric vehicles. *IEEE Transactions on Vehicular Technology*, 66(1), 57–70.
7. Adam, H. (2014). *Automated topology synthesis and optimization of hybrid electric vehicle powertrains*. Master's thesis, University of Waterloo.
8. Liu, J., & Peng, H. (2008). Modeling and control of a power-split hybrid vehicle. *IEEE Transactions on Control Systems Technology*, 16(6), 1242–1251.
9. Zhang, X., Peng, H., Sun, J., & Li, S. (2014). Automated modeling and mode screening for exhaustive search of double-planetary-gear power split hybrid powertrains. *Dynamic Systems and Control Conference*, 46186, V001T15A002.
10. Yang, Y., Hu, X., Pei, H., & Peng, Z. (2016). Comparison of power-split and parallel hybrid powertrain architectures with a single electric machine: Dynamic programming approach. *Applied Energy*, 168, 683–690.
11. Zhu, F., Chen, L., & Yin, C. (2012). Design and analysis of a novel multimode transmission for a HEV using a single electric machine. *IEEE Transactions on Vehicular Technology*, 62(3), 1097–1110.

7 Dedicated Thermal Propulsion Systems for Electrified Passenger Vehicles

Yanfei Li

7.1 BACKGROUND

The goals of carbon peaking and carbon neutrality and increasingly strict regulations on fuel economy and emissions pose huge challenges for passenger vehicles. It is predicted that in the transportation section, more than 50% of the passenger cars will still be equipped with internal combustion engines (ICEs); thus, it poses a great challenge for the ICEs to realize the goals and how to improve thermal efficiency. As a result, carbon-neutral ICEs have become the main motivation for ICE research and development. Herein, the main objective of this chapter is to present a preliminary overview of the thermal propulsion systems for electrified passenger cars in recent years, with an emphasis on the IC engines, including the recent advances in IC engine techniques, and the trend for the future development to overcome the challenges. Additionally, some potential techniques will also be presented, including gasoline compression ignition engines, hydrogen and fuel/engine co-optimization.

The adoption of two power sources in HEVs provides more degrees of freedom to design the powertrain, and different configuration can be found in the market. Their advantages and disadvantages can be can be found in recent reviews [1]. To date, using advanced artificial intelligence algorithms to realize the rapid component sizing of HEV powertrains becomes viable, considering the facts, like part performance, cost, lifetime, and so on [2, 3]. However, there are some limits for the engine specification selection. Nowadays, in order to achieve better fuel economy, more and more advanced and complicated technologies have been applied, and the engine operation characteristics are varied even given engines with the same displacement. Recent studies have demonstrated that the engine operations and the fuel savings are quite different when the HEV architectures are varied [4]. Therefore, how to co-design the powertrain and engine is still an important concern for the automotive industry. In spite of the different requirement on engine performance where the powertrain

 DOI: 10.1201/9781003302827-7

configuration varies, some common points can be achieved for the development of dedicated IC engines for HEV.

In contrast to the engines developed for traditional vehicles, the operation region for dedicated IC engines for HEVs is much narrower with the power assisted from the motor, which could significantly improve the engine part design. The engine low-end torque might be ignored, and this will provide the potential to further raise compression ratio; the low-engine efficiency zone (part loads) due to the pumping loss can also be suppressed. In addition, dedicated IC engines may not need to take into account the performance at the high-speed zone (e.g., >4500 rpm), and this will help improve the camshaft profile and reduce friction loss. But it should be noted that for different HEV configurations, some of the engine operation points might have to be located in the low-efficiency zone, due to the rigid connection between the engine and wheels.

Apart from the difference in the engine operation zone between the hybrid engines and traditional engines as mentioned before, one of the main differences is the intermittent work of the hybrid engines, or say frequent engine start/stop, which could raise much higher requirements for the engine aftertreatment systems.

Thus, for hybrid engines, in order to reduce the emissions during the start/stop, the requirements on the thermal management and dynamics of the aftertreatment systems become higher. Also, the intermittent engine operation could also lead to low oil temperature. It will influence the oil performance, such as emulsification and oil dilution, and finally influence the lubrication. Furthermore, considering the addition of a set of power source (torque motor), it is necessary to reduce the package volume and mass of the engine in order to fulfill the limited cabin capacity. For better drivability, dynamic response of the engine is also demanded.

7.2 ADVANCED ENGINE TECHNOLOGIES

High-efficiency and clean combustion are not only desirable for traditional engines but also core requirements on hybrid engines. Increasing compression ratio is the key technology to improve engine efficiency, and the combustion knocking caused by high compression ratios is the key obstacle. In the following, the main technologies used or to be used in dedicated IC engines will be introduced.

7.2.1 Fuel Injection

Generally, for hybrid gasoline engines, two type of fuel supply systems are used in the market, port fuel injection (PFI) and direct injection (DI). DI are normally used in combination with turbocharging in order to achieve better efficiency and higher power output. For engines equipped with the PFI system, they can adopt higher compression ratios. Furthermore, it will also significantly reduce the system costs. Nowadays, the Naturally-inspired engines equipped with PFI systems have achieved the peak brake thermal efficiency of 43.02%, the highest among the engines in the market. Nevertheless, DI is still the mainstream technology globally.

The fuel spray process plays an important role in influencing DI engine performance. It not only determines the air/fuel mixture quality and the subsequent combustion but also contributes to the emissions very much, especially for PN emissions. For DI fuel supply systems, increasing injection pressure could effectively reduce the Sauter mean diameter and enhance the fuel/air mixture quality to realize high-efficiency combustion and lower PN emissions. Nowadays, the maximum injection pressure has reached 50 MPa for the engines launched into the market. Injector nozzle design is another key issue to improve the mixture formation by well organizing the fuel distribution. For instance, Toyota optimized the injector to enhance fuel radial dispersion and reduce the fuel impingement into the cylinder wall [5].

In recent years, flash boiling attracted plenty of attention, not only its frequent occurrence in engine operations but also its huge potential to improve atomization quality, even with a much lower injection pressure than the injection pressure in modern DI engines. However, for the multi-hole injectors, which are widely used, the occurrence of flash boiling may cause spray collapse due to the stronger jet-to-jet interactions. The change in the original jet orientation could deteriorate the mixture formation. Thus, in order to utilize the advances of flash boiling in improving the atomization and mitigate the side effect of spay collapse, massive studies have been conducted to understand its influencing factors and collapse mechanism, and some potential ways are proposed to promote the application in engines. The recent studies have shown that the flash boiling also contributes to the injector tip wetting, which could produce the tip deposit and PN drift, that is, much higher PN emissions compared to clean injectors. Nowadays, adding detergent is an effective way to prohibit the formation of tip deposit.

Injection strategy is another important influence on fuel economy and emissions. It will affect the fuel distribution and the interactions of high-speed jets, and charging air may also promote the mixture. Early injection is conducive to fuel impingement due to the low ambient pressure, while late injection can lead to the less homogeneity of the fuel/air mixture. Via multiple injection strategy, the fuel impingement on the cylinder wall can be reduced, mitigating the pool firing. Furthermore, by co-ordinating the injection timing and tumble motion, the sprays have the potential to enhance the tumble.

7.2.2 Air Motion Organization

The intake port determines the intake flow. With the increase in the flow intensity, the combustion will become more stable and faster. With the increase in power density, the combustion duration will be significantly shortened, helping reduce the heat transfer loss. Furthermore, more challenges are raised with the wide application of exhaust gas recirculation (EGR). The introduction of exhaust gas can lower the flame propagation and deteriorate combustion stability. The enhancement of intake air motion can offset this disadvantage. However, with the increase to the tumble ratio, the port flow co-efficient could go lower. This will be more prominent for Miller cycle, and this is the reason why engines with Miller cycle normally adopt turbocharging [6].

For modern hybrid engines, Atkinson and Miller cycles are used instead of the traditional Otto cycle to improve engine efficiency. The essence of Atkinson and Miller cycles is to increase the expansion ratio by decreasing the effective compression ratio, which can be realized by late intake valve closing (Atkinson cycle/ LIVC) or early intake valve closing (Miller cycle/(EIVC). Both of them could dramatically reduce pumping loss, but there are also some differences between the two cycles during the engine operations. In the low-speed high-load conditions, EIVC could contribute to the charge expansion, and lower charge temperature can be obtained, which will be of help achieving a better combustion phasing. For Atkinson cycle, the partial charge will be pushed back into the intake port, and this will cause the higher intake charge temperature, which will negatively influence the knocking. However, it should be noted that EIVC can strongly affect the charge motion due to the shorter valve lift than Miller and Otto cycles. The tumble intensity can be largely reduced, and the decay time is much longer, which may affect combustion stability and the tolerance to exhaust gas. Thus, in order to guarantee the engine power output, the engines adopting Miller cycle are normally supercharged. Additionally, intake valve masking is also adopted to enhance the tumble inside the chamber.

EGR was initially adopted in diesel engines as an effective measure to reduce NOx. With the spread of DI in gasoline engines, EGR has been commonly used in combination with other technologies (e.g., high compression ratio, turbocharging). In addition to the NOx reduction, the adoption of EGR can also reduce throttling loss at part loads. With the ever-increasing demand on fuel economy, EGR is also found to be a good way to replace fuel enrichment and knocking inhibition. There are several types of EGR (e.g., hot EGR, cooled EGR, high-pressure EGR and low-pressure EGR) depending on the positions where the exhaust gas is taken and where the exhaust gas is introduced in the intake pipe [7]. Roth et al. [8] investigated different EGR on the engine map and reported that different EGRs have their own applicability. Nowadays, low-pressure EGR is the mostly investigated and used in practical engines. Furthermore, for low-pressure EGR, Roth et al. [9] also proposed the concepts of clean and dirty EGR, representing the exhaust gas induction positions before and after the pre-catalyst. With the dirty EGR, in addition to the better optimization of EGR pipes, it can also have the following advantages: 1) further reduction in the throttling loss, 2) the larger the pressure drop, the higher the EGR rate, 3) better cool performance due to the lack of the exothermic reactions in the pre-catalyst, and 4) the residual hydrocarbon can be reintroduced into the combustion chamber.

Water injection by PFI or DI can broaden the combustion knocking boundary by reducing the in-cylinder temperature; thus, it has the potential in increase compression ratio and combustion phasing optimization. It can be realized by both PFI and DI. In the PFI mode, water is injected in the intake port or manifold and enters the combustion chamber together with the fresh charge. Thus, the water distribution inside the chamber is homogeneous. But due to the water impingement, the water consumption is relatively higher than that of DI. The main advantages of water DI are the flexibility in controlling the water amount and distribution inside the chamber by adjusting the injection timing and water/fuel ratios over different operating conditions. Generally, the injection strategy should avoid the water film formation inside the chamber and be vaporised before the start of combustion. Improper injection strategy could cause extinguishment, oil emulsification, higher coefficient of variation, and emission deterioration.

7.2.3 Combustion Organization

For the timing being, stochiometric combustion is normally adopted for gasoline engines launched into the market due to the low-cost three-way catalysts (TWCs). Lean burn is another technology that can significantly raise the engine thermal efficiency. The lean mixture can increase the specific heat, the low-temperature combustion can reduce the heat transfer loss to the cylinder wall and has the potential to further increase the compression ratio, and more air sucked into the chamber can reduce the throttling loss. The lean burn or ultra-lean burn in the engines brings about challenges for the ignition systems. Therefore, innovative ignition systems were intensively investigated in recent years [10]. The prechamber ignition has attracted plenty of attention. By igniting the fuel/air mixture inside the prechamber, the flame will travel through the holes in the prechamber and produce several high-speed reacting jets, which will then ignite the mixture in the main chamber and finally realize the multi-point ignition, shortening the combustion duration. Generally, the traditional spark ignition can be termed as single-point ignition, and prechamber can be termed as spatial ignition. Additionally, the prechamber ignition can also be divided into two categories: passive and active prechamber ignition. For the former one, the ignitable mixture in the prechamber comes from the main chamber, while for the latter one, the mixture is obtained by the fuel supply system inside the prechamber and can be gaseous or liquid fuel. The prechamber ignition is believed to be promising for lean burn. Still, intensive work is still being investigated. With the adoptions of the advanced combustion technologies, the brake thermal efficiency larger than 50% has been demonstrated for the time being [11].

7.2.4 Fuel/Engine Co-Optimization

Previous technologies to enhance brake thermal efficiency improve the hardware or control strategy to boost the air/fuel mixture and combustion. In fact, the physiochemical properties of the fuel are important influences on the efficiency. With the advancement of engine technologies (e.g., turbocharging, high compression ratio, EGR, and so on), the traditional indices for gasoline should be further evaluated to maximize the engine efficiency and optimize the emissions, especially in the context of HEV, narrower engine operation range, and the utilization of biofuels blended with gasoline. Thus, how to realize the fuel-engine co-optimization becomes important by integrating the oil refining and engine manufacturing, and intensive studies have been conducted. The most systematic investigation is sponsored by US Department of Energy (DoE) Co-Optima program, aiming to identify the critical fuel properties for maximizing engine efficiency and emissions performance, in combination with the advanced combustion modes. Then, six fuel properties (i.e., research octane number, octane sensitivity, latent heat of vaporization, laminar flame speed, particulate matter index, and catalyst light-off temperature) were proposed and combined into a unified merit function to evaluate the potential for efficiency increase of fuels with conventional and non-conventional compositions.

7.2.5 Others

Apart from the technologies as mentioned previously, there are still a lot of potential ways to improve thermal efficiency and emissions in terms of combustion modes and engine operations.

Engine displacement is one of the key parameters in determining engine power. Turbocharging and downsizing were believed to be effective ways to increase the thermal efficiency because they can improve the thermal efficiency at the part loads. However, with enhanced turbocharging, the knocking tendency at low-speed high-load conditions becomes more severe and might bring in super-knock. Thus, engine rightsizing becomes one of the key objectives during engine R&D. It is a compromise among compression ratio, the degree of turbocharging, Atkinson/Miller cycles, and other technologies in pursuing the optimal fuel economy. In the context of HEVs, the selection of engine displacement is quite complicated and is relevant with the engine technologies used and the powertrain.

Long stroke (i.e., large stroke/bore diameter ratio) technology can also increase the thermal efficiency mainly. For a given displacement and compression ratio, larger stroke/bore diameter ratio can obtain smaller volume/surface area ratio, mitigating the heat transfer loss. It can also reduce the distance of flame propagation and shorten the combustion duration. Then, the thermal efficiency can be improved. Larger S/B can increase the piston speed, but this can be ignored in the context of HEV application because the high-speed rotation speed may not exist. Another disadvantage for large S/B is the increase in friction loss. Therefore, there is a threshold valve for the S/B to obtain the optimal fuel economy.

Cylinder deactivation is another technology worth considering, and its contribution to fuel consumption is realized primarily by the reduction in pumping losses, improved combustion, and reduced oxygen saturation of catalysts during deceleration fuel cut events [12]. Additionally, cylinder deactivation can also smooth the output torque fluctuation when both engine and motor works. Variable compression ratio (VCR), by varying the clearance volume at the top dead center, can mitigate the limits of the knocking boundary in traditional engines with a fixed compression ratio. Thus, in the non-knocking-limited region, higher compression ratio can be used to achieve high efficiency. In order to realize the compact design of ICEs, some innovative designs, e.g., the integration of start-starter-generation with engine flywheel [13], free piston engines were also proposed. The new combustion modes, like gasoline compression ignition (GCI), also have great potential to contribute to dedicated IC engines for HEV. But still there is a lot of work to do before the launch into the market in terms of the reliability and cost.

Recently, hydrogen and ammonia have become hot topics due to pressure from the goal of carbon peaking and carbon neutrality. Hydrogen-fuelled IC engines can continue thermal propulsion with zero tank-to-wheel CO_2 emissions. Also, it can be fed H_2 with lower purity as well as has much lower production costs. Meanwhile, for the research and development of H_2-fuelled engines, the combustion system should be significantly improved due to the different air/H_2 mixing process and the combustion characteristics from those fuelled with gasoline. However, due to the difficulty in the storage and transportation and the lower energy density, ammonia, as a H_2 carrier fuel and also the representative of hydrogen economy 2.0, may become the viable alternative. Still we must confront the following challenges: combustion and emissions. First, on combustion, ammonia has much lower flame propagation speed and high auto-ignition temperature, causing the combustion instability and cold start problems. Duel fuel can be possible ways to solve the two issues but may increase the cost in fuel supply systems. Next, regarding emissions, although ammonia is carbon-free, combustion production can include NOx and residual $NH_{3,}$ and this raises the question that dedicated aftertreatment systems for ammonia becomes inevitable.

7.3 HEV EMISSIONS

HEVs have demonstrated their ability to reduce fuel consumption. However, emission challenges continue due to intermittent operations, where the majority of emissions due to the low exhaust temperature and rich combustion come from. Huang et al. [14] compared two pairs of hybrid and conventional vehicles of the same model for RDE tests and found that the fuel economy of the HEVs can be improved by up to 49%, but there was no reduction in HC emissions and consistently higher CO in contrast the conventional ICE only vehicles. Yang et al. [15] compared two HEVs (equipped with PFI and GDI systems, respectively) with the engine-only counterparts and reported that the PN emissions are obviously higher for HEVs. Furthermore, the GDI-HEV also had higher PN emissions than PFI-HEV. Suarez-Bertoa et al. [16] tested two PHEVs under WLTC conditions with ambient temperature equal to 23 °C and –7 °C and reported the modern PHEV could emit similar or even higher levels of PN and NOx than the conventional gasoline and diesel vehicles. Furthermore, for the blended PHEV, once the vehicle needs higher power out during the charge depleting state, IC engines would be engaged with high power command in a short time. This process is normally termed as high-power cold starts. The research from the California Air Resources Board reported that high-power cold starts could cause significantly higher pollution than conventional vehicles. Connectivity and automation are believed to be a transformative way to raise fuel economy by the optimization of vehicle velocity profile and energy management strategy. However, the study by Amini et al. [17] reported that eco-driving could lower the thermal response, leading to the increase in HC emissions due to the delayed light off of TWC.

For RDE emissions, the most challenging impact is the driving style. Although the main power requirements are relatively stable, the huge difference in engine dynamics can be found in terms of engine torque, engine speed, and engine power, causing the emission species spikes and enhancing the trend of engine right-sizing [18]. It also poses the challenges for the aftertreatment systems, e.g., fast light-off, better oxygen storage capacity under high space velocity, higher PN filtration and more intelligent strategy for GPF regeneration. Recently, in order to address the poor performance of TWC during low temperature conditions, electric heating catalysts also have been intensively studied [19]. Thus, in spite of the electrified powertrain achieving better fuel economy from different ways (e.g., engine start/stop, brake energy regeneration), the energy control strategies play an important role in determining emissions. The key for emission control is thermal management. Advanced thermal management should be taken into consideration for EMS design to realize the optimization of both fuel economy and emissions, not only the engines and their aftertreatment systems but also the vehicle level.

7.4 SUMMARY AND OUTLOOK

The goals of carbon peaking and carbon neutrality and the increasingly strict regulations on fuel economy and emissions bring in great challenges for IC engines. Electrified passenger vehicles will co-exist with electric vehicles in the mid- and long- terms. To address the challenges, intensive efforts have been paid to improve fuel economy and emission performance. For engines in the context of electrified passenger vehicles, the advancement in air organization, fuel delivery, and combustion organization has been widely applied. The challenges also promote fuel diversification of IC engines (e.g., hydrogen, ammonia) and much closer collaboration between oil refineries and engine manufacturers to realize the fuel/engine co-optimization and application of advanced combustion modes. However, for the further development of dedicated ICEs for HEV, there are still necessary work to do, including advanced control strategy integrating the power management and thermal management, dedicated key parts (e.g., injection, turbocharging and aftertreatment systems) and material compatibility with advanced fuels.

REFERENCES

1. Zhuang W, Li S, Zhang X, Kum D, Song Z, Yin G, et al. A survey of powertrain configuration studies on hybrid electric vehicles. *Applied Energy* 2020;262:114553.
2. Anselma PG, Biswas A, Bruck L, Amirfarhangi Bonab S, Lempert A, Roeleveld J, et al. *Accelerated Sizing of a Power Split Electrified Powertrain*. SAE International; 2020.
3. Zhou Q, He Y, Zhao D, Li J, Li Y, Williams H, et al. Modified particle swarm optimization with chaotic attraction strategy for modular design of hybrid powertrains. *IEEE Transactions on Transportation Electrification* 2020;7.
4. Vijayagopal R, Nieto Prada D, Costanzo V. *Powertrain Choices for Emerging Engine Technologies*. SAE International; 2020.

5. Saitoh S, Shibata H, Ookuma M, Shigenaga M. *Evolution of Gasoline Direct Injection System for Reduction of Real Mode Emission*. SAE International; 2019.
6. Schnorpfeil S-J, Fuchs B, Baum E. Targets and parameters for combustion system layout. In: Liebl J, Beidl C, eds. *Internationaler Motorenkongress 2017*. Springer Fachmedien; 2017:341–56.
7. Wei H, Zhu T, Shu G, Tan L, Wang Y. Gasoline engine exhaust gas recirculation—A review. *Applied Energy* 2012;99:534–44.
8. Roth D, Sauerstein R, Becker M, Meilinger R. Application of hybrid EGR systems to turbocharged GDI engines. *MTZ Worldwide* 2010;71(4):12–17.
9. Roth DB, Keller P, Becker M. *Requirements of External EGR Systems for Dual Cam Phaser Turbo GDI Engines*. SAE International; 2010.
10. Zheng M, Chen G, Tjong J, Li L, Yu S, Yu X, Yang Z. *Spark-Based Advanced Ignition Control for Future Diluted Gasoline Engines*. 4th International Conference of Ignition Systems for Gasoline Engines, Berlin, Germany; 2019.
11. Nagasawa T, Okura Y, Yamada R, Sato S, Kosaka H, Yokomori T, et al. Thermal efficiency improvement of super-lean burn spark ignition engine by stratified water insulation on piston top surface. *International Journal of Engine Research* 2020. https://doi.org/10.1177/1468087420908164.
12. Wilcutts M, Nagashima M, Eisazadeh-Far K, Younkins M, Confer K. *Electrified Dynamic Skip Fire (eDSF): Design and Benefits*. SAE International; 2018.
13. Hu Y, Yang F, Du L, Zhang J, Ouyang M. A novel method to actively damp the vibration of the hybrid powertrain by utilizing a flywheel integrated-starter-generator. *IEEE Access* 2020;8:147045–58.
14. Huang Y, Surawski NC, Organ B, Zhou JL, Tang OHH, Chan EFC. Fuel consumption and emissions performance under real driving: Comparison between hybrid and conventional vehicles. *Science of the Total Environment* 2019;659:275–82.
15. Yang Z, Ge Y, Thomas D, Wang X, Su S, Li H, et al. Real driving particle number (PN) emissions from China-6 compliant PFI and GDI hybrid electrical vehicles. *Atmospheric Environment* 2019;199:70–9.
16. Suarez-Bertoa R, Pavlovic J, Trentadue G, Otura-Garcia M, Tansini A, Ciuffo B, et al. Effect of low ambient temperature on emissions and electric range of plug-in hybrid electric vehicles. *ACS Omega* 2019;4(2):3159–68.
17. Amini MR, Feng Y, Wang H, Kolmanovsky I, Sun J. Thermal responses of connected HEVs engine and aftertreatment systems to eco-driving. *2019 IEEE Conference on Control Technology and Applications (CCTA)*. 2019:724–9.
18. Fraidl G, Kapus P, Vidmar K. *The Gasoline Engine and RDE Challenges and Prospects*. Springer Fachmedien; 2016:257–83.
19. Okajima T, Sone R, Yan X, Inoue R, Sivakumar S, Shingyouchi H, et al. *Exhaust Purification Performance Enhancement by Early Activation of Three Way Catalysts for Gasoline Engines Used in Hybrid Electric Vehicles*. SAE International; 2019.

8 Thermal Management in E-Mobility Systems Based on Heat Pump Air Conditioner

Fundamentals and Key Developments

Zhi Li, Xiaoli Yu and Ruicheng Jiang

8.1 INTRODUCTION

E-mobility systems have experienced giant advances in recent years since the carbon neutrality strategy has received broad consensus around the world [1]. Especially in transport sectors, electric vehicles (EVs) are replacing conventional fuel vehicles at an unprecedented speed due to their superiority in high energy-efficiency and low carbon emissions, and the share of EVs has grown significantly in the past ten years [2, 3]. For fuel vehicles, the heating load of vehicle cabins under cold winters can be totally satisfied by the waste heat of internal combustion engines [4], but that is not the case for EVs since the electricity consumed by heating devices is provided by batteries. It is predicted that there will be more than two billion EVs on the road by 2035 [5], and an efficient heating method remains a major bottleneck for further development of EVs. Positive temperature coefficient (PTC) electrical heaters used to be the main heating method for EVs due to their simple structure and low cost. Previous results demonstrated that the energy consumption to meet the heat load by sole PTC accounted for more than 30% of the total energy during the driving process of EVs [6, 7].

In order to improve the thermal efficiency of heating under cold conditions, heat pump air conditioners (HPAC) have been gradually instituting PTC in EVs because HPACs possess the advantages of high efficiency and can effectively prolong the driving mileages of EVs [8]. However, PTC is also used to provide back-up heating capacity when HPAC cannot satisfy the heating load under extremely cold conditions. Meanwhile, the heating capacity and coefficient of performance (COP) of HPAC decline significantly with the decrease of the ambient temperature; thus

DOI: 10.1201/9781003302827-10

TABLE 8.1
Heating Methods Used by Different Car Company in Recent Years

Car company	Brand	Year to market	Heating method
BMW	i3	2013	PTC
Renault	zeo	2013	PTC
Nissan	Leaf	2013	PTC
Volkswagen	New Golf	2017	PTC, HPAC+WHR
Jaguar	I-Pace	2018	PTC, HPAC+WHR
SAIC	Ei5, MARVEL X	2018	PTC
NIO	ES6	2018	PTC, HPAC+WHR
Tesla	Model Y	2019	PTC, HPAC+WHR
Geely	Geometry C	2020	PTC
ZEEKER	001	2021	PTC, HPAC+WHR
XPENG	P5	2021	PTC, HPAC+WHR

the energy efficiency of HPAC needs to be further improved [9]. Considering the waste heat generated by batteries, electric motors and motor controllers, HPAC integrated with a waste heat recovery (WHR) unit has been proposed to utilize this waste heat and achieve a higher COP in recent years, and these novel combined systems have been applied in some popular EVs as shown in Table 8.1.

Although heat pump air conditioners have better comprehensive performance than conventional heating methods, there are also some challenges hindering the further development and application of HPAC. On the one hand, HPAC generally adopt organic refrigerants (R134a and R407c) as working fluids, and the leak of working fluids will cause severe greenhouse effect [10], especially when the number of EVs grows. As a result, developing novel environmentally friendly working fluids such as CO_2 and R1234yf is quite important. On the other hand, operating conditions of HPAC cannot always be maintained under steady state since the ambient temperature and heat load of EV cabins are variable. Proper control strategies are indispensable for the safe and efficient operation of HPAC for EVs. Furthermore, there is an apparent trend that the thermal management of HPAC will be combined with the temperature control of batteries, motors, and motor controllers, forming an integrated intelligent thermal management system in EVs [11]. The integrated thermal management system with intelligent control strategies can achieve higher energy efficiency and lower power consumption.

Considering these giant advances and existing challenges of HPAC for EVs, this chapter aims to give an overview of HPAC for EVs. Firstly, different typical HPAC system protypes are concluded from the aspect of energy and environmental performance, including the conventional air source heat pump air conditioner systems, HPAC with WHR systems, and CO_2-based HPAC systems. Then the intelligent thermal management of single HPAC systems and integrated HPAC-batteries-motors systems are summarized. With the information and knowledge concluded in this chapter, the development and optimization of future HPAC systems can be expedited.

8.2 KEY DEVELOPMENTS OF HEAT PUMP AIR CONDITIONERS

The air source heat pump air conditioner is the typical system used in electric vehicles. In recent years, this system has been modified for the sake of higher energy efficiency and better environmental performance. The main solution to improve the energy efficiency of air source HPAC is to recover waste heat from batteries, electric motors, and motor controllers, and many studies focus on the design and optimization of novel HPAC-WHR combined systems. Developing more environmental working fluids is also an important issue for HPAC systems, especially adopting CO_2 as the working fluid has attracted lots of attention in the last few years. The high operating pressure of CO_2-based HPAC systems requires more efforts to improve the system design and optimization. Therefore, this section will give an overview of different typical HPAC systems.

8.2.1 AIR SOURCE HEAT PUMP AIR CONDITIONER SYSTEMS

The schematic diagram of the air source heat pump air conditioner is shown in Figure 8.1. The liquid working fluid absorbs the heat from air and becomes vapor, then the vapor working fluid is compressed as high-temperature vapor. Later, the high-temperature vapor condenses at the condenser and the released heat is transferred to the cabin by the fan. Finally, the exhausted liquid working fluid is throttled by the expansion valve and starts a new cycle. The air source HPAC can achieve a relatively comfortable thermal environment of the cabin under very cold ambient conditions, but it consumes large amounts of energy derived from batteries. To avoid the shortage of heat capacity, PTC heaters are widely applied as auxiliary heating in EVs [12]. HPAC systems possess superiorities in simple system configuration and high safety. According to the second law of thermodynamics,

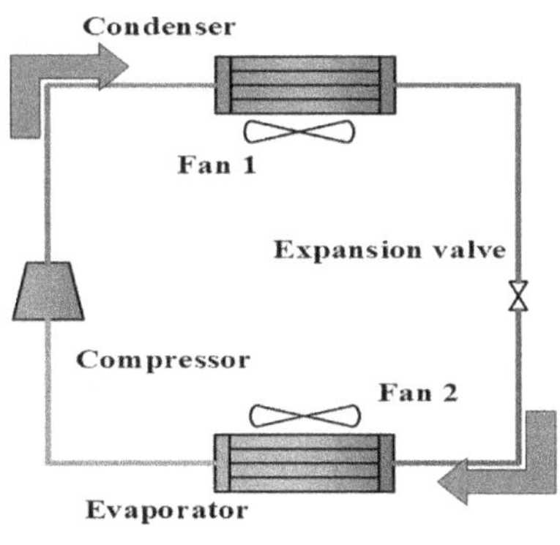

FIGURE 8.1 Basic layout of air source heat pump system.

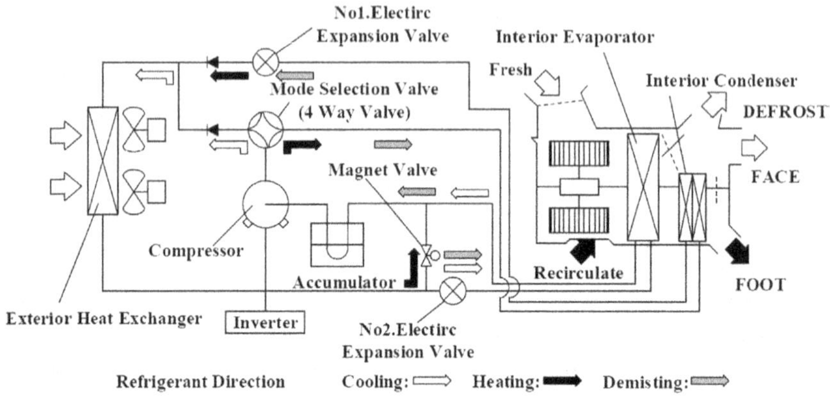

FIGURE 8.2 Schematic diagram of R134a HPAC for EVs.

the COP of HPAC can exceed 1, and it can be further improved by system optimization and control.

Air source HPAC systems generally adopt R134a as the working fluid. Suzuki and Katsuya [13] first proposed the protype of R134a HPAC for EVs, and the schematic diagram is shown in Figure 8.2. The proposed system provided multi functions including heating, cooling, demisting, and dehumidification. The experimental results indicated that the heat capacity and COP could attain as large as 2.3 kW and 2.3 under the ambient temperature of −10 °C. Working fluids play an important role in heating performance of HPAC systems. Wang et al. [14] compared the heating performance of R134a and R407C HPAC systems for EVs under the ambient temperature of –10 °C, and the performance comparison is depicted in Figure 8.3. The results showed that the heating capacity and compressor power of R407C system increased compared to R134a system, and the R407C system achieved a heating COP of 2.3. In addition to working fluids, operating conditions also affect the heat performance of HPAC systems. Lee et al. [15] investigated the steady state and start-up performance characteristics of air source HPAC with R134a as working fluid for cabin heating in a passenger EV. Although the experimental results illustrated that the heat COP and capacity could be as high as 3.3 and 3.1 kW at the ambient temperature of –10 °C, the observed heating and transient performance of HPAC could be insufficient to satisfy the heating load of EV cabin, which implied the importance of hybrid heat methods or further improving the energy efficiency of HPAC under cold conditions.

8.2.2 Heat Pump Air Conditioner Systems with Waste Heat Recovery

As the ambient temperature decreases, the heating capacity and COP of the HPAC decrease significantly. One of the most important solutions is to utilize the waste heat from the motors, motor controllers, and batteries. Heat pump air conditioners combining waste heat recovery (HPAC-WHR) systems are conducive to improving

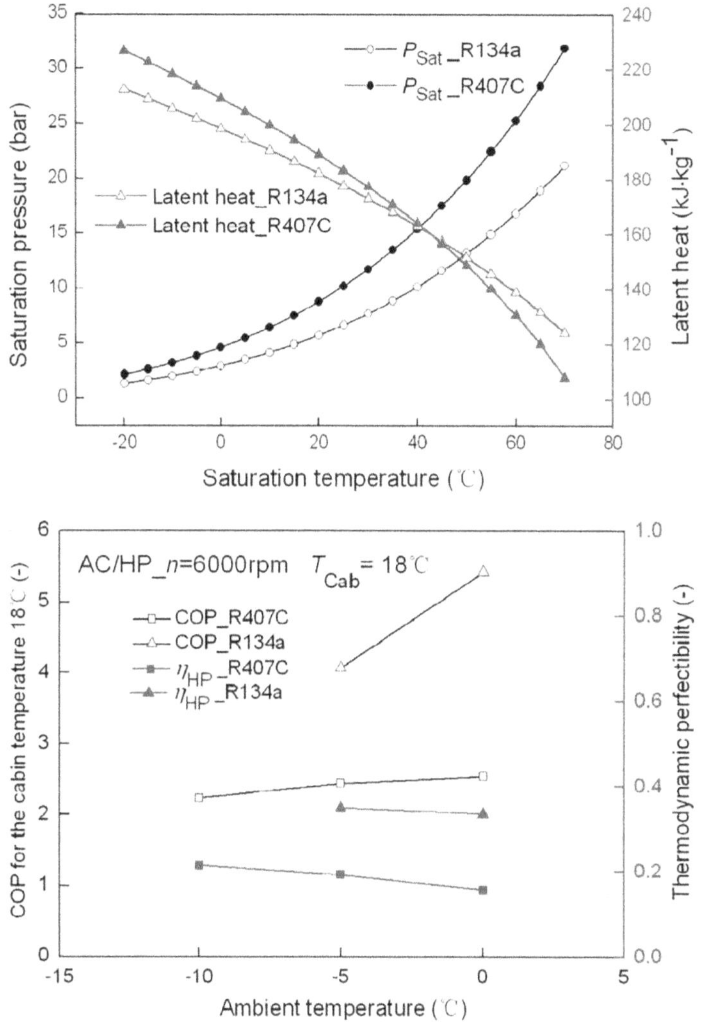

FIGURE 8.3 Performance comparison of HPAC using R134a and R407C: (a) variations of saturation pressure and latent heat with saturation temperature; (b) COP and thermodynamic perfectibility.

the energy efficiency of EVs and have advantages in cost and technical maturity. Ahn et al. [16–18] conducted a series of studies to investigate the performance of HPAC with waste heat recovery. In 2014, Ahn et al. [16] proposed a dual-source R134a HPAC system including both air and waste heat for EVs. The experiment results indicated that the dual-source system outperformed the air source–only and waste heat–only HPAC systems in the heating mode, and the results are shown in Figure 8.4. However, the heating performance of the dual-source system was

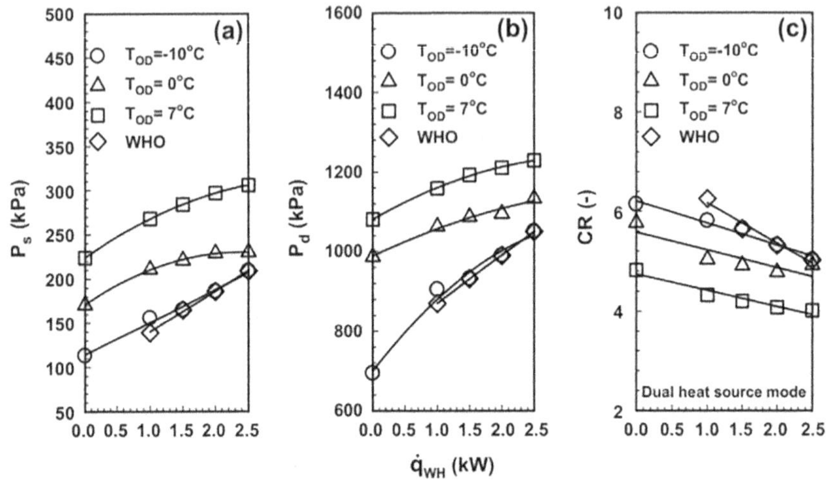

FIGURE 8.4 Variations of (a) suction pressure, (b) discharge pressure, and (c) compression ratio with waste heat amount and outdoor air temperature in the dual heat source mode.

greatly dependent on the amount of waste heat when the ambient temperature was −10 °C, since the amount of heat absorbed from the ambient air was inappreciable under low ambient temperature. In 2015, Ahn et al. [17] proposed a dual-evaporator HPAC using waste heat recovery from the dehumidifying process. The result showed that COP was 62% higher than that of the conventional HPAC system at the indoor air temperature of 13 °C. In 2016, Ahn et al. [18] investigated the performance improvement of a dehumidifying HPAC system using an additional waste heat source in EVs. The heating capacity and COP of the dual-source dehumidifying HPAC was found to increase by 75.8% and 5.2% respectively compared with the air source dehumidifying HPAC.

When combining HPAC and WHR, there are different kinds of layouts since waste heat can be recovered at different locations of HPAC for further improvement of heating performance. Therefore, different HPAC-WHR systems have been proposed and investigated for EVs in recent years. These layouts can be divided into three types, namely evaporator-side system, condenser-side system and refrigerant-side system. Figure 8.5 shows the schematic diagram of the HPAC system that the waste heat is recovered at the evaporator side. The waste heat of the motor and motor controller is first absorbed by the coolant of the WHR unit, then the waste heat is dissipated into the ambient air at the radiator located ahead of the evaporator of HPAC. Then the high-temperature air is pulled in to evaporate the working fluid of HPAC by the fan, obtaining a higher temperature working fluid of HPAC and achieving the aim of waste heat recovery. Finally, the working fluid in a vapor state is compressed by the compressor and releases heat to the cabin air. Qian et al. [19] investigated the heating performance of an evaporator-side HPAC-WHR system for EV. It turned out that the heating capacity

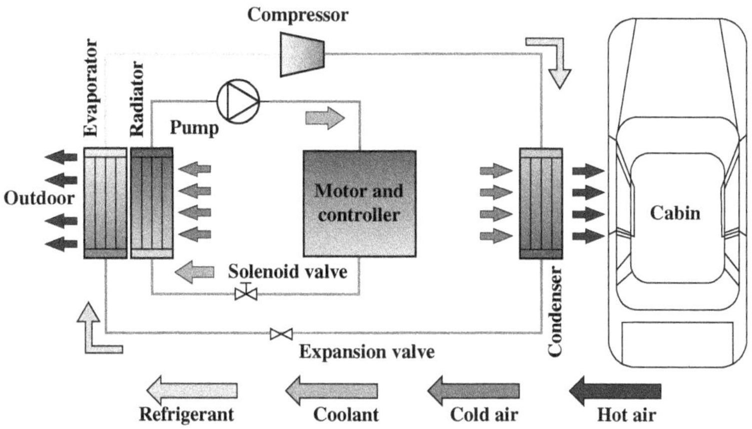

FIGURE 8.5 HPAC-WHR system with waste heat recovered at the evaporator side.

and COP were increased by 24.2% and 10.8% respectively compared with the air source HPAC system.

Figure 8.6 depicts the schematic diagram of the HPAC system where the waste heat is recovered at the condenser side. The waste heat of the motor and motor controller is first absorbed by the coolant of the WHR unit, then the waste heat is dissipated into the cabin air at the radiator located ahead of the condenser of HPAC. Later, the cabin air flowing through the condenser is heated by the working fluid of HPAC. Finally, it is further heated by the working fluid in the radiator of the WHR unit when pulled into the radiator by the fan, achieving the aim of waste heat recovery. Yokoyama et al. [7] studied the energy performance of a condenser-side HPAC-WHR system for EV. The results indicated that the power consumption could be reduced from 850 W to 580 W to satisfy the heat capacity of 2 kW when the ambient temperature was below 0 °C.

Figure 8.7 presents the schematic diagram of the HPAC system where the waste heat is recovered at the refrigerant side. This system is more complex than the previous two systems since the working fluid of HPAC is split into two parts to recover waste heat from the motor and motor controller. During the working process of HPAC, one part of the working fluid evaporates at the evaporator by absorbing heat from ambient air, while the other part of the working fluid evaporates at the radiator by absorbing heat from the high-temperature coolant of the WHR unit. Then these two parts of working fluid converge at the inlet of the compressor. Finally, the working fluid releases heat into the cabin air after the compression process, achieving the aim of waste heat recovery. Kowsky et al. [20] designed a refrigerant-side HPAC-WHR system for heating the cabin air and batteries of EVs. The results demonstrated that the proposed HPAC-WHR system could achieve a COP of 2.3 to meet the requirement of a heating capacity of 6.5 kW when the ambient temperature was below –10 °C, while the corresponding COP were 0.95 and 1.3 for only PTC mode and HPAC combining PTC mode, respectively.

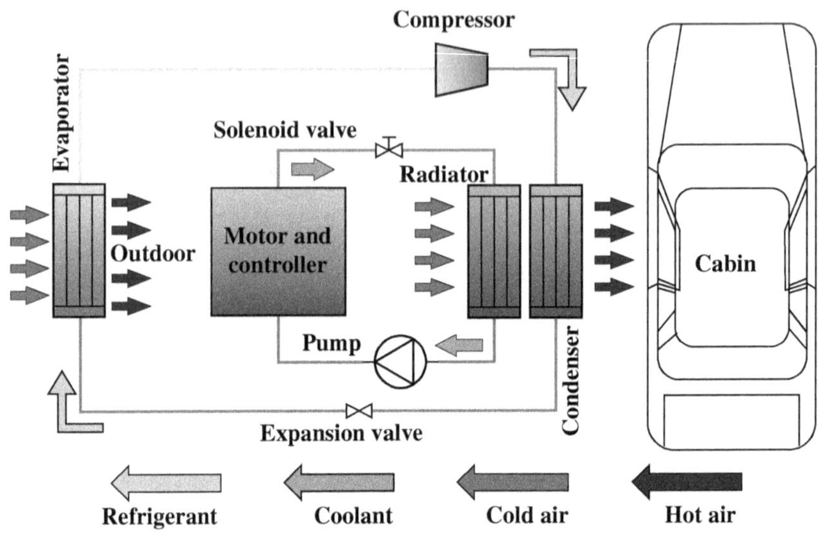

FIGURE 8.6 HPAC-WHR system with waste heat recovered at the condenser side.

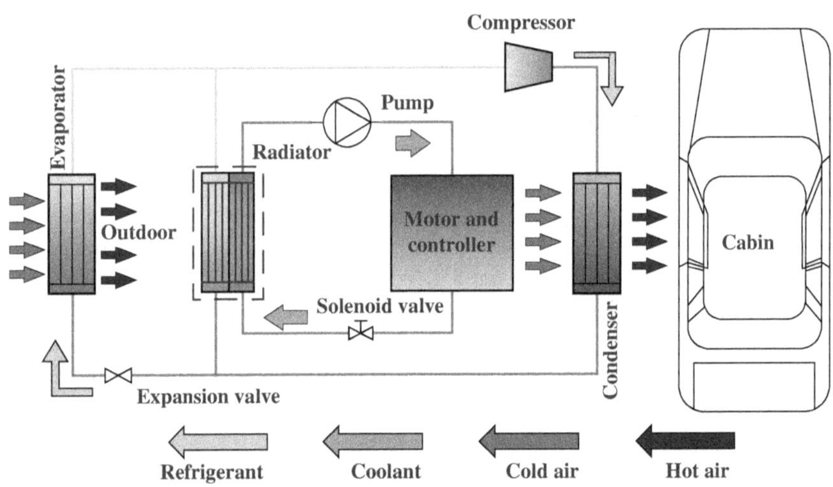

FIGURE 8.7 HPAC-WHR system with waste heat recovered at the refrigerant side.

The energy performance of these three typical HPAC-WHR systems cannot be compared due to the different operating conditions in the present studies. To have a deeper insight into the advantages and disadvantages of these three HPAC-WHR systems and provide theoretical guidelines for the future design of a more efficient HPAC system, it is important to investigate the unified evaluation criteria for HPAC-WHR systems. As for the performance evaluation of HPAC-WHR systems,

COP is widely used to evaluate energy performance. Compared with traditional HPAC systems, HPAC-WHR systems incorporate some additional components, including the radiator(s) for releasing waste heat, solenoid valves, coolant pump, pipeline, etc. These additional components can affect the performance of HPAC, which has not been considered and investigated in previous studies. On the one hand, the operations of the pump and fan in the WHR subsystem need additional electricity provided by batteries, and the additional components can increase the total weight of the EV, causing increased power consumption during the driving process. On the other hand, these additional components can cause power consumption that has not been considered in conventional COP. To reflect the effects of the WHR subsystem on the energy performance of HPAC, the previous addition power consumptions have to be considered when calculating the COP of HPAC-WHR systems. Except for energy performance reflected by the COP indicator, economic and environmental performance of the HPAC-WHR system are also important for engineering design and applications. The cost of HPAC-WHR systems determines whether they can be accepted by the market, while the good environmental performance is urgent to meet increasingly rigorous carbon emission regulations.

8.2.3 CO_2-Based Heat Pump Air Conditioners

Passenger vehicles generally use R134a as the working fluid, while commercial vehicles mostly adopt R407C as the working fluid. Either R134a or R407C HPAC systems, there is a bottleneck that the heat capacity of systems descends significantly when the ambient temperature is lower than –10 °C [21]. Although using compressors with larger capacity and expansion tanks can improve the COP of HPAC systems by 10% under low ambient temperature [22, 23], the system configurations become more complicated and are not suitable for engineering applications. Therefore, developing novel working fluids to replace conventional R134a and R407C has been an important issue and has attracted increasing attention around the world with the fast development of EVs [24, 25].

As a kind of natural working fluid, the ODP of CO_2 is 0 while its GWP is only 1, therefore, it is environmentally friendly in spite of the leakage sometimes [26]. In addition, the CO_2-based trans-critical cycle possesses a wide operating temperature range and has superiority in heating performance [27]. Therefore, CO_2 has been extensively investigated as a working fluid of HPAC in EVs in the past few years, and it is very promising to be the replacement of R134a and R407C. The layout of CO_2 HPAC is identical to that of conventional R134a HPAC, but the operating temperature and pressure of CO_2 HPAC is quite different. For a CO_2-based HPAC of EVs, the temperature of the heat releasing process is higher than the critical point of CO_2, and CO_2 only releases sensible heat to the cabin air. Since there is no phase change during the heat releasing process, the condenser in a conventional HPAC is called a gas cooler in CO_2-based HPAC. Trans-critical CO_2 cycle is theoretically less efficient than vapor compression cycle under the same conditions, and the reasons lie in the following two aspects. On the one hand, the average temperature of CO_2 at the heat releasing process is high and the heat loss is also large in this process. On the other hand, the pressure difference of CO_2 at the inlet and outlet of expansion device is large, leading to more exergy destruction [28]. Although the pressure difference of the trans-critical CO_2 cycle is large, its pressure ratio is only about 3 while the pressure ratio of the compression cycle for other refrigerants reaches about 8, so the compressor of the trans-critical CO_2 cycle has higher efficiency and has a higher clearance volume and smaller size.

Considering existing blocks of CO_2-based HPAC, present studies focus on the optimization of system layouts and operating parameters to improve the heating capacity and COP. In the trans-critical CO_2 HPAC, the pressure difference through the throttling device can reach about 6 MPa, which is much higher than that of traditional HPAC using other refrigerants and causes the largest irreversible loss in the entire trans-critical CO_2 cycle. Reducing the throttling loss or recovering the expansion work is the key to improving the efficiency of CO_2-based HPAC [29]. Therefore, an internal heat exchanger (IHX) is generally considered in a CO2-based HPAC, and a typical system layout and T-S diagram of a trans-critical CO_2 system with and without internal heat exchanger is depicted in Figure 8.8. IHX is used for the heat transfer between high-pressure CO_2 after the gas cooler and low-pressure CO_2 before suction. The supercritical CO_2 gas flowing out of the gas cooler is subjected to isobaric cooling in the IHX, which reduces the enthalpy and dryness of CO_2 at the evaporator inlet, and the enthalpy at the evaporator inlet and outlet increases, leading to better heat transfer performance of the refrigerant in the evaporator. Studies conducted by Aprea et al. [30] and Llopis et al. [31] demonstrated that the introduction of IHX could always improve the performance of trans-critical CO_2-based HPAC, and comparison of COP with and without IHX is shown in Figure 8.9. It should be pointed out that the effectiveness of IHX is also dependent to the evaporating pressure [32].

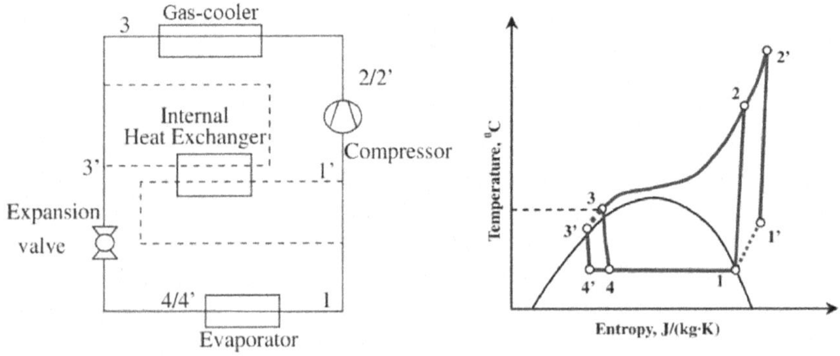

FIGURE 8.8 The layout and T-S diagram of a trans-critical CO_2 system with and without IHX [33].

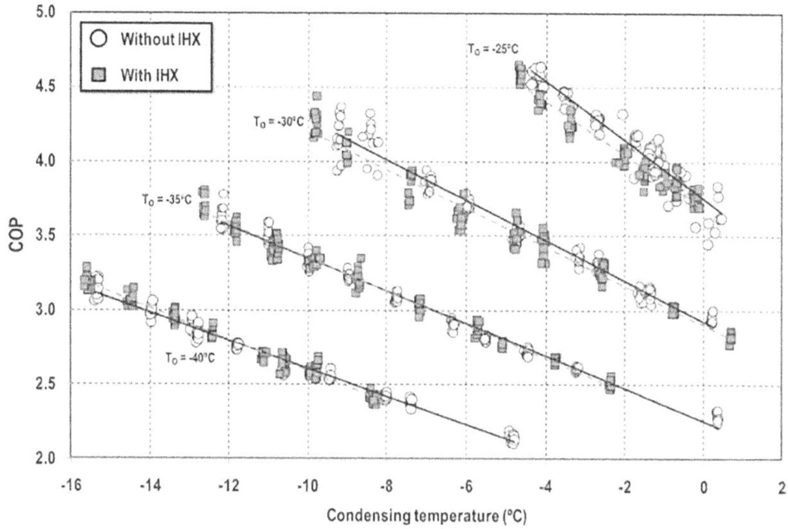

FIGURE 8.9 Comparison of COP with and without IHX under different ambient temperature [31].

Except for IHX, using the technology of adding gas and increasing enthalpy is also important, and these kinds of technologies can significantly improve the performance of trans-critical CO_2-based HPAC in cold regions [34]. Economizer and flash tank are two commonly used technologies of adding gas and increasing enthalpy, and these two systems are depicted in Figure 8.10 and Figure 8.11, respectively. In the economizer system, the CO_2 flows out of the gas cooler and then passes through the economizer and the regenerator and is divided into two paths: one path is partially throttled to the intermediate pressure (state point 4) through the expansion valve 1 and then enters the economizer for heat exchange. It is injected into the compressor through the air supply port. The other CO_2 is completely throttled by expansion valve 2 after passing through the economizer, and then it enters the evaporator to evaporate and absorb heat, later it flows out through the regenerator, and finally it enters the gas-liquid separator. The outgoing CO_2 enters the compressor for first-stage compression and is mixed with the gas entering from the air supply port to complete the second-stage compression to become high-temperature and high-pressure CO_2. In the system with flasher, after the CO_2 flows out of the gas cooler (state point 3), it enters the flasher for flashing after being throttled by the expansion valve (state point 4), and the gaseous CO_2 (state point 6) mixed with partially compressed CO_2 (state point 7 and state point 8). Liquid CO_2 (state point 5) enters the evaporator after being throttled by throttle valve 2 (state point 9).

Heo et al. [35] studied the heating performance of the HPAC system using a flash evaporator, and the system diagram is shown in Figure 8.12. The experimental results showed that the refrigerant flow rate of the system increased at –15 °C. The COP and heating capacity increased by 10% and 25% respectively compared

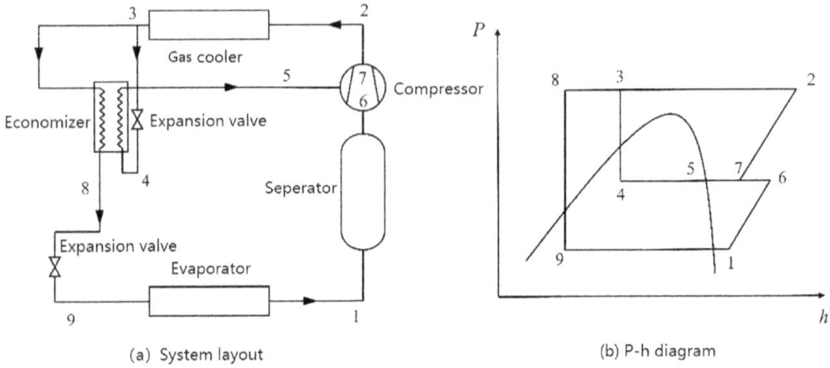

FIGURE 8.10 Vapor injection with economizer for trans-critical CO_2 HPAC.

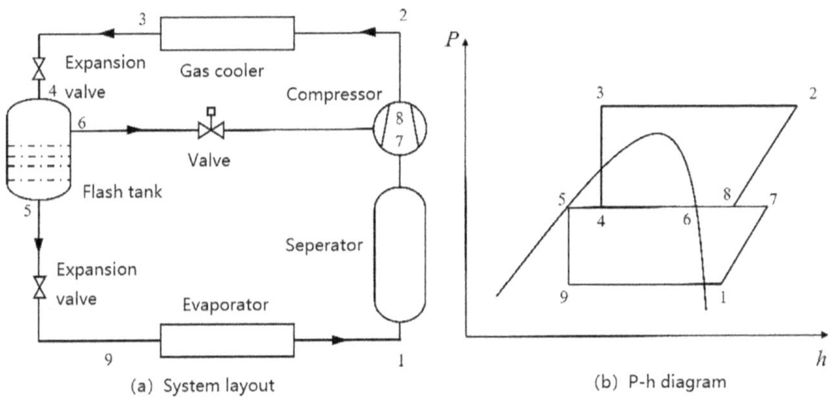

FIGURE 8.11 Vapor injection with flash tank for trans-critical CO_2 HPAC.

with the ordinary single-stage heat pump system. Similarly, experimental results conducted by Nicholas et al. [36] indicated that 8–12% increase of working fluid mass flowrate and 4% decrease of enthalpy difference through gas cooler could be observed, resulting in a slight improvement of heating capacity and COP without consuming more compressor power. In recent years, different kinds of modifications have been conducted based on the economizer system and flash-tank system, and these novel systems have been theoretically and experimentally proved to have obvious enhancement in COP and heating capacity [37–39].

8.3 THERMAL MANAGEMENT OF HEAT PUMP AIR CONDITIONERS

Heat pump air conditioner system is an important part for the thermal management of EVs. The operation of HPAC determines the quality of cabin air, and it

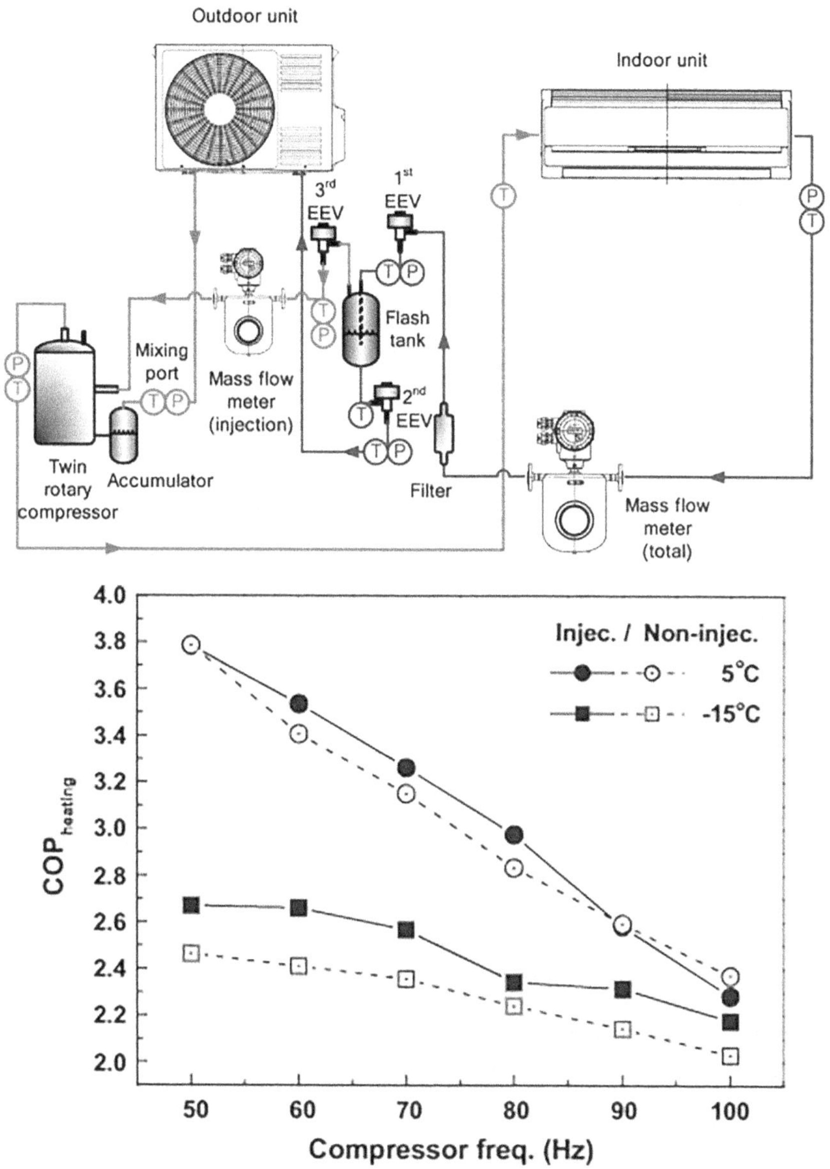

FIGURE 8.12 The proposed system and performance comparison [35]. (a) System diagram (b) Comparison of COP.

relies on thermal management strategies. In terms of thermal management objectives for EVs, both comfort and efficiency need to be considered when designing thermal management strategies since the mileage of EVs is largely related to the power consumption of HPAC. With the fast development of EVs and urgent requirements for energy efficiency, thermal management is not only subjected to HPAC but also needs to consider the integrated system including HPAC, batteries, motors, and motor controllers. Designing and optimizing intelligent thermal management strategies can help improve the comfort of passengers and the energy efficiency of EVs.

8.3.1 Thermal Management of Individual HPAC

The thermal management of HPAC relies on sensors, actuators, and controllers. The sensors mostly include temperature and pressure transmitters. Actuators mainly include electric compressor and electric expansion valves, as well as some auxiliary device fans and water pumps. The generally used control methods involve proportion integration differentiation (PID), model predictive control (MPC), and intelligent control methods. PID is a classic control method and has been widely used in industries. The control parameters can be obtained by Ziegler/Nichols methods [40]. PID control is a relatively flexible and high-performance control method. However, HPAC of EVs is a nonlinear system, and it is extremely hard to achieve nonlinear control of complex system parameters by using a classical PID controller [41]. Fuzzy control has advantages in relatively strong robustness and rapid response, and it does not need a specific mathematical model; therefore, it is also widely used since its control rules and parameters can be obtained from real engineering experiences or optimized by simulation operations [42]. Hu et al. [43] developed an efficient-cost advanced control strategy to boost the operating performance of an air source HPAC based on MPC method. By using the proposed control-oriented model and objective function, the compressor speed and water mass flowrate can be optimized in real time, leading to the minimum power consumption of the compressor and water pump in the real operating conditions. Wang et al. [44] proposed a novel method to predict the heating capacity and COP of R134a HPAC in EVs based on a support vector regression estimator. The procedure of this method is as follows: data standardization, train/test split, feature selection, model optimization, and performance prediction. The simulation results demonstrated that the developed method can predict the heating capacity and COP within the error of 8.25% and 8.33% compared with experimental results, respectively. Tang et al. [45] proposed a novel self-adaptive control strategy using dimensionless artificial neural network for frost prevention and retardation in an air source HPAC in cold winter.

The preceding studies achieved proper thermal management of individual HPAC system by various controllers, considering the different input, output, and other variables, and the effects to different extent were proved theoretically and experimentally. In summary, traditional HPAC controllers, including on/off controller and PID controller, have been widely used in HPAC systems of EVs due to their simple control logic and low cost in real engineering applications. However, an on/off controller can cause large fluctuations of compressor speed and cabin temperature, while a PID controller has relatively poor robustness to adapt to the variation of ambient temperature and may cause uncertain results.

Fuzzy controller is less able to assure the control precision. Although a Fuzzy-PID controller can ensure the control precision, the cost is not appreciable to the total HPAC system at present. Intelligent controllers including MPC and ANN-based controllers have more accurate control precision and better robustness for HPAC of EVs, but the obvious disadvantages of intelligent controllers are their quite high cost and calculation resources because they operate in online mode and require massive computing power to calculate the real-time feedback of the HPAC system, and they are not proper for real-world engineering at this stage.

8.3.2 Thermal Management of Integrated System

The integrated thermal management system of EVs can include batteries, motors, motor controllers, and HPAC, and these key components should work under proper temperature by using control strategies. The thermal management of the integrated system is an important method to improve the mileage of EVs. Ensuring that batteries under a suitable temperature range can provide a stable and reliable power for EVs, and the performance improvement of the HPAC system can save more energy, thereby increasing the mileage of the car. At present, the integrated thermal system of EVs is built based on HPAC, and the HPAC is connected to the batteries, motors, and controllers by using additional heat exchangers to recover waste heat, achieving the comfortable cabin environmental and thermal management of these power devices.

Tian et al. [46] proposed an integrated thermal management system considering the cabin comfort, battery cooling, and waste heat recovery of motors, and the system diagram is depicted in Figure 8.13. In this system, the evaporator and condenser of HPAC are coupled to the cooling cycle of batteries and motors by adding addition parallel heat exchangers, and operations of these heat exchangers are switched by valves according to different seasons. The results indicated that cooling capacity required for batteries is reduced by 26.3%~32.1%, while the waste heat recovery rate of motors is within the range of 18.7%~45.2%. In addition, recovering the waste heat of motors improved the heating COP by 25.6% and mileage by 31.7% compared with those of the PTC method. Yang et al. [47] designed an integrated thermal management system for EVs and conducted a simulation study, shown in Figure 8.14. Results demonstrated that the designed integrated thermal management system can provide suitable temperature environments for the cabin, batteries, and motors. Especially in the winter, the overall power consumption is reduced 16.4%, and the mileage is improved by 18.3% compared with PTC heating.

8.4 OUTLOOK OF HEAT PUMP AIR CONDITIONERS IN EVs

Heat pump air conditioners seem a more promising solution to satisfy the heating requirement of electric vehicles compared with only PTC in cold conditions, due to its better performance in energy efficiency and vehicle mileage. Although giant advances have been achieved to propel the applications of HPAC in EVs, there are

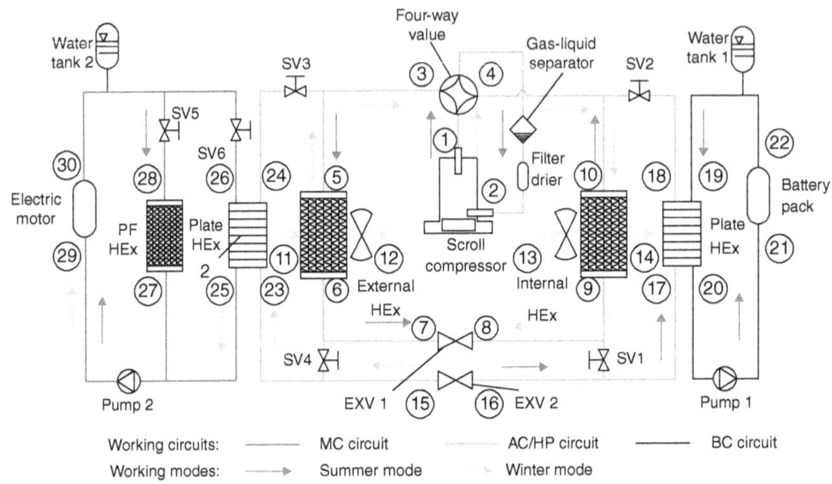

FIGURE 8.13 Schematic diagram of integrated thermal management system [46].

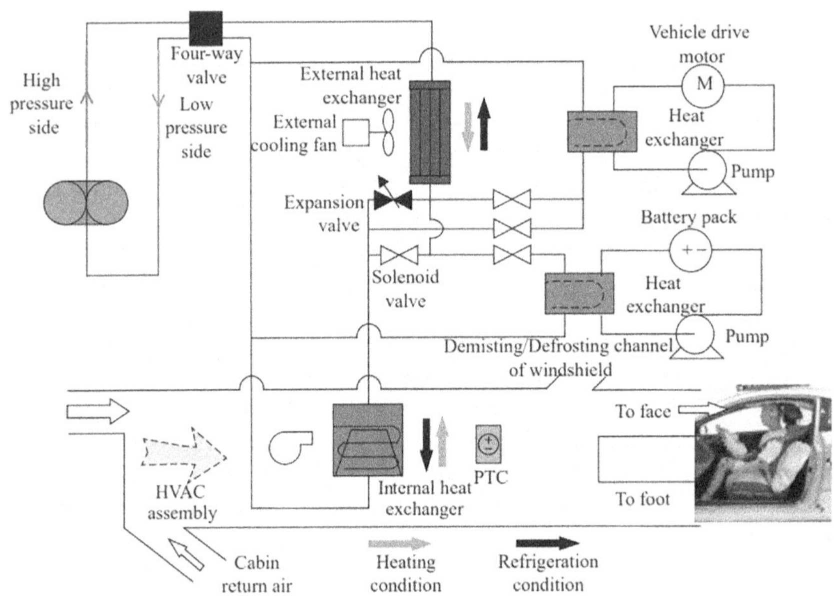

FIGURE 8.14 Schematic diagram of integrated thermal management system [47].

still some key issues to be addressed for the further improvement of performance, and these outlooks are summarized as follows.

- Conventional working fluids used in HPAC including R134a and R407C are less able to meet the increasingly rigorous requirements of energy and environmental performance. As a consequence, developing novel working fluids is an important direction, including CO_2, R1234yf, R290, and R410A, for example. Among these substitute working fluids, CO_2 is a representative one since it is a kind of natural working fluid and the CO_2-based trans-critical cycle has good heating performance. Meanwhile, mixed working fluids using CO_2 and other refrigerants can improve the cooling performance of HPAC.
- Heat pump air conditioners have better comprehensive performance than PTC, and one of the main reasons is that HPAC can improve energy efficiency by absorbing heat from ambient air or the high temperature coolant of EVs. This implies the importance of designing integrated systems to combine HPAC and waste heat recovery units for further enhancement of energy efficiency. Recovering waste heat from power devices such as batteries, motors, and controllers can be considered. There are different layouts to integrate HPAC with waste heat recovery units, the matching between the heat source and working fluid (thermodynamic cycles) needs to be specially focused.
- To ensure the performance of HPAC under different conditions, thermal management is quite important. Developing proper thermal management strategies is the key issue for a thermal management system. The integrated thermal management system with cost-effective control strategies is a future direction since the integrated system can achieve a better overall energy efficiency and longer mileage for EVs. An integrated thermal management system has more sensors and actuators compared to an individual HPAC system as well as more objectives; therefore, intelligent thermal management methods deserve more investigation.
- Conventional air source HPAC systems have very simple structures. COP is generally used as the evaluation method, and thermal management method is also relatively monotonous. For integrated systems, more key components and auxiliary electric devices are added. To comprehensively reflect the energy, environmental, and economic performance, new evaluation criteria should be developed in the future before designing corresponding thermal management methods and control strategies.

8.5 CONCLUSIONS

This chapter overviews the fundamentals and key development of heat pump air conditioners for heating and corresponding thermal management methods in electric vehicles. HPAC is a substitute technology to PTC due to its high energy efficiency, but further developments and applications of HPAC are subject to energy efficiency, working fluid, and thermal management. To achieve higher COP, integrating with HPAC and waste heat recovery of batteries, motors, or controllers is promising. Meanwhile, adopting novel working fluids such as CO_2 can also enhance COP as well as environmental performance, and the system layout needs to be re-designed to adapt to the new working fluids. In an integrated system, the global demand on the thermal management of HPAC and other power devices is inevitable, and proper control strategies to maintain all the objectives under the designed temperature environment deserve more efforts in the future.

ACKNOWLEDGEMENTS

This chapter is greatly supported by the National Natural Science Foundation of China under grant No. 51976176.

REFERENCES

1. L. Prause, K. Dietz, Just mobility futures: Challenges for e-mobility transitions from a global perspective, *Futures* (2022) 102987.
2. Q. Zhou, J. Li, B. Shuai, H. Williams, Y. He, Z. Li, H. Xu, F. Yan, Multi-step reinforcement learning for model-free predictive energy management of an electrified off-highway vehicle, *Applied Energy*, 255 (2019) 113755.
3. Q. Zhou, W. Zhang, S. Cash, O. Olatunbosun, H. Xu, G. Lu, Intelligent sizing of a series hybrid electric power-train system based on chaos-enhanced accelerated particle swarm optimization, *Applied Energy*, 189 (2017) 588–601.
4. Z. Qi, Advances on air conditioning and heat pump system in electric vehicles—a review, *Renewable and Sustainable Energy Reviews*, 38 (2014) 754–764.
5. M. Bentrcia, M. Alshatewi, H. Omar, Developments of vapor-compression systems for vehicle air-conditioning: A review, *Advances in Mechanical Engineering*, 9 (2017) 1687814017717186.
6. S. Bellocchi, G. Leo Guizzi, M. Manno, M. Salvatori, A. Zaccagnini, Reversible heat pump HVAC system with regenerative heat exchanger for electric vehicles: Analysis of its impact on driving range, *Applied Thermal Engineering*, 129 (2018) 290–305.
7. A. Yokoyama, Thermal management system for electric vehicles, *SAE International Journal of Materials and Manufacturing*, 4 (2011) 1277–1285.
8. Q. Tu, C. Deng, X. Gong, L. Wei, L. Zhang, X. Wu, C. Ji, L. Boban, V. Soldo, Characteristics of active refrigerant injection based on external heat source and its effects on performance of an air source heat pump, *Sustainable Energy Technologies and Assessments*, 45 (2021) 101164.
9. Q. Ning, G. He, G. Xiong, W. Sun, H. Song, Operation strategy and performance investigation of a high-efficiency multifunctional two-stage vapor compression heat pump air conditioning system for electric vehicles in severe cold regions, *Sustainable Energy Technologies and Assessments*, 48 (2021) 101617.
10. Y. Huang, X. Wu, J. Jing, Research on the electric vehicle heat pump air conditioning system based on R290 refrigerant, *Energy Reports*, 8 (2022) 447–455.
11. A. Ibrahim, F. Jiang, The electric vehicle energy management: An overview of the energy system and related modeling and simulation, *Renewable and Sustainable Energy Reviews*, 144 (2021) 111049.
12. H. Zou, W. Wang, G. Zhang, F. Qin, C. Tian, Y. Yan, Experimental investigation on an integrated thermal management system with heat pipe heat exchanger for electric vehicle, *Energy Conversion and Management*, 118 (2016) 88–95.

13. T. Suzuki, K. Ishii, *Air Conditioning System for Electric Vehicle*, SAE International, 1996.
14. Z. Wang, M. Wei, F. Peng, H. Liu, C. Guo, G. Tian, Experimental evaluation of an integrated electric vehicle AC/HP system operating with R134a and R407C, *Applied Thermal Engineering*, 100 (2016) 1179–1188.
15. H.-S. Lee, M.-Y. Lee, Steady state and start-up performance characteristics of air source heat pump for cabin heating in an electric passenger vehicle, *International Journal of Refrigeration*, 69 (2016) 232–242.
16. J.H. Ahn, H. Kang, H.S. Lee, H.W. Jung, C. Baek, Y. Kim, Heating performance characteristics of a dual source heat pump using air and waste heat in electric vehicles, *Applied Energy*, 119 (2014) 1–9.
17. J.H. Ahn, H. Kang, H.S. Lee, Y. Kim, Performance characteristics of a dual-evaporator heat pump system for effective dehumidifying and heating of a cabin in electric vehicles, *Applied Energy*, 146 (2015) 29–37.
18. J.H. Ahn, J.S. Lee, C. Baek, Y. Kim, Performance improvement of a dehumidifying heat pump using an additional waste heat source in electric vehicles with low occupancy, *Energy*, 115 (2016) 67–75.
19. C. Qian, B. Gu, Z. Tian, L. Ma, L. Yang, Performance analysis of dual source heat pump in electric vehicles, *Shanghai Jiaotong Daxue Xuebao/Journal of Shanghai Jiaotong University*, 50 (2016) 569–574.
20. C. Kowsky, E. Wolfe, L. Leitzel, F. Oddi, Unitary HPAC system, *SAE International Journal of Passenger Cars-Mechanical Systems*, 5 (2012) 1016–1025.
21. Z. Wang, M. Wei, C. Guo, M. Zhao, Enhance the heating performance of an electric vehicle AC/HP system under low temperature, *Energy Procedia*, 105 (2017) 2384–2389.
22. J. Jung, Y. Jeon, H. Lee, Y. Kim, Numerical study of the effects of injection-port design on the heating performance of an R134a heat pump with vapor injection used in electric vehicles, *Applied Thermal Engineering*, 127 (2017) 800–811.
23. W. Li, R. Liu, Q. Liu, Y. Liu, D. Wang, S. Xia, J. Shi, J. Chen, Upstream and downstream injection effects on R134a economized vapor injection heat pump system at low temperatures for electric vehicles, *International Journal of Refrigeration*, 120 (2020) 1–11.
24. J. Luo, K. Yang, Z. Zhao, G. Chen, Q. Wang, Experimental investigations on the performance of a single-stage compound air-source heat pump using CO2/R600a in cold regions, *Applied Thermal Engineering*, 205 (2022) 118050.
25. A. Wang, F. Cao, J. Fang, F. Jia, X. Yin, X. Wang, Research on efficient defrosting control logic for transcritical CO2 electric vehicle heat pump air-conditioning system, *International Journal of Refrigeration*, 138 (2022) 13–22.
26. H. Wang, Y. Song, Y. Qiao, S. Li, F. Cao, Rational assessment and selection of air source heat pump system operating with CO2 and R407C for electric bus, *Renewable Energy*, 182 (2022) 86–101.
27. J. Ko, K. Thu, T. Miyazaki, Transient analysis of an electric vehicle air-conditioning system using CO2 for start-up and cabin pull-down operations, *Applied Thermal Engineering*, 190 (2021) 116825.
28. B. Yu, J. Yang, D. Wang, J. Shi, J. Chen, An updated review of recent advances on modified technologies in transcritical CO2 refrigeration cycle, *Energy*, 189 (2019) 116147.
29. F. Cao, Y. Song, M. Li, Review on development of air source transcritical CO2 heat pump systems using direct-heated type and recirculating-heated type, *International Journal of Refrigeration*, 104 (2019) 455–475.
30. C. Aprea, A. Maiorino, An experimental evaluation of the transcritical CO2 refrigerator performances using an internal heat exchanger, *International Journal of Refrigeration*, 31 (2008) 1006–1011.

31. R. Llopis, C. Sanz-Kock, R. Cabello, D. Sánchez, E. Torrella, Experimental evaluation of an internal heat exchanger in a CO2 subcritical refrigeration cycle with gas-cooler, *Applied Thermal Engineering*, 80 (2015) 31–41.
32. E. Torrella, D. Sánchez, R. Llopis, R. Cabello, Energetic evaluation of an internal heat exchanger in a CO2 transcritical refrigeration plant using experimental data, *International Journal of Refrigeration*, 34 (2011) 40–49.
33. F.Z. Zhang, P.X. Jiang, Y.S. Lin, Y.W. Zhang, Efficiencies of subcritical and transcritical CO2 inverse cycles with and without an internal heat exchanger, *Applied Thermal Engineering*, 31 (2011) 432–438.
34. S. Lu, R. Liang, J. Zhang, C. Zhou, Performance improvement of solar photovoltaic/thermal heat pump system in winter by employing vapor injection cycle, *Applied Thermal Engineering*, 155 (2019) 135–146.
35. J. Heo, M.W. Jeong, Y. Kim, Effects of flash tank vapor injection on the heating performance of an inverter-driven heat pump for cold regions, *International Journal of Refrigeration*, 33 (2010) 848–855.
36. N. Fernandez, Y. Hwang, R. Radermacher, Comparison of CO2 heat pump water heater performance with baseline cycle and two high COP cycles, *International Journal of Refrigeration*, 33 (2010) 635–644.
37. C. Baek, J. Heo, J. Jung, H. Cho, Y. Kim, Effects of the cylinder volume ratio of a twin rotary compressor on the heating performance of a vapor injection CO2 cycle, *Applied Thermal Engineering*, 67 (2014) 89–96.
38. C. Baek, J. Heo, J. Jung, E. Lee, Y. Kim, Effects of vapor injection techniques on the heating performance of a CO2 heat pump at low ambient temperatures, *International Journal of Refrigeration*, 43 (2014) 26–35.
39. C. Baek, J. Heo, J. Jung, H. Cho, Y. Kim, Performance characteristics of a two-stage CO2 heat pump water heater adopting a sub-cooler vapor injection cycle at various operating conditions, *Energy*, 77 (2014) 570–578.
40. G. Huang, S. Wang, X. Xu, A robust model predictive control strategy for improving the control performance of air-conditioning systems, *Energy Conversion and Management*, 50 (2009) 2650–2658.
41. H. Khayyam, S. Nahavandi, E. Hu, A. Kouzani, A. Chonka, J. Abawajy, V. Marano, S. Davis, Intelligent energy management control of vehicle air conditioning via look-ahead system, *Applied Thermal Engineering*, 31 (2011) 3147–3160.
42. X. Huang, K. Li, Y. Xie, B. Liu, J. Liu, Z. Liu, L. Mou, A novel multistage constant compressor speed control strategy of electric vehicle air conditioning system based on genetic algorithm, *Energy*, 241 (2022) 122903.
43. W. Wang, B. Hu, R.Z. Wang, M. Luo, G. Zhang, B. Xiang, Model predictive control for the performance improvement of air source heat pump heating system via variable water temperature difference, *International Journal of Refrigeration*, 138 (2022) 169–179.
44. Y. Wang, W. Li, Z. Zhang, J. Shi, J. Chen, Performance evaluation and prediction for electric vehicle heat pump using machine learning method, *Applied Thermal Engineering*, 159 (2019) 113901.
45. J. Tang, C. Herman, G. Gong, A novel self-adaptive control strategy of frost prevention and retardation for air source heat pumps in winter conditions, *Applied Mathematical Modelling*, 83 (2020) 284–300.
46. Z. Tian, W. Gan, X. Zhang, B. Gu, L. Yang, Investigation on an integrated thermal management system with battery cooling and motor waste heat recovery for electric vehicle, *Applied Thermal Engineering*, 136 (2018) 16–27.
47. X.L. Yang, Z.H. Ma, L. Yang, Thermal management system of electric vehicle based on heat pump, *Journal of Central South University*, 47 (2016) 2855.

Part III

Prediction and Planning of Electric Vehicles in Real-World Driving

9 Driver Behavior Prediction and Driver-Oriented Control of Electric Vehicles

Ji Li, Quan Zhou, Yanhong Wu and Hongming Xu

9.1 WHAT IS DRIVING BEHAVIOR?

The term "driving behavior" refers to the set of deliberate and subconscious actions and reactions exhibited by an individual while operating a motor vehicle. It encompasses a broad spectrum of activities, including but not limited to steering, braking, accelerating, and decision-making.

In recent years, there has been an increasing interest in studying driving behavior and its effects on road safety and traffic management. The factors that shape driver behavior are diverse and complex, ranging from age, experience, and gender, to attitudes, emotions, and the surrounding driving conditions [1]. These internal and external factors can have a significant impact on a driver's ability to assess risk and make safe driving decisions, and they can vary from one situation to the next, even for the same driver.

Driver behavior can also be categorized according to driver skills and styles, including prudence (aggressive versus cautious), stability (unstable versus stable), conflict proneness (risk-taking versus risk-avoidance), skillfulness (non-skilled versus skilled), and self-discipline (law-abiding versus frequent rule violators) [2]. One notable area of research that has attracted considerable attention is the use of physiological measures, such as heart rate and brain wave patterns via an electroencephalogram (EEG), to monitor and predict driving behavior. By analyzing changes in physiological signals, researchers have been able to detect patterns of behavior that are suggestive of hazardous driving or fatigue. Predicting and controlling driving behavior can be challenging. Some literature describes driving behavior using various biological signal data, including electrocardiogram (ECG), EEG, electromyogram (EMG), and electrodermal activity (EDA) [3–5]. In general, drivers with more favorable attitudes towards a behavior, with a stronger endorsement for the behavior from important others and with higher levels of perceived behavioral control will likely form stronger intentions to engage in the behavior (see Figure 9.1).

DOI: 10.1201/9781003302827-12

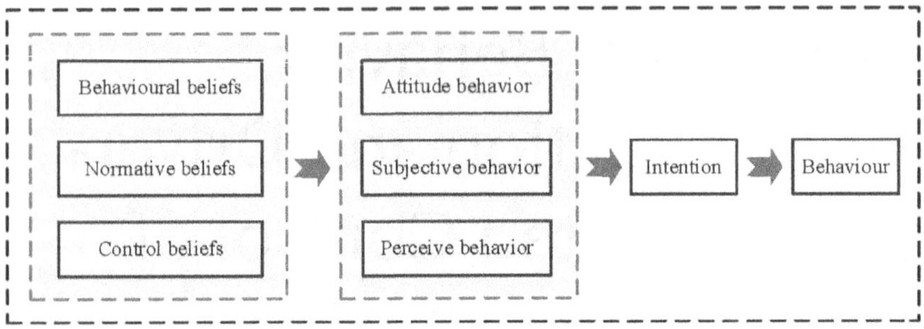

FIGURE 9.1 The framework of planned behavior [6].

9.2 WHY STUDY DRIVING BEHAVIOR?

Effective traffic management requires understanding and addressing driver behavior as it plays a crucial role in road safety. Human error accounts for many serious traffic incidents with a massive portion of them attributed to distracted driving. This poses a growing challenge for the traffic system, as it must ensure the safety of both its drivers and the public while balancing the cost implications.

9.2.1 Safety

Since the 1970s, human drivers have been the subject of intensive research in various aspects, with most of the existing studies focused on driver behaviors, attention, intention, fatigue, and cognitive and neural functions, among others. These studies aim to gain a better understanding of the driver's psychological and physiological state to improve driving safety and assist in driving tasks.

The previous data presents a 2020 analysis of traffic accidents caused by various driving behaviors. As shown in Figure 9.2, the statistics show that the leading causes of road accidents were pedestrians not following traffic rules, excessive speeding, and improper lane usage. Drunk driving, violation of traffic signal lights, and fatigued driving were also significant factors. The number of traffic accidents caused by drivers has been listed in Figure 9.3. These figures underline the importance of safe and responsible driving practices, such as obeying traffic laws and signals, staying within designated lanes, and avoiding driving under the influence or while tired. Drivers should remain alert and cautious, especially in high-risk areas, to ensure the safety of themselves and other road users. By implementing these practices, we can reduce the number of accidents and make our roads safer.

The analysis of driver behavior in traffic accidents is crucial for enhancing traffic safety for autonomous driving vehicles. Despite the anticipated safety benefits of autonomous vehicles, there is still a need to address human factors that contribute to accidents, such as driver error and unsafe driving practices. The investigation of driver behavior in traffic accidents can provide insights into the factors that lead to accidents, such as distraction, fatigue, and impairment. By identifying these factors, researchers can develop effective interventions and safety measures

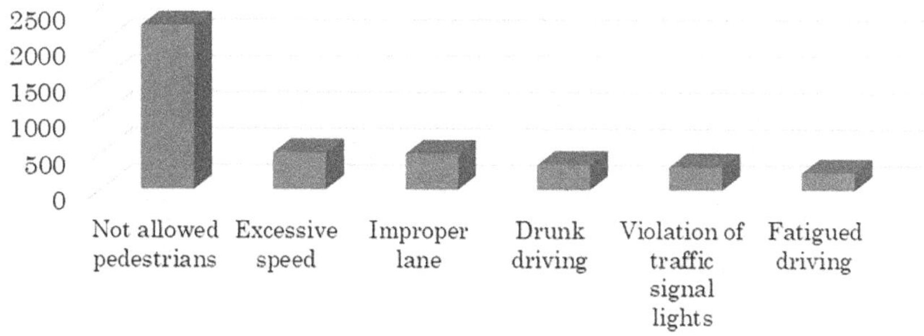

FIGURE 9.2 The number of traffic accidents caused by drivers in 2020 [7].

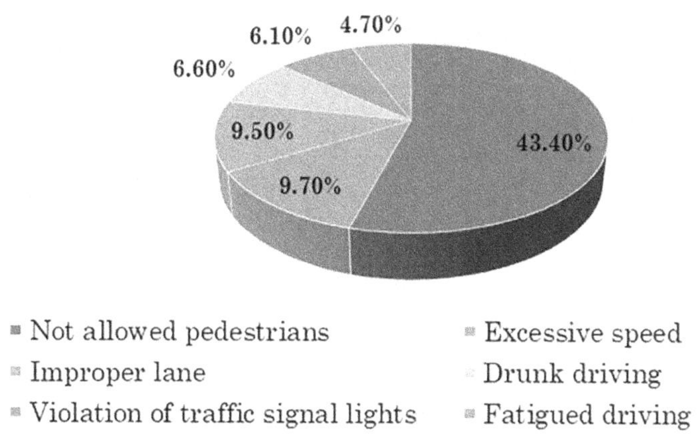

FIGURE 9.3 The probability of traffic accidents caused by drivers in 2020 [7].

that address the root causes of accidents. It is essential to analyze driver behavior in traffic accidents, especially in the context of autonomous driving vehicles. By understanding the role of driver behavior in accidents involving these vehicles, researchers and policymakers can implement appropriate measures to improve traffic safety and prevent accidents. Recent studies have investigated the effects of several factors on driver behavior, including the impact of smartphone use on driving performance [8], the effects of cognitive load on driver behavior [9], and the relationship between driver fatigue and driving performance [10]. These studies provide vital information that can be used to develop effective interventions to promote safe driving practices and reduce the number of accidents on our roads.

9.2.2 Autonomous Driving

Driver behavior research in the automotive industry has become more important and serious. One of the most significant goals is to improve road safety by reducing the number of accidents caused by driver misunderstanding. Research in

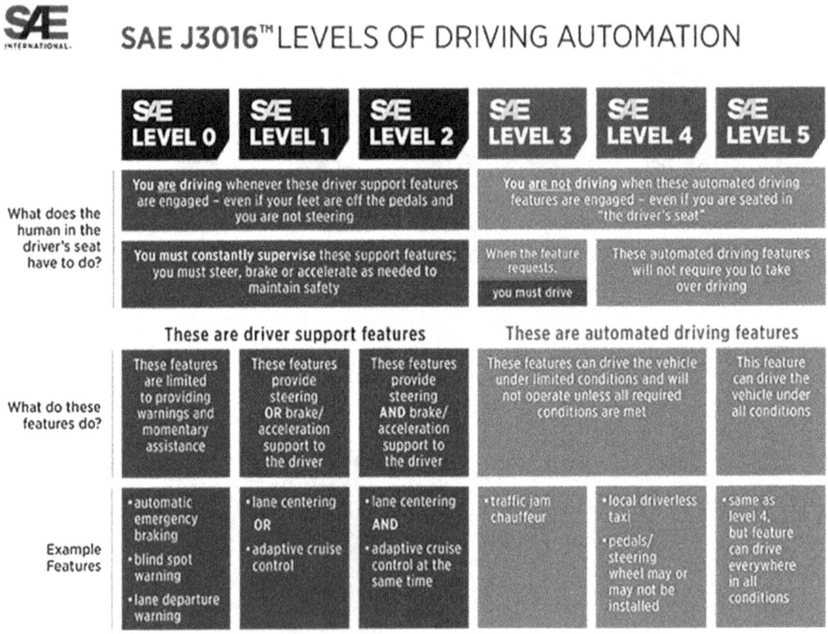

FIGURE 9.4 SAE levels of driving automation [11].

driver behavior is used to develop advanced driver assistance systems (ADAS) and autonomous vehicles that can anticipate and react to driver behavior on the road. This research can also inform the design of vehicles, road infrastructure, and traffic laws. Research on autonomous vehicles still relies on driver behavior. As shown in Figure 9.4. The Society of Automotive Engineers (SAE) autonomous driving classification system includes six levels, ranging from Level 0 (no automation) to Level 5 (full automation). One principal factor that distinguishes Level 2 (partial automation) from Level 3 (conditional automation) is the role of the human driver. In Level 2, the driver is still responsible for monitoring the driving environment and taking over when necessary. In contrast, Level 3 vehicles can handle most driving tasks but require the driver to be able to intervene when prompted. The importance of driver behavior in Level 3 automation cannot be overstated, as the driver's ability to respond quickly and effectively to a takeover request can mean the difference between a safe and an unsafe driving experience. As the industry moves towards higher levels of automation, it is becoming increasingly clear that understanding and optimizing driver behavior will be critical to ensuring the safety and effectiveness of autonomous vehicles on our roads.

9.2.3 Energy Economy

Another aspect of driver behavior research is to help improve fuel efficiency and reduce emissions [12] as shown in Figure 9.5. By understanding how drivers interact

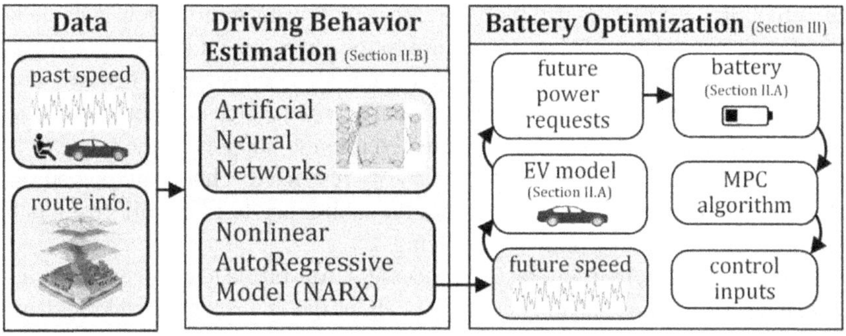

FIGURE 9.5 The framework of energy management based on driving behavior [13].

with their vehicles, manufacturers can design more fuel-efficient cars and develop a system to encourage more eco-friendly driving behaviors. There is one commercial case study from the package delivery company UPS when it implemented a system called "ORION," which stands for "On-Road Integrated Optimization and Navigation." This system uses a combination of data analysis, GPS tracking, and driver behavior research to optimize the routes of delivery trucks. By analyzing data on traffic patterns, delivery addresses, and driver behaviors, ORION generates an optimal delivery route for each truck, reducing the number of miles driven and minimizing the amount of time spent idling or stuck in traffic. The system has reportedly saved UPS over 10 million gallons of fuel and reduced CO_2 emissions by around 100,000 metric tons per year. Advances in technology and research methods have enabled researchers to gain a better understanding of the complex nature of driver behavior, and ongoing research in this area will continue to provide valuable insights into how we can improve driver performance and safety.

9.3 RECENT DEVELOPMENT IN DRIVING BEHAVIOR

The study of driver behavior has undergone significant progress in recent years, driven by the principles of human-machine interaction. This research has followed a comprehensive approach that involves analyzing driving tasks, understanding cognitive processes, and designing experiments. Advanced equipment, such as driving simulators and eye-tracking devices, has allowed for a better understanding of how drivers interact with their environment. Overall, the advancement in driver behavior research has provided valuable insights into improving road safety and creating more efficient transportation systems.

9.3.1 Paradigm of Driver-Vehicle Interaction

Due to the shortage of computing power in the past, the electronic control unit can only provide a limited development platform for vehicles, resulting in vehicle supervisory control (VSC) with simple control logic. Figure 9.6 shows an interactive process of conventional VSC, in which state information is environmental condition and the driver's response to the environment that will be mixed and sent to a conventional VSC for operating energy management. However, conventional VSC cannot extract valuable references from a large amount of unknown information. Due to having no reliable information to support it, the predictivity and adaptability of the system will be seriously restricted. This could be one reason vehicles with conventional VSC pass the laboratory test but still give a deficient performance during real-world driving.

In the past decade, their interactive processes have started to change. The industry realized that at this stage, uncertainties about the environment and the driver have been the main reason hindering the overall performance of vehicles. These external uncertainties need to be defined and quantified for clean and safe driving. However, hybrid electric vehicles (HEVs) designed with extra degrees of freedom could increase the difficulty of quantifying external uncertainties and magnify the effect on energy consumption. In this case, there is an urgent need for modern energy management schemes for HEVs that effectively deal with uncertainty. In this work, the author has investigated if there is a conceivable way to improve energy management efficiency by considering the impact of human drivers on state-of-the-art VSC.

Differing from the conventional VSC, the uncertainty of drivers is considered in the design of these VSC. As independent sub-models, they work in parallel with the vehicle powertrain. As shown in Figure 9.7, a driver model has been moved from a conventional VSC to make a new driver-oriented VSC, that is, a man-machine system. Uncertainty about drivers can be quantified in the new system and state information becomes more abundant. Thus, this form of interaction makes it possible to reduce external uncertainties. Although the transfer of a driver model may bring new design problems into the VSC, its significance is to break the structural framework of the conventional system and show driving behavior characteristics' effects on HEV energy management finally working towards a new level of human-machine fusion.

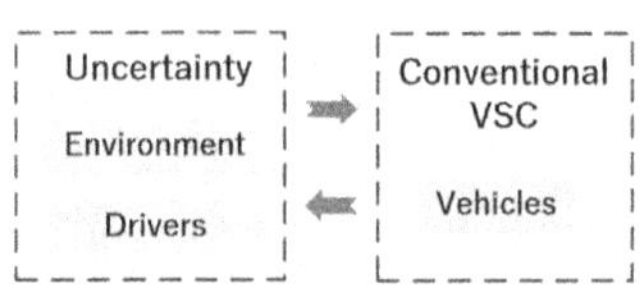

FIGURE 9.6 Interactive process of conventional VSC.

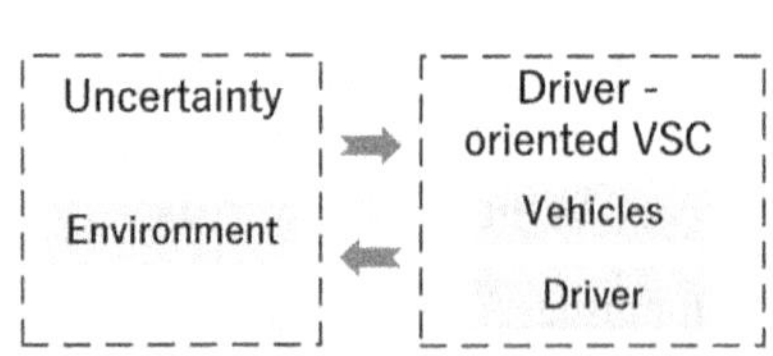

FIGURE 9.7 Interactive process of driver-oriented VSC.

9.3.2 Research Technical Routes

To guide the development of the driver-oriented supervisory control methodology, research technical routes are designed to clarify work contents and the development process. As illustrated in Figure 9.8, the driver-oriented VSC with additional driver dynamics delivers new opportunities to improve the performance of four critical elements of the vehicle system. They are adaptability (ability to adapt to unknown driving conditions), global optimality (ability to find the global optimal solution), synergy (synergistic promotion for multiple evaluation indexes), and predictability (accuracy and length of prediction) [14]. For the new driver-oriented VSC, the author plans to break them off from the following four phases individually.

In Phase 1: control rule design optimization, a novel approach of using personalized non-stationary inference is proposed to increase the robustness of the rule-based control system. The difference between the current system to the previous ones in the literature is that it introduces real-time driving behavior monitoring to increase the robustness of VSC.

In Phase 2: control frame design optimization, the concept of the driver-identified supervisory control system is introduced, which forms a novel architecture of adaptive energy management for HEVs. As a man-machine system, the proposed system can accurately identify the human driver from natural operating signals and provide driver-identified globally optimal control policies as opposed to mere control actions.

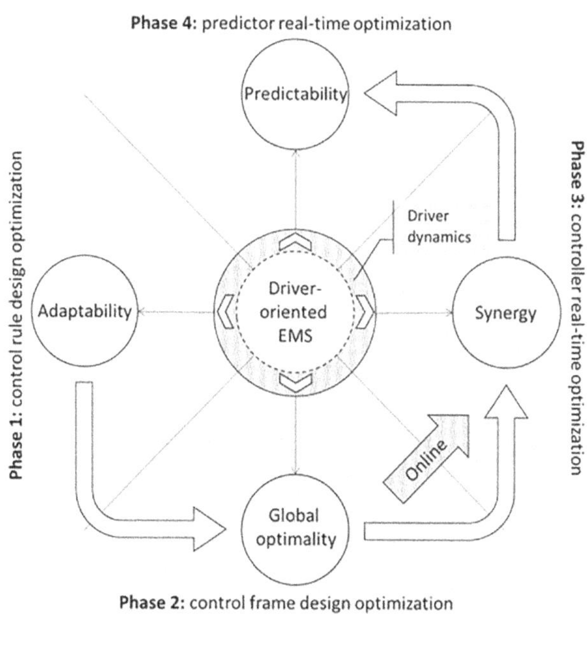

FIGURE 9.8 Roadmap for driver-oriented VSC performance.

Starting from the work in Phase 3, all work is upgraded to an online level, which means the process of optimization and control will be simultaneously carried out during real-world driving. In Phase 3: controller real-time optimization, a novel back-to-back competitive learning mechanism is proposed. This mechanism allows continuous competition between two fuzzy logic controllers during real-world driving.

In Phase 4: predictor real-time optimization, an online predictive control strategy is investigated, resulting in a novel online optimization methodology named dual-loop online intelligent programming, which is proposed for velocity prediction and energy-flow control.

9.3.3 Experimental Equipment

Using a suitable driving simulator is widely acknowledged to reduce system development costs and shorten the development cycle. An effective simulator allows for the study of driver behavior and testing of system performance in hazardous conditions or situations that are difficult or impossible to replicate in the real world [15]. By utilizing a simulator, researchers and developers can thoroughly analyze and evaluate various scenarios and conditions, making it possible to improve system functionality and safety measures. Moreover, the use of a simulator can also reduce the risk of harm or injury to individuals, vehicles, and infrastructure that would otherwise be involved in real-world testing. Therefore, driving simulators have become an essential tool for the development and testing of ADAS and autonomous driving technologies.

Eye-tracking devices have been increasingly used in driver behavior research for the identification and analysis of visual attention and eye movements in Figure 9.9. By recording eye movements, researchers can gain insights into the cognitive processes involved in driving and the factors that impact driver performance and safety [16]. The use of eye-tracking devices allows for objective and precise measurement of driver behavior and can provide valuable information for the development of interventions aimed at improving driving safety.

FIGURE 9.9 Driver's eye–tracking devices [17].

FIGURE 9.10 Electroencephalography (EEG) devices [19].

The use of EEG devices has become increasingly prevalent in driver behavior research for the identification and analysis of cognitive processes involved in Figure 9.10. By measuring electrical activity in the brain, EEG can provide insights into attention, perception, and decision-making processes and help to identify factors that impact driver performance and safety [18]. The use of EEG devices allows for objective and precise measurement of driver behavior and can inform the development of interventions aimed at improving driving safety.

Electromyography (EMG) devices (see Figure 9.11) are commonly used in driver behavior research to measure and analyze muscle activity involved in driving. By recording muscle activity, researchers can gain insights into the cognitive processes involved in driving and the factors that impact driver performance and safety. The use of EMG devices allows for objective and precise measurement of driver behavior, which can provide valuable information for the development of interventions aimed at improving driving safety. EMG devices can be used to monitor muscle activity related to fatigue, distraction, and other factors that may impact driving performance.

The driving simulator platform is used for generating real-world cycles. As Figure 9.12 shows, data collection is conducted in the cockpit package (supported by a Thrustmaster T500RS) with the same scale HEV model with an automatic gearbox. This is to make sure the driving characteristics exhibited by drivers are under the same constraints and their results are comparable. With respect to real-world road conditions, the roadmap model used with reconstructed traffic simulates a cyclic undivided highway with uphill, downhill, curved, and straight roads, and it is provided by IPG CarMaker. It is developed specifically for testing passenger cars and light-duty vehicles. Users can accurately model real-world test

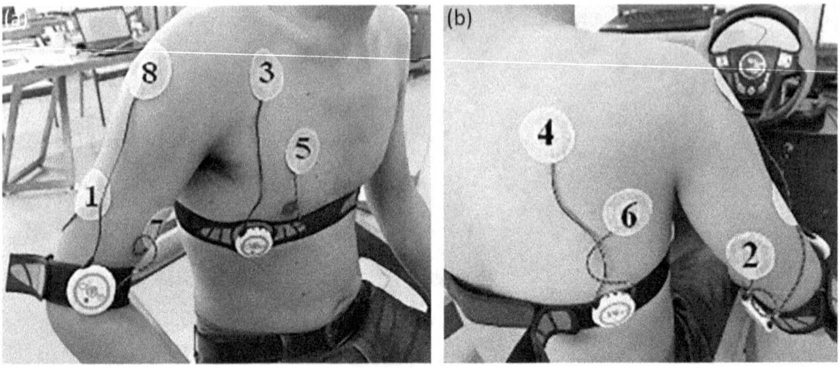

FIGURE 9.11 Electromyography (EMG) devices [20].

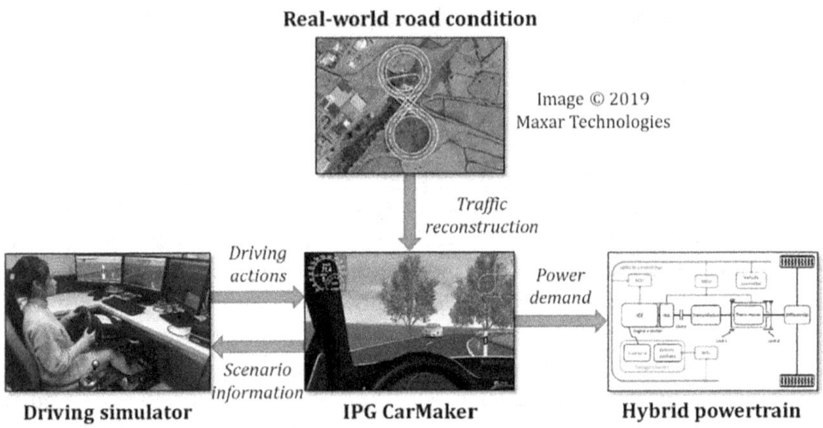

FIGURE 9.12 Driving simulator platform.

scenarios, including the entire surrounding environment, in the virtual world. To reduce the impact of different traffic and road conditions on human drivers, they are restricted to the same cycling road conditions and required to follow the speed limits, stop signs, traffic lights, and other traffic regulations. It should be noted that the driver's pedal behavior might be dependent on the vehicle, the pedal-to-torque map, and even the physical pedal resistance feedback.

9.4 TYPICAL APPLICATIONS

Today, benefiting from informatics' expansion, plenty of optimization methods via information fusion for hybrid vehicles are appearing. As the primary decision-maker of modern vehicles, the human driver plays a significant role in driving safety as well as in eco-driving. Therefore, a vehicle control strategy that seeks a highly optimized performance, which requires optimizing the system composed of the vehicle and the driver, needs to explicitly consider driver behavior [21]. The main challenge is how to exploit driver-related data to precisely describe driving behavior and establish a relationship with the vehicle's system. Therefore, a relevant literature survey is carried out from three categories: a) driving behavior monitoring, b) driving behavior modeling, and c) driver-oriented control method.

9.4.1 Driving Behavior Monitoring

Driving behavior monitoring (DBM) refers to the process of using various sensing technologies and machine learning algorithms to analyze driver behavior and identify patterns that can inform decisions related to driving safety. DBM involves the collection of data related to driver behavior, such as speed, acceleration, and steering patterns, and the use of algorithms to analyze this data and identify any unsafe behaviors or patterns.

The importance of DBM lies in its potential to improve driving safety by identifying risky behavior and providing timely alerts to drivers as well as informing the development of interventions aimed at improving driving behavior. By detecting and addressing unsafe driving behaviors, DBM has the potential to reduce the incidence of accidents caused by human error, which is a leading cause of traffic accidents worldwide.

Under the impetus of global research programs, neuroscience is gradually developing its research on human attention and cognition. As one of the essential functions of the human brain, cognitive function is mainly responsible for complex brain activities such as perception, memory, judgment, reasoning, and problem-solving. Currently, the most used method for studying cognitive science is the analysis of EEG signals. The brain signal produced by the scalp's surface point variation caused by the activity of brain neurons reflects the brain's functional state. By effectively extracting EEG information and capturing the changes in scalp surface potentials, the brain's activity and function can be explored more profoundly [22]. The cognitive workload is the result of the interaction between work motivation, task demands, cognition, and behavior when people perform tasks [23]. Generally, when task demands are high, ore cognitive workload is required [24]. During driving, drivers are often subject to various external factors, such as the interference of cognitive workload, which can directly affect the driver's judgment and cause serious impacts on driving safety.

The driver's gaze behavior is the main visual feature. Drivers identify effective traffic information by looking at the external traffic environment and making judgments to operate. During the process of visual attention, the driver's attentional resources will shift in space. The multi-resource theory model proposed by Wickens et al. [25] is shown in Figure 9.13. When multiple tasks have the same demands in the same or multiple dimensions, these tasks compete, resulting in reduced time sharing and weakened corresponding performance levels. While driving, a certain percentage of the driver's visual channel resources will be occupied by secondary tasks such as texting or operating the in-car system, which will compete directly

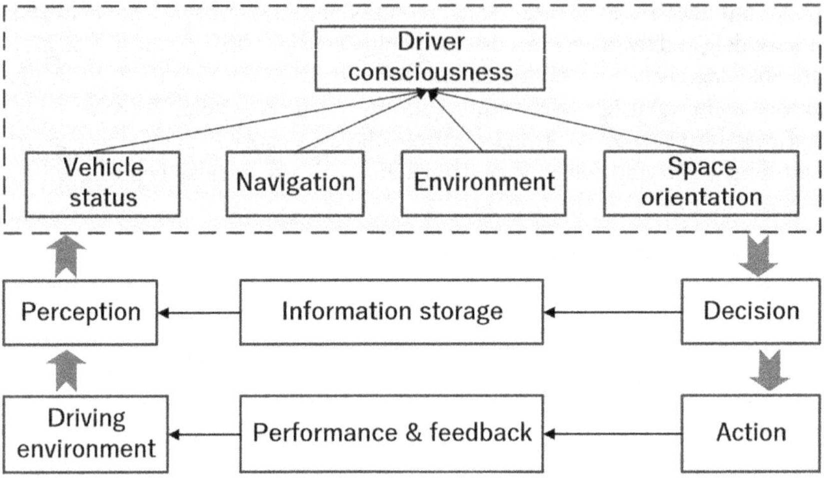

FIGURE 9.13 Driver perception and decision-making theory.

with the primary driving task. When drivers are under a high workload, attentional decline may occur, resulting in decreased driving performance. Ma et al. [26] studied the entire process of the driver's perception of the environment, identification of external traffic information, and finally decision-making and action implementation. They also compared the effects of answering the phone and using the in-car system on attention through experiments. The results showed that attentional loss caused by answering the phone made drivers unable to have a complete and accurate understanding of the driving state, resulting in a decrease in situational awareness and decreased driving performance.

In fact, driver behavior recognition algorithms have attracted considerable attention in the fields of automotive and energy management. With the development of advanced sensor technology, machine learning algorithms, and Big Data analysis techniques, a variety of driver behavior recognition algorithms have been proposed in recent years. These algorithms are mainly used to improve vehicle safety, reduce fuel consumption, and optimize energy management. In addition, they have potential applications in intelligent transportation systems and autonomous driving. Some of the recent studies in this area include the use of convolutional neural networks (CNNs) for driving event detection [27], fuzzy clustering and hidden Markov models for driving pattern recognition [28], and dynamic time warping for driver identification and verification [29]. Moreover, driver behavior recognition algorithms have also been integrated with VSC to achieve optimal energy efficiency [30].

9.4.2 Driving Behavior Modeling

Driving behavior differs among individuals. They differ in the way they press the gas and brake pedals, the way they turn the steering wheel, and how far away they keep when following a vehicle. Consequently, energy management is anticipated to be tailored for each driver according to individual driving behaviors. To realize this goal, one way is to assist each driver by controlling a vehicle based on a driver model representing the typical driving patterns of the target driver [31]. Driver models for individual drivers or a subset of drivers classified based on their driving behaviors can be trained in offline or online mode. A vehicle controller needs to choose an appropriate driver model for supporting the target driver, by distinguishing the driver or assigning the model that fits driving behaviors.

Various definitions of a driver's driving style are caused by many influencing factors and possible explanations for the driver's reaction. Martinez et al. summarized some of these variables listed in Figure 9.14. Given the large numbers and that most factors are difficult to measure, it is unreasonable to expect control over all of them. A simplified method of actual driving style identification focuses on its effect on a single variable, for example, fuel consumption average speed or range. However, these indicators may be oversimplified to assess distinct levels of traffic congestion and cannot represent a complete driving situation.

Generally, the driving behavior can be treated as an extent within (−1, 1); with −1, 0 and 1 being mild, normal, and aggressive, respectively [32]. This criterion is usually formulated based on the relative fuel consumption or overall efficiency rather than the driver's level of aggressiveness. Manzoni et al. [33] used an estimated value of the fuel consumed during the trip and compared it with a benchmarked value to calculate the percentage of excessive consumption, indicating

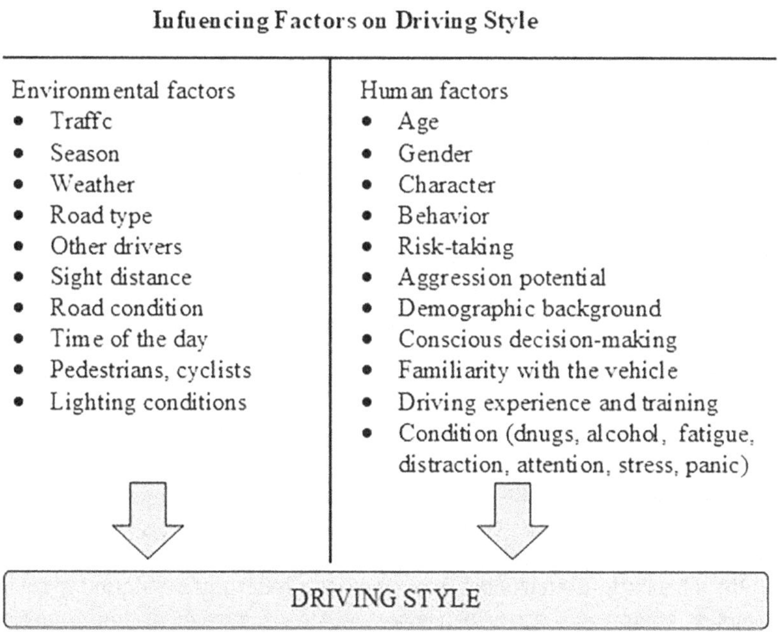

FIGURE 9.14 Influencing factors on driving style [32].

additional costs. Corti et al. [34] used an energy-oriented cost function to evaluate the driver's driving style, which estimated the excessive consumption of power. However, such classification methods based on whether discrete classes or continuous indexing are suitable for HEV energy management need to be further investigated.

In terms of driving style recognition algorithms, Murphey et al. [35] presented a practical approach based on counting aggressive maneuvers, i.e., jerk. The final score is based on a percentage, where 'stays calm' is in the range below 50%, and 'stays aggressive' is in the range above 100%, otherwise between 50% and 100%. Besides, the use of the rules-based (RB) algorithm unifies simplicity, is easy to explain and implement, but limits the number of parameters that can be managed. Larger sets of variables generate unnecessarily complex rules that can be substituted by fuzzy logic (FL) maps. Syed et al. proposed an FL algorithm to evaluate the optimal operation of the pedals in HEVs [36]. The algorithm can monitor the operation of the gas pedal and brake pedal and then can calculate the appropriate correction value and produce tactile feedback to the driver. The author claims that fuel consumption can be reduced by at least 3.5% with the mildest driving in the simulated environment without compromising vehicle performance. Although the RB and FL algorithms unify the acceptable results by achieving simplicity, the quality of the classification is closely related to the choice of threshold.

The threshold definition in the RB algorithm decides the robustness of the results and requires a lot of data analysis. Unsupervised algorithms do not need to understand the underlying process. In the work of Miyajima et al., a Gaussian mixture model was implemented based on the analysis of car-following behavior and pedal operation spectrum [37]. The car-following identifiability of the model is 69%, and the classification rate of the pedal spectrum analysis in the simulated environment is 89.6% and in the field test is 76.8%. The unsupervised algorithm has proved its applicability to driving style identification. However, the output needs to be guided based on the number of interpretations and clusters. In addition, the classification performance may be worse than the supervised algorithm. Supervised algorithms represent understanding the driving style of the data used for training. Augustynowicz applied an Elman neural network to identify driving behavior by speed and accelerator pedal position [38]. The Markov model has also successfully achieved driving style recognition. Guardiola et al. defended the advantages of combining the Markov model with the Monte Carlo application, [39], thereby generating random patterns based on previous data in the driving style representation. Pentland and Liu defended the adaptability of the dynamic Markov model to driving style, which was supported by the fact that it is best to capture human movements through a series of control steps. To address the real-world problem regarding chaotic time series prediction, a driver-centric velocity prediction model is proposed for enhancing vehicle intelligent control and advanced driver assistance [40]. This multi-dimensional fuzzy predictor is designed and used to capture the complex dynamics of driver behavior and vehicle movement, improving prediction accuracy and enabling more effective control strategies.

9.4.3 Driver-Oriented Control Method

In the work of Zhang et al., a novel method for driving blocks classification is proposed to classify the driving pattern into diverse groups, as opposed to using the existing unbroken driving cycles [41]. One concern is that as the driving cycle grows, more driving blocks need to be classified. Langari et al. designed a driving condition recognition component that uses long-term and short-term numerical characteristics of the driving cycle to evaluate the driving conditions, the driving style of the driver, and the operation mode of HEVs. Another alternative method involves a self-reporting driving behavior questionnaire designed by Zhang et al. for offline cluster analysis [42]. Martinelli et al. proposed a machine learning–based method to continuously characterize the driver by data analysis for built-in vehicle sensors [43]. They found that with the features such as cold start-up, cruising down the motorway, and idling in heavy traffic, they were able to discriminate the car owner by an impostor. Adaptability can be also added to the vehicle in another form, which is the driving style recognition employed by Yang et al. [44] and Tian et al. [45]. Differing from the work of Yang et al., the nearest-neighbor method needs labeled learning materials for training purposes as opposed to statistical classification. An interesting piece of work by Gu et al. proposed a pedestrian-aware engine management strategy that considers the environmental effects of the vehicles on pedestrians outside of the vehicles. The strategy helps VSC switch to EV mode when a cloud server informs them that the density of pedestrians has become higher [34]. As Gu et al. stated, an accelerator pedal opening and its change rate are considered as inputs of a driving style recognizer, to define a driving style factor for optimal control of plug-in HEVs [46]. Driver models using fuzzy sets can be well integrated with energy management. However, how to establish the mapped relationship between driving style and control actions is still a big challenge [47].

A highly dynamic driver model not only is dependent on historical driving data but also must consider the current driving data. It utilizes up-to-date driving segments of the driver to model, thus ensuring an accurate judgment of prediction trends. Through real-time or regularly updating driver models, their reliability can be relatively guaranteed. The Markov decision process, also called stochastic dynamic programming, is mainly used to model and solve dynamic decision-making problems. In the work of Guo et al. [48], an onboard learning algorithm for Markov chain models is engaged to produce transition probability matrices of power requirements. Furthermore, fuzzy encoding technology is applied for Markov chain models to add in ports of continuous intervals in reinforcement learning-based energy management [49]. Cairano et al. [50] developed a self-learning stochastic (model predictive control (MPC) for driver-oriented predictive control of an HEV, where using quadratic programming, larger state dimension models than in stochastic dynamic programming, can be reconfigured in real-time to adapt to changes in driving behaviors. Bichi et al. [51] used a linear filtering algorithm to estimate a transformation possibility matrix. The driver model is learned in real-time, permitting the control algorithm to adapt to various drivers and drivers' behaviors. Obviously, the price of the improvement is mainly to sacrifice computing resources in exchange for high-precision state information. As indicated by Moore's Law, it is anticipated this relation can be gradually improved and performed on the actual onboard controller of HEVs for real-time energy saving soon.

9.5 SUMMARY AND OUTLOOK

This chapter provides a comprehensive overview of the research conducted on driver behavior and its applications for vehicle supervisory control. Beginning with the definition of driver behavior, its significance in the context of energy management is clarified. The historical development of driver-vehicle interaction is discussed as the different research paths have been taken. It highlights the importance of integrating driver behavior into the control strategies for optimizing vehicle performance and reducing energy consumption. The chapter also delves into the specific applications of driver behavior in vehicle supervisory control, such as predictive control, eco-driving assistance system, and intelligent route planning. It explores the role of advanced technologies, including machine learning and Big Data analytics, in developing accurate models for predicting energy consumption and optimizing control strategies.

Generally, this is a comprehensive guide that may serve researchers, engineers, and practitioners interested in developing more intelligent and sustainable transportation systems. The importance of integrating driver behavior into the design and development of vehicle control systems is clear, which has enormous potential to reduce energy consumption and mitigate the impact of transportation on the environment.

ACKNOWLEDGEMENT

This work is partly derived from the author's PhD thesis [32]. The author would like to thank his supervisor Professor Hongming Xu, who has offered timely support and guidance whenever needed.

REFERENCES

1. Nauto. (n.d.). *What is driver behavior?* Retrieved from www.nauto.com/glossary/what-is-driver-behavior
2. Habibifar, N., & Salmanzadeh, H. (2022). Relationship between driving styles and biological behavior of drivers in negative emotional state. *Transportation Research Part F: Traffic Psychology and Behaviour*, 85, 245–258.
3. Meseguer, J. E., Calafate, C. T., & Cano, J. C. (2018). On the correlation between heart rate and driving style in real driving scenarios. *Mobile Networks and Applications*, 23, 128–135.
4. Yang, L., Ma, R., Zhang, H. M., Guan, W., & Jiang, S. (2018). Driving behavior recognition using EEG data from a simulated car-following experiment. *Accident Analysis & Prevention*, 116, 30–40.

5. Yan, F., Liu, M., Ding, C., Wang, Y., & Yan, L. (2019). Driving style recognition based on electroencephalography data from a simulated driving experiment. *Frontiers in Psychology*, 10, 1254.
6. Ajzen, I. (2015). Consumer attitudes and behavior: the theory of planned behavior applied to food consumption decisions. *Italian Review of Agricultural Economics*, 70(2), 121–138.
7. Lv, C., Cao, D., Zhao, Y., Auger, D. J., Sullman, M., Wang, H., Dutka, L. M., Skrypchuk, L., & Mouzakitis, A. (2017). Analysis of autopilot disengagements occurring during autonomous vehicle testing. *IEEE/CAA Journal of Automatica Sinica*, 5(1), 58–68.
8. Guo, Y., Wu, Y., Xiong, S., Xu, X., & Huang, H. (2020). Investigating the relationship between driver fatigue and driving performance: A driving simulator study. *Accident Analysis & Prevention*, 142, 105572.
9. Jiang, Y., Wang, L., Xie, J., & Dong, R. (2021). Investigating the effects of cognitive load on driving performance using naturalistic driving data. *Accident Analysis & Prevention*, 155, 106105.
10. Wang, Y., Li, Y., Zhang, J., & Wu, C. (2021). Investigating the impact of smartphone use on driver behavior: A naturalistic driving study. *Accident Analysis & Prevention*, 159, 106336.
11. Society of Automotive Engineers International. (2021). *Taxonomy and definitions for terms related to driving automation system for on-road motor vehicles* (SAE J3016-202103). Retrieved from www.sae.org/standards/content/j3016_202103/
12. Holland, C., Levis, J., Nuggehalli, R., Santilli, B., & Winters, J. (2017). UPS optimizes delivery routes. *Interfaces*, 47(1), 8–23.
13. Vatanparvar, K., Faezi, S., Burago, I., Levorato, M., & Al Faruque, M. A. (2018). Extended range electric vehicle with driving behavior estimation in energy management. *IEEE transactions on Smart Grid*, 10(3), 2959–2968.
14. Apollo. (n.d.). *Apollo white paper*. Retrieved from https://developer.apollo.auto/platform/whitepaper_cn.html
15. Gietelink, O., Ploeg, J., De Schutter, B., & Verhaegen, M. (2006). Development of advanced driver assistance system with vehicle hardware-in-the-loop simulations. *Vehicle System Dynamics*, 44(7), 569–590.
16. Pradhan, A. K., Hammoud, R., & Chute, P. (2019). Eye-tracking technology in driver behavior research: A review. *Journal of Safety Research*, 68, 105–118.
17. Goaltouch. (n.d.). *Product detail*. Retrieved from www.goaltouch.com/ProductDetail140.html
18. Di Stasi, L. L., Catena, A., & Cañas, J. J. (2019). Electroencephalography in driving research: A critical review of applications, methods, and metrics. *Neuroscience & Biobehavioral Reviews*, 103, 382–397.
19. Cao, J., Huang, W., Xu, Z., & Cao, Y. (2020). Driving simulation with electroencephalogram (EEG) measurements and video recording. *Frontiers in Human Neuroscience*, 14, 587287. https://doi.org/10.3389/fnhum.2020.587287
20. Wu, Y., Wei, H., Chen, X., Xu, J., & Rahul, S. (2020). Adaptive authority allocation of human-automation shared control for autonomous vehicle. *International Journal of Automotive Technology*, 21, 541–553.
21. Martinez, C. M., Heucke, M., Wang, F., Gao, B., & Cao, D. (2018). Driving style recognition for intelligent vehicle control and advanced driver assistance: A survey. *IEEE Transactions on Intelligent Transportation Systems*, 19(3), 666–676. https://doi.org/10.1109/TITS.2017.2706978
22. Jung, T. P., Makeig, S., Stensmo, M., & Sejnowski, T. J. (1997). Estimating alertness from the EEG power spectrum. *IEEE Transactions on Biomedical Engineering*, 44(1), 60–69.

23. Moray, D. N. (1977). *Mental workload: Its theory and measurement*. Springer Science and Business Media LLC.
24. Salas, E., & Maurino, D. E. (2010). *Human factors in aviation* (Second Edition). Academic Press.
25. Wickens, C. D., & Damos, D. L. (1991). Processing resources and attention. In *Multiple-task performance* (pp. 3–34). CRC Press.
26. Ou, Y. K., & Liu, Y. C. (2012). Effects of sign design features and training on comprehension of traffic signs in Taiwanese and Vietnamese user groups. *International Journal of Industrial Ergonomics*, 42(1), 1–7.
27. Zhang, C., Wang, J., & Wang, L. (2020). A driving event detection algorithm based on convolutional neural networks. *IEEE Access*, 8, 23066–23075.
28. Li, H., Zhang, Y., & Li, W. (2021). Driving pattern recognition based on fuzzy clustering and hidden Markov models. *IEEE Transactions on Intelligent Transportation System*, 22(1), 89–101.
29. Ahsan, A. R., & Sajeev, A. S. M. (2020). Driver identification and verification using dynamic time warping of in-vehicle sensor data. *Journal of Advanced Transportation*, 2020, Article ID 8841175.
30. Jiang, Z., Lu, Y., & Chen, W. (2021). Integrating driver behavior recognition with energy management for electric vehicles. *IEEE Transactions on Industrial Informatics*, 17(6), 4044–4053.
31. Miyajima, B. C., Tokunaga, Y., & Kikuchi, S. (2007). Driver modeling based on driving behavior and its evaluation in driver identification. *Proceedings of the IEEE*, 95(2), 427–437.
32. Li, J. (2021). *Driver-oriented intelligent control methodology for series-parallel hybrid electric vehicles* (Doctoral dissertation, University of Birmingham).
33. Manzoni, V., Corti, A., De Luca, P., & Savaresi, S. M. (2010). Driving style estimation via inertial measurements. In *IEEE conference on intelligent transportation system, proceedings, ITSC* (pp. 777–782). IEEE.
34. Corti, A., Ongini, C., Tanelli, M., & Savaresi, S. M. (2013). Quantitative driving style estimation for energy-oriented applications in road vehicles. In *Proceedings—2013 IEEE international conference on system, man, and cybernetics, SMC 2013* (pp. 3710–3715). IEEE.
35. Murphey, Y. L., Milton, R., & Kiliaris, L. (2009). Driver's style classification using jerk analysis. In *2009 IEEE workshop on computational intelligence in vehicles and vehicular system, CIVVS 2009—proceedings* (pp. 23–28). IEEE.
36. Syed, F. U., Filev, D., & Ying, H. (2007). Fuzzy rule-based driver advisory system for fuel economy improvement in a hybrid electric vehicle. In *Annual conference of the North American fuzzy information processing society—NAFIPS* (pp. 178–183). IEEE.
37. Neubauer, J., & Wood, E. (2014). Thru-life impacts of driver aggression, climate, cabin thermal management, and battery thermal management on battery electric vehicle utility. *Journal of Power Sources*, 259, 262–275. https://doi.org/10.1016/j.jpowsour.2014.02.083
38. Augustynowicz, A. (2009). Preliminary classification of driving style with objective rank method. *International Journal of Automotive Technology*, 10, 607–610. https://doi.org/10.1007/s12239-009-0071-8
39. Pentland, A., & Andrew, L. (1999). Modeling and prediction of driver behavior. *Neural Computation*, 11(1), 229–242.
40. Li, J., Zhou, Q., He, X., & Xu, H. (2022). Driver-centric velocity prediction with multidimensional fuzzy granulation. *IEEE/CAA Journal of Automatica Sinica*, 10(2), 547–549.

41. Zhang, S., & Xiong, R. (2015). Adaptive energy management of a plug-in hybrid electric vehicle based on driving pattern recognition and dynamic programming. *Applied Energy*, 155, 68–78.
42. Zhang, Y., Chu, L., Fu, Z., Xu, N., Guo, C., Zhang, X., Chen, Z., & Wang P. (2017). Optimal energy management strategy for parallel plug-in hybrid electric vehicle based on driving behavior analysis and real-time traffic information prediction. *Mechatronics*, 46, 177–192. https://doi.org/10.1016/j.mechatronics.2017.08.008
43. Martinelli, F., Mercaldo, F., Orlando, A., Nardone, V., Santone, A., & Sangaiah, A. K. (2018). Driver behavior characterization for driving style recognition in vehicle system. *Computers and Electrical Engineering*, 1–16. https://doi.org/10.1016/j.compeleceng.2017.12.050
44. Yang, S., Wang, W., Zhang, F., Hu, Y., & Xi, J. (2018). Driving-style-oriented adaptive equivalent consumption minimization strategies for HEVs. *IEEE Transactions on Vehicular Technology*, 67(10), 9249–9261. https://doi.org/10.1109/TVT.2018.2855146
45. Tian, X., Cai, Y., Sun, X., Zhu, Z., & Xu, Y. (2019). An adaptive ECMS with driving style recognition for energy optimization of parallel hybrid electric buses. *Energy*, 189, 116151.
46. Gu, Y., Liu, M., Naoum-Sawaya, J., Crisostomi, E., Russo, G., & Shorten, R. (2018). Pedestrian-aware engine management strategies for plug-in hybrid electric vehicles. *IEEE Transactions on Intelligent Transportation System*, 19(1), 92–101.
47. Li, J., Zhou, Q., He, Y., Williams, H., Xu, H., & Lu, G. (2021). Distributed cooperative energy management system of connected hybrid electric vehicles with personalized non-stationary inference. *IEEE Transactions on Transportation Electrification*, 8(2), 2996–3007.
48. Guo, Q., Zhao, Z., Shen, P., Zhan, X., & Li, J. (2019). Adaptive optimal control based on driving style recognition for plug-in hybrid electric vehicle. *Energy*, 186, 115824.
49. Liu, T., Wang, B., & Yang, C. (2018). Online Markov Chain-based energy management for a hybrid tracked vehicle with speedy Q-learning. *Energy*, 160, 544–555. https://doi.org/10.1016/j.energy.2018.07.022
50. Di Cairano, S., Bernardini, D., Bemporad, A., & Kolmanovsky, I. V. (2014). Stochastic MPC with learning for driver-predictive vehicle control and its application to HEV energy management. *IEEE Transactions on Control System Technology*, 22(3), 1018–1031.
51. Bichi, M., Ripaccioli, G., Di Cairano, S., Bernardini, D., Bemporad, A., & Kolmanovsky, I. V. (2010). Stochastic model predictive control with driver behavior learning for improved powertrain control. In *Proceedings of the IEEE conference on decision and control* (pp. 6077–6082). IEEE. https://doi.org/10.1109/CDC.2010.5717791

10 Global Control Optimizations of Electrified Vehicles

Nan Xu, Yan Sui, Yan Kong, Qiao Liu and Fenglai Yue

10.1 INTRODUCTION

Conserving fossil energy and rationally using renewable energy are inevitable demands for the sustainable development of human society [1]. Therefore, low-carbon travel tools are people's future vision and realistic requirements, and it is also one of the main development directions of the world's automobile industry [2]. In order to cope with the increasingly severe contradiction between oil and energy supply and demand, environmental pollution, and the pressure of carbon dioxide emissions, governments around the world have proposed corresponding strategies and laws for the energy-saving and emission reduction of automobiles. In order to effectively achieve the goal of the dual-carbon strategy [3], the China Association of Automobile Manufacturers, together with relevant research institutions and experts, has formulated a low-carbon strategy for the automobile industry to "achieve carbon peaking in 2028, near zero emissions in 2050, and carbon neutrality in 2060" [4]. From January 2020 onwards, according to the EU 2019/631, the average CO2 emissions of newly launched passenger cars shall not exceed 95 g/km, and the average CO2 emissions of light commercial vehicles shall not exceed 147 g/km. Once this target is exceeded, a fine of 95 euros per gram per vehicle [5]. Encouraged by the Zero Emission Vehicle Act and the Governor's Order implemented in California, more than 400,000 hybrid vehicles have been driven on the road [6].

The multi-energy source vehicle represented by PHEV is a focus of development direction in the automobile industry. Involved vehicle energy management technology has become a focus of automobile research and development due to its outstanding performance in the field of energy-saving and emission reduction. The rule-based strategy [7–9], as the most basic strategy, is widely applied due to its simplicity and practicality. However, it cannot obtain the globally optimal solution. To achieve better performance in the fuel economy of MEVs, optimization-based energy management [10–12] control strategies have been developed, which mainly include instantaneous optimization [13, 14] and global optimization

 DOI: 10.1201/9781003302827-13

[15, 16]. Dynamic programming (DP) [17–19], as a typical global optimization method, can obtain the theoretical optimal fuel economy, which provides a benchmark [20, 21] for assessing the optimality of other energy management strategies.

However, there are four main challenges in the practical application of DP on MEVs.

The first challenge is the standardization problem of the DP control strategy, which mainly refers to establishing a unified state space model, matching feasible work modes and driving conditions, and building a unified global optimization framework. For a fixed vehicle model, the selection of state variables and control variables is slightly different. For example, for a single-axis, series-parallel, plug-in hybrid electric bus, the battery state of charge (SOC) is chosen as the only state variable, and engine torque (T_e) and motor torque (T_m) are chosen as the independent control variables according to Wang XM [22]. For Peng JK [23] and He HW [24], the DP model has three control variables: T_e, n_e (engine speed), and T_m.

To solve the standardization problems of DP, Zhou W established a unified state space model of DP based on the work modes of electric vehicles/hybrid electric vehicles [25]. Xu Nan concluded the main steps and technical routes of the DP control strategy [26]. To achieve superior performance in real-time applications, Zhang YJ proposed a novel hierarchical control framework for PHEVs [27], which includes an energy utilization plan by iterative dynamic programming and energy utilization management by model predictive control (MPC). The majority of the research standardizes the DP problem from the perspective of the powertrain configuration or cooperative control strategy. Because energy conversions are related to information scenarios and vehicle configurations, the overlap of these factors complicates the DP solution process. To quickly obtain an optimal solution, this chapter will organically integrate the vehicles, information, and energy to standardize the DP control processes more effectively.

The second challenge is the real-time application problem of DP on MEVs. On the one hand, the implementation of the DP control strategy relies on prior knowledge of the driving power demand; that is, the entire trip information should be acquired in advance. To avoid the adverse effects of optimal results against unknown cycles, the DP method is usually combined with the driving pattern recognition method, MPC [28], velocity prediction, or rule extraction. Based on driving pattern recognition and DP, Zhang S proposed an adaptive energy management of PHEVs [29], which recognizes driving patterns by a fuzzy logic controller. Under the MPC framework, Guo JQ [30] proposes an adaptive energy management strategy with dynamically updated traffic information. The short-term velocity is predicted by a deep neural network, and DP is applied to calculate the optimal energy distribution at each MPC control step. Analogously, in Ref. [31], according to the optimal results of the DP strategy, Peng JK calibrated a rule-based control strategy to ensure timely control. Three-segment control rules are extracted from the DP results and then a load-adaptive rule-based control strategy is proposed according to Liu C [32]. Li MC [33] analyzes the offline optimal control rules of DP under different driving modes and uses a random forest method for learning.

Against the background of intelligent transportation systems (ITSs) [34] and automotive Big Data [35, 36], basic information (vehicle state, road conditions, and traffic conditions) and derived data (road capacity, driving style or driver behavior) are available, which provide an information basis for global optimization energy management. With the prior trip information, the global optimal energy distribution can be realized.

Yet, with the increase in the number of control variables, DP suffers from the "curse of dimension" [37] and can only be implemented offline owing to its time consumption and tremendous memory. For basic DP, after meshing the state variable (grids), the cost-to-go function is calculated backward based on the state transition equation, then a forward calculation is performed to search for the optimal control consequence. However, this results in a heavy computational burden. In addition, tremendous memory is reflected in the storage of the fuel matrix, which involves three different approaches: point to point, a highly sparse matrix, and a dense matrix with column by column [38]. To ensure the optimality of EMS while improving real-time performance, an effective method is to generate SOC reference trajectory [39] by fully considering current and future road-traffic information. Based on the traffic information, Lei ZZ leveraged a simplified DP [40] to determine the optimal SOC trajectory with fast calculation speed. In Ref. [41], Astarita V obtained the reference SOC trajectory only by the average velocity information of traffic flow, which is smoothed by a one-order low filter for the floating car data. If the battery SOC is restrained to fluctuate near a reference trajectory [42], the exploring region can be significantly shrunk. Thus, under the deterministic trip information, we will propose a fast DP by narrowing the exploring region as much as possible to ensure the optimal performance with satisfactory calculation speed.

TABLE 10.1
Nomenclature

MEVs	Multi-energy source vehicles
PHEVs	Plug-in hybrid electric vehicles
SOC	State of charge
MPC	Model predictive control
GPSs	Global positioning systems
GISs	Geographical information systems
DDP	Deterministic dynamic programming
HEVs	Hybrid electric vehicles
EMS	Energy management strategy
DP	Dynamic programming
ADP	Adaptive dynamic programming
ITSs	Intelligent transportation systems
IPE	Information layer-physical layer-energy layer
ADHDP	Action-dependent heuristic dynamic programming

With the rapid development of artificial intelligence algorithms and data processing technology, neurodynamic programming [43], deep learning [44] reinforcement learning [45], and adaptive dynamic programming (ADP) [46, 47], as data-driven [48] control strategies, show superior performance due to online implementation and approximation of DP results. Due to the strong self-learning ability and adaptability of neural networks, the ADP method can adaptively approximate optimal control and optimal cost function in DP optimizing problems to improve the real-time performance of global optimization energy management. It is regarded as an effective method to improve the vehicle economy in the case of uncertain operating conditions. Thence, aiming to the stochastic trip information, we will utilize ADP methods to improve real-time performance while ensuring the global optimality.

The third challenge is the accuracy of the results, which mainly refers to acquiring comprehensive and accurate trip information and reducing the cumulative errors caused by interpolation leakage [49]. In real-world driving, traffic congestion, upgrades or downgrades, and traffic lights are inevitable, which can significantly affect the power demand. Thanks to intelligent transportation systems (ITSs), geographical information systems (GISs) [50], and global positioning systems (GPSs) [51], the acquired trip information becomes more realistic and accurate. In Ref. [52] by Gong QM, based on traffic information, DP was applied to reinforce charge-depletion control, which showed a significant improvement in fuel economy compared with rule-based control. In particular, road grade has a large influence on the battery charging and discharging processes; thus, it significantly affects vehicle energy consumption. Guo JQ proposed a DP method that considers road grades [53]. Considering that the future route is unknown, Zeng XR modeled the road grade as a Markov chain in [54]. However, the majority of the literature only accounts for vehicle speed and slope, without involving slip rate and gross weight. To obtain more accurate driving power demand, we will propose full-factor driving data to model trip information, including vehicle speed, slope, and slip rate.

During the backward calculation, if the terminal state is not on discrete grids, linear interpolation will be utilized to calculate the optimal cost-to-go function, which causes accumulative errors. The basic solution is to reduce the discrete interval as much as possible; however, it will increase the computational burden. To address the interpolation leakage issue, Zhou W proposed a level-set DP algorithm [49], which uses a level-set function to describe the backward reachable space. To avoid interpolation calculations, Zhou W designed a state variable filter to transform the standard linear grid into a nonlinear grid [25]. Aiming to obtain gridded cost-to-go, Larsson V [55] used a local approximation (linear and quadratic spline) and derived an analytic solution for the optimal torque split decision at each point, which significantly reduced the memory storage requirements. An additional module, such as an approximate cost function or the filtration of invalid states, is required in the preceding methods, which increases the algorithmic complexity. In summary, minimizing the cumulative errors caused by interpolation issues and balancing the calculation accuracy and computational burden are key points.

The last challenge is to satisfy drivability, which mainly refers to restrictions on gear shifting and engine start-stop. For vehicles with automatic manual transmission, frequent gear shifting will not only accelerate the internal wear of the transmission but also make the driver feel tired. Furthermore, frequent starting and stopping of the engine will worsen the fuel economy. To obtain the optimal engine on-off command, intelligent algorithm, such as a genetic algorithm incorporated with quadratic programming [56], is developed. Compared with DP, it can significantly reduce the computational time, but the price to pay is additional model approximations (and heuristics) for discrete decision variables [57].

To fit for practical conditions, engine frequent start-stop problem, frequent shift problem [58], and excessive transient torque response should be considered in a DP optimizing process to ensure safety, drivability, and comfort. However, a real-time EMS, including engine start-stop and gear selection in an optimal control framework, is rarely found in previous literatures. Two main approaches can be found in several previous studies. On the one hand, a penalty function proportional to the number of gear shifts or engine starts/stops is added to the fuel consumption criterion [59, 60]. On the other hand, an additional augmented cost function [61] is introduced with multi-criterion cost, which integrates fuel consumption and drivability constraints. Multi-criterion performance indexes often use weighting factors to adjust the tradeoff between each individual criterion [62]. The selection of penalty factors should take the coefficient of fuel consumption as a reference and ensure that the battery SOC is within the preset range. To embody the consideration of vehicle drivability, Fan LK added a penalty function containing the engine start/stop and gear shifting to the cost function [63]. Unfortunately, once penalty factors are selected inappropriately, it will have a significant influence on the optimal results. To avoid the weighting factors having adverse effects on the optimal results, we will address these constraints from a global perspective. That is, relevant constraints will be regarded as restrictions to filter out unsatisfactory control sequences.

To solve these problems of DP on MEVs, energy management technology based on information and intelligent control stands out with its outstanding performance. This type of energy management system obtains all-factor working condition information and realizes the power control of the power components through the utilization of the vehicle power system components and then achieves a reasonable distribution of energy from different energy sources, to reduce fuel consumption and reduce emissions. In other words, working condition information, vehicle power system, and energy distribution are the three elements of energy management research, and they interact and complement each other to achieve optimal vehicle energy consumption. Therefore, a framework of "Information Layer—Physical Layer—Energy Layer—Dynamic Programming" (IPE-DP) is proposed for global optimal energy management on MEVs. The global optimization framework organically integrates vehicles, information, and energy, which realizes the unity of different information scenarios, different vehicle configurations, and energy conversions.

10.2 THE GLOBAL OPTIMIZATION FRAMEWORK OF "INFORMATION LAYER—PHYSICAL LAYER—ENERGY LAYER—DYNAMIC PROGRAMMING" (IPE-DP)

Essentially, global optimization energy management of the vehicle is based on the available trip information to produce the global optimal energy distribution by making full use of the characteristics of the vehicle. Based on hierarchical thinking, a framework of "Information Layer—Physical Layer—Energy Layer—Dynamic Programming" (IPE-DP) [64] is proposed, which reveals the energy-saving mechanism of global energy management.

The framework consists of three main layers, namely, the information layer, physical layer, and energy layer, and two interface layers, the interface layer between the information layer and physical layer and the interface layer between the physical layer and energy layer; there is an application layer in the end. A schematic diagram is shown in Figure 10.1.

According to the correlation between the information, physical, and energy layers, the proposed framework has the following properties:

1) If the trip information in the information layer and the vehicle parameters and operation modes in the physical layer are available, then the corresponding optimal fuel economy of the vehicle can be obtained in the energy layer.

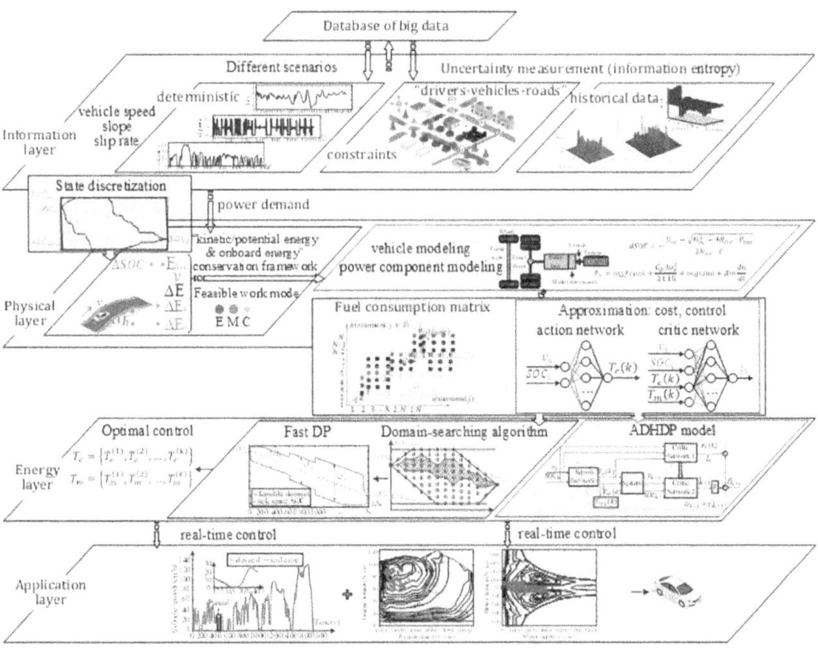

FIGURE 10.1 Schematic diagram of the overall framework (IPE-DP).

2) If the trip information in the information layer and the related energy constraints in the energy layer are available, the optimal powertrain configuration and the operation modes can be matched.
3) If vehicle parameters and work modes in the physical layer and related energy constraints in the energy layer are available, the corresponding economical driving profile can be obtained as well as the optimal traffic flow control for traffic signal facilities.

10.3 THE INFORMATION LAYER—ACQUISITION OF FULL-FACTOR TRIP INFORMATION

To acquire more accurate power demand, full-factor trip information is provided in the information layer. According to the uncertainty of available information, the acquisition of full-factor trip information is realized from three scenarios [65]:

1) Deterministic information
   For a MEV with a fixed line, trip information can be fully understood in advance with ITSs, GISs and GPSs. Simultaneously, the energy saving potential of a certain vehicle configuration can be explored, which provides a benchmark for assessing the optimality of other energy management strategies.
2) Information supported by historical data
   Supported by historical driving data, the state transition probability matrix can be obtained to reflect the distribution of the trip information.
3) Information with constraints
   The following subsection will introduce the acquisition of the trip information with constraints or historical data.

10.3.1 Information with Constraints

Considering that the entropy will be utilized to measure the uncertainty of the trip information, if the influencing factors are not independent, the joint entropy must be used to calculate the information entropy. However, the joint probability density between certain factors is difficult to obtain or even impossible to obtain. Thus, constraints about the drivers, vehicles, and roads are reflected by independent factors to facilitate subsequent work.

10.3.1.1 Factors about "Drivers–Vehicles–Roads"

By comprehensively considering the constraints from the drivers' driving style, dynamic performances of the vehicle, traffic flow, and road conditions, the factors that can characterize the corresponding performance are shown in Figure 10.2.

If only the constraints on the trip information are available, the possible value of the trip information at each moment (or location, the same as below) can be limited by constraints about the drivers, vehicles, or roads, or combination constraints.

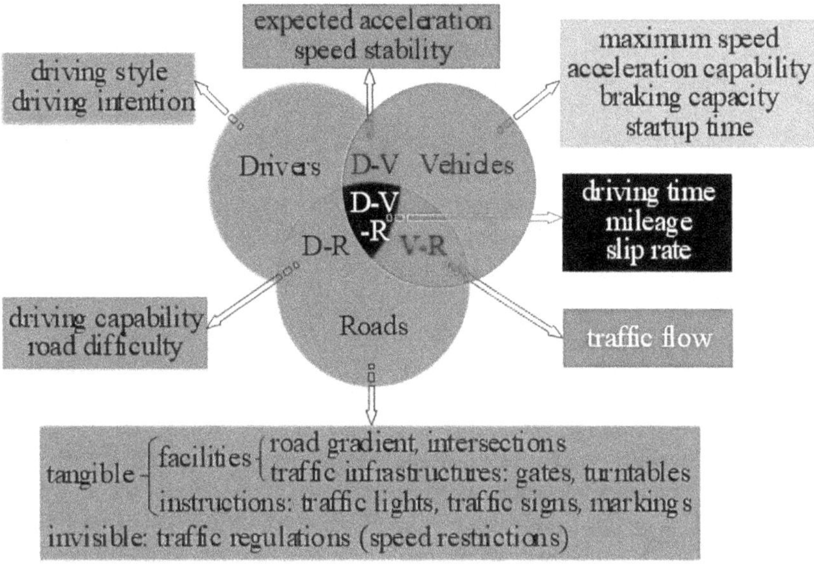

FIGURE 10.2 Factors about "drivers–vehicles–roads."

10.3.1.2 Constraints about "Drivers–Vehicles–Roads"

The factors mentioned before can form corresponding constraints individually or in combination. From the aspects of drivers, vehicles, and roads, the constraints are divided into the following categories:

1) *Constraints about roads*

   On the one hand, tangible constraints refer to constraints imposed by concrete objects, which include facilities and instructions.

   Facilities include mainly slopes and traffic infrastructure (such as gates, intersections and turntables), which are formed based on natural terrain conditions.

   - Slopes and intersections

     Assuming that different routes correspond to different slopes, the slope on a fixed route is available. For the fixed road, the slope is determined. However, at the intersection, once the driving direction changes, the slope changes accordingly.
   - Traffic infrastructure

     At the gates, the vehicle speed reduces to 0. For turntables with traffic lights, the driving rules are related to the external and internal signal lights.

     From the perspective of indicative signs, the instructions include mainly the traffic lights, traffic signs, and markings.

- Traffic lights
  When the signal light is red or yellow, the vehicle speed reduces to 0 before the stop line. Otherwise, the vehicle passes through the signal light at the normal speed. If the position of the traffic lights can be acquired in advance, the entire cycle can be split into multiple segments.
- Traffic signs
  In the traffic field, warning signs, prohibition signs, and mandatory signs are all traffic signs. For roads with speed limit signs, vehicles should follow the speed restriction.
- Markings
  The markings impose the restrictions on the vehicle speed, driving direction, and lane changing conditions (overtaking). For instance, when encountering a white diamond pattern, the vehicle must slow down and pay attention to pedestrians crossing the road [66].

  However, intangible constraints refer to traffic regulations, which are mainly reflected in speed restrictions. If there is no speed limit sign, aiming at different roads, different countries or regions have different traffic laws, which results in different speed restrictions. For example, according to traffic regulations in China, for roads without center lines, urban roads have a speed limit of $30 km / h$, and highways have a speed limit of $40 km / h$ [67].

  Based on these rules, there exists a maximum or minimum speed limit when the vehicle is driving on a certain road.

2) *Constraints about vehicles*
   For a fixed vehicle, the average speed is mainly affected by the maximum speed, acceleration capability, and braking capacity. The maximum speed depends on the maximum power that power components can provide.

   The acceleration capability is commonly reflected in the starting acceleration time and overtaking acceleration time. According to the longitudinal dynamic equation, the maximum acceleration (a_{dm}) can be determined.

   The braking capacity is mainly related to brake performance, tires, and road adhesion coefficient. The maximum deceleration (a_{bm}) is related to the braking intensity (z): $a_{bm} = zg$.

   Because a kinematic fragment usually contains idling, acceleration, uniform speed, and deceleration, the startup time of the vehicle is accounted for to characterize the vehicle's dynamic performance.
3) *Constraints about the combination of roads and vehicles*
   When multiple vehicles are driving in a certain area, the vehicle speed will be affected by the traffic flow, which reflects the heavy traffic or smooth traffic.

   In different time periods, based on the space speed survey, the speed distribution histogram and cumulative frequency curve of the velocity can be obtained. Generally, the speed distribution agrees with the normal distribution. That is, it satisfies the following probability density function:

$$p(v) = \frac{1}{\sqrt{2\pi}\sigma_s} e^{-\frac{(v-\bar{v}_s)^2}{2\sigma_s^2}} \tag{10.1}$$

where $\bar{v}_s$ is the space mean speed and σ_s is the standard deviation (variance), which reflects the discretization degree of the speed distribution.

According to the cumulative frequency curve of the vehicle speed, the 85th percentile speed (85% of all vehicles are driving below this speed) [68] is regarded as the maximum speed limit to ensure driving safety. The 15th percentile speed (15% of all vehicles drive below this speed) [68] is regarded as the minimum speed limit to reduce congestion.

The standard deviation of the speed distribution has the following approximate relationship with the difference between the 85th percentile speed and the 15th percentile speed [68]:

$$S = \sigma \approx \frac{v_{85\%} - v_{15\%}}{2.07} \tag{10.2}$$

That is,

$$\begin{cases} \int_{v_{min}}^{v_{max}} \frac{1}{\sqrt{2\pi}\sigma_s} e^{-\frac{(v-\bar{\bar{v}}_s)^2}{2\sigma_s^2}} dv = 70\% \\ \sigma_s \approx \frac{v_{max} - v_{min}}{2.07} \end{cases} \tag{10.3}$$

Based on this information, the constraints on traffic flow are transformed into speed restrictions.

4) *Constraints about the combination of drivers and vehicles*
For different driving styles, the expected acceleration (deceleration) and speed stability will be different. Taking the maximum acceleration (deceleration) as the baselines, the expected acceleration (a_{de}) and expected deceleration (a_{be}) can be determined by:

$$\begin{cases} a_{de}^{l} = \beta_l \cdot a_{dm} \\ a_{be}^{l} = \beta_l \cdot a_{dm} \end{cases} \tag{10.4}$$

where β represents the weighting factors of different driving styles, and the subscript l indicates the driving style, which corresponds to 0, 1, or 2.

Furthermore, for different driving styles, the speed stability is different, which corresponds to the discretization degree of the speed distribution (σ_s).

5) *Constraints about the combination of drivers and roads*
Based on the driving mileage and working seniority, the professional level of drivers is divided into five levels: beginner, intermediate, advanced, technician, and senior technician. Based on the driver professional classification, driving ability can be divided into five grades, which are recorded as 1, 2, 3, 4, and 5. The higher the value is, the higher the driving ability.

Due to the different types, properties, and functions of roads, it is not possible to classify all roads with a single standard. Currently, roads are classified first and then graded based on technical standards. According to the characteristics and application, the roads are divided into five levels: urban roads, highways, mine roads, forest roads, and country roads. Based on the road types, the road difficulty can be divided into five grades, which are recorded as 1, 2, 3, 4, and 5. The higher the value is, the greater the road difficulty.

By comprehensively considering the road difficulty and driving ability, feasible routes can be screened out. The filtering rules can be formulated as

$$route = \begin{cases} 0, if\ p > q \\ 1, if\ p \leq q \end{cases} \tag{10.5}$$

where p represents the road difficulty level of this road, and$q$ represents the level of driving ability. The value 1 means that the route is feasible; otherwise, it is not feasible.

6) *Constraints about joint constraints*
Based on the driving ability and road difficulty, feasible routes can be selected out. From the starting point to the destination, the driving time and mileage of each route are different.

Moreover, different routes correspond to different adhesion coefficients. Based on the road adhesion coefficient and speed range, the range of the slip rate of each feasible route can be determined.

10.3.1.3 Limitations of Trip Information

Based on the driving time, mileage, and slip rate of each route, the unique route and lane can be selected according to different objective functions. The objectives mainly involve the shortest driving time, shortest mileage, and so forth.

From the starting point to the destination, all possible routes in this district and the corresponding absolute altitude can be obtained by Google Maps. By setting the reference point, the slope of each route can be acquired, which composes the sets $\Theta_1 = \{\theta_1^1, \theta_1^2, \ldots, \theta_1^n\}, \Theta_2 = \{\theta_2^1, \theta_2^2, \ldots, \theta_2^n\}, \ldots, \Theta_L = \{\theta_L^1, \theta_L^2, \ldots, \theta_L^n\}$. Where, θ_i^k is the slope at the kst position (or moment, the same as below) of route i, and Θ_L is the slope set of the Lth route.

The mileage on each route composes the set $\Omega_L = \{l_1, l_2, \cdots, l_L\}$. Based on the speed limits, the driving time on each route is determined based on the average speed, which composes the set $\Omega_T = \{t_1, t_2, \cdots, t_L\}$.

By minimizing the driving time or mileage, a unique route can be determined. Correspondingly, the slope is fixed.

Once the route is determined, by integrating constraints about "drivers–vehicles–roads," the speed profile (with position or time coordinates, the same as below) from the starting point to the destination can be obtained. This profile represents the maximum and minimum speed at each moment.

As shown in Figure 10.3 (a), constraints are represented by arrows, and its effective scope is represented by a line segment (the same as below). With the imposed constraints, the corresponding speed-distance curve ($v-x$) can be determined, which can be converted into the speed-time curve ($v-t$), as shown in the Figure 10.3 (b).

Simultaneously, once the route is determined, the pavement type is fixed, that is, the road adhesion coefficient is determined. By discretizing the driving speed at a certain interval, the slip rate can be determined.

In summary, the speed profile, slip rate, and slope are obtained, which lay a foundation for determining the work mode in the physical layer.

* ① There exists the traffic light, gate, or turntable at point A and point B, and speed restrictions of each section are v_{m1}, v_{m2}, v_{m1};

② There exist the constraint about restricted traffic flow between point C and point D, and the corresponding speed is limited to $\left[v_{fm}, v_{fn}\right]$.

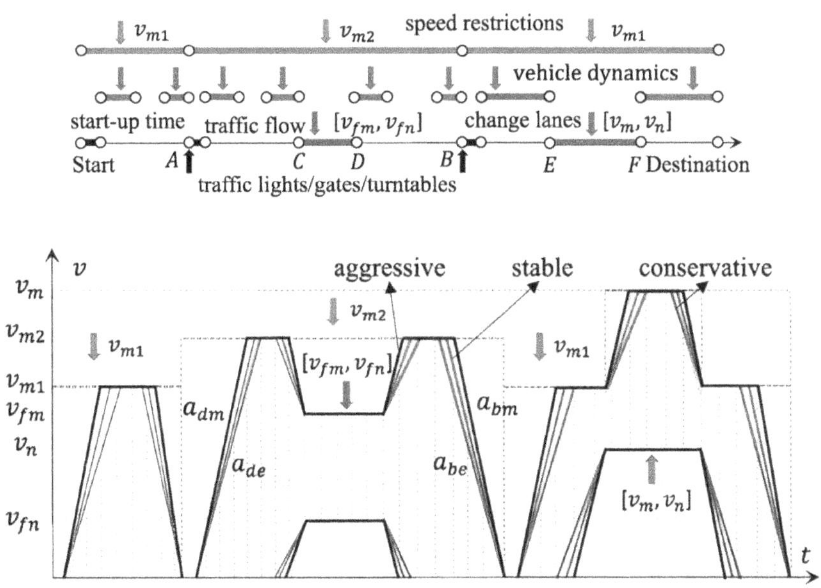

FIGURE 10.3 The schematic diagram of related constraints and speed profile. (a) Schematic diagram of the imposed constraints on a certain route (b) The speed limits.

*Different slopes represent different driving styles.

③ There exists the lane changing (or acceleration behavior) between point E and point F, and the speed is limited to $[v_m, v_n]$.
④ The constraint about the combination of vehicle dynamics and roads (speed limits, traffic flow, overtake/accelerate).

10.3.2 Information Supported by Historical Data

Within a certain time, the vehicle speed and route selection (at multiple junctions) conform to the Markov property. Supported by historical driving data acquired from large databases, the state transition probability matrix can be obtained to reflect the distribution of the trip information. Generally, the relative altitude and slope can be regarded as the state spaces to generate the state transition matrix of the slope. With respect to the vehicle speed, due to the diversity of driving cycles, multiple state transition probability matrices can be generated to improve the prediction accuracy.

Once the pavement type is determined, the road adhesion coefficient α is determined accordingly. With the predicted speed, slip rate can be calculated by the magic formula of the tire [69].

10.4 The Interface Layer Between the Information Layer and Physical Layer—Discretization of the State Space

Because the DP problem is a numerical solution, it is necessary to discrete the state space, including battery SOC. The discretization process primarily includes the following steps.

10.4.1 Boundary of State Feasible Domain

The available trip information is taken into account to limit the maximum charging/discharging current at each moment (or geographic location, the same below), which can significantly narrow the state exploring region. Meanwhile, it provides ideas for the development of rapid DP strategy.

Once the vehicle speed and powertrain configuration are fixed, the motor speed is determined accordingly. Based on the external characteristic of the motor [70], the maximum allowable motor power (P_{m_line}) can be determined, which is less than the fixed maximum motor power (P_{m_max}). Correspondingly, the maximum charging/discharging current (I_{m_line}) at each moment can be determined, that is, $I_{m\_line} = 1000P_{m\_line} / U_{op}$.

By considering the trip information, power demand, and external characteristics of the motor, the transition range of each state point are restricted by the maximum charging/discharging current limit. Meanwhile, the boundary of state feasible domain can be determined as well as the highest/lowest SOC at each moment.

10.4.2 Grid Interval

To ensure equivalent state transition (SOC) between adjacent moments and no drastic change of the number of state points at each moment, state feasible domain is discretized uniformly.

In order to balance the contradiction between calculation accuracy and computational burden, the maximum allowable motor and the required power (P_{req}) at each moment are taken into account to determine allowable battery power, that is, $P_{bat}(k) = g(P_{req}(k), P_{m\_line}(k))$, where the function $g(\cdot)$ is described in Table 10.2.

Then, the maximum discrete interval at each moment can be determined by:

$$\Delta SOC_m(k) = -\frac{U_o(k) - \sqrt{U_o^2(k) - 4R_{int}(k)\cdot P_{bat}(k)}}{2R_{int}(k)\cdot C} \tag{10.6}$$

where U_o, P_{bat} and R_{int} are the SOC, open-circuit voltage, electric power, and internal resistance of the battery, respectively.

Based on the preceding infomration, an approximate discrete interval (ΔSOC) can be determined.

10.4.3 Feasible Domain and the Number of State Points

The initial SOC (SOC_0) and terminal SOC (SOC_f) are regarded as the baselines. Based on the maximum SOC, minimum SOC and baselines, state feasible domain is divided into multiple areas. A schematic diagram is shown in Figure 10.4.

TABLE 10.2
The Function $g(\cdot)$

If $P_{req}(k) > 0$, $P_{bat}(k) = \{0, \pm min\{P_{req}(k), P_{m\_line}(k)\}\}$

If $P_{req}(k) < 0$, $P_{bat}(k) = -min\{P_{req}(k), P_{m\_line}(k)\}$

If $P_{req}(k) = 0$, $P_{bat}(k) = \{0, -P_{m\_line}(k)\}$

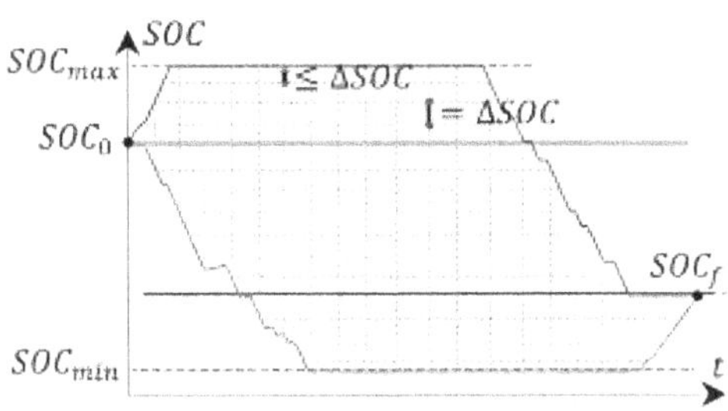

FIGURE 10.4 The division of state feasible domain.

It should be noted that the upper and lower boundaries still need to be discretized when the discrete interval is less than ΔSOC. In each area, the number of discrete points can be calculated based on the highest/lowest SOC (at each moment) and the initial/terminal SOC. Then, the total number of discrete points at each moment can be determined.

10.4.4 Reachable State Set

Starting from the first state point, state points at each moment are numbered sequentially from top to bottom. According to the initial/terminal SOC, the highest/lowest SOC and the number of state points, the SOC value corresponding to each state point can be calculated, which is stored at the corresponding position of SOC matrix. It is defined as $SOC(i,k)$, which represents the SOC value of the ith state point at the moment k.

10.5 THE PHYSICAL LAYER—VEHICLE MODELING AND FEASIBLE OPERATING MODE DETERMINATION

As the physical subject, the modeling of vehicle longitudinal dynamics and power components (engine, motor, and battery) is completed in the physical layer. According to the driving equation, the required power can be determined. With the available trip information, power demand and state matrix, the work modes between any two reachable state points are determined in the physical layer based on the conservation framework of "kinetic/potential energy and onboard energy."

10.5.1 System Architecture and Vehicle Modeling

10.5.1.1 Vehicle Powertrain

Multi-energy source vehicles usually consist of engine, drive motor, power battery, and electronic control system, which includes a vehicle control unit, a battery management system, and so forth.

1) Hybrid vehicle with single electric machine (SEM)
   According to the position of SEM, the hybrid configurations can be divided into five categories, as illustrated in Figure 10.5, namely P_0 (belt driven starter/generator), P_1 (SEM mounted on crankshaft), P_2 (SEM mounted on the gearbox input), P_3 (SEM mounted on the gearbox output), and P_4 (SEM mounted on the driving axle) [71].

   Generally, the engine state can be on or off; the motor state can be off, electric or generate; and the clutch state can be engaging or detaching. For a fixed vehicle configuration, once the operating states of the engine and the motor are determined, the clutch state is determined accordingly. By combining the possible operating states of various components, the feasible work modes can be determined.

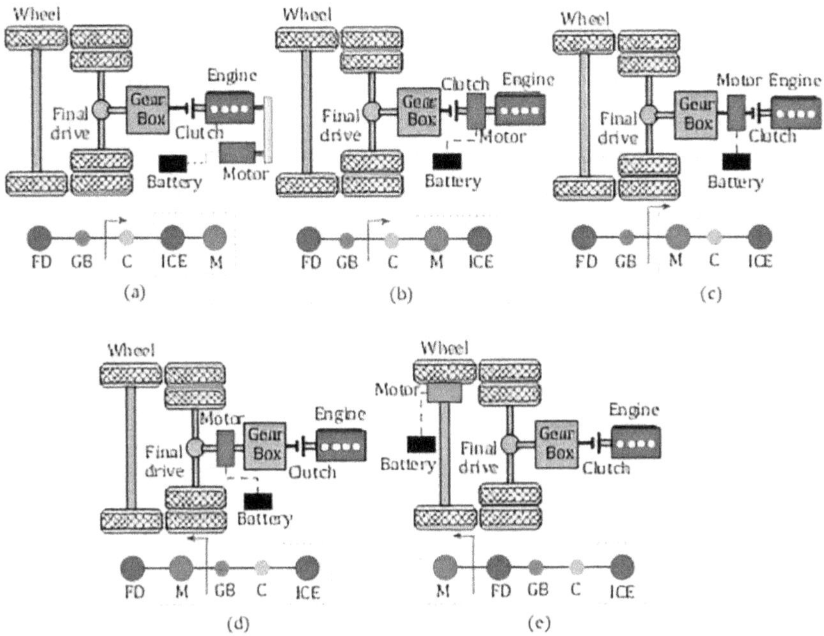

FIGURE 10.5 Topology of the vehicle with SEM. (a) P_0, (b) P_1, (c) P_2, (d) P_3, (e) P_4.

*** FD-the final drive, M-the motor, GB-gear box, C-clutch, ICE-the engine, B-the battery.**

The feasible work modes of each configuration are summarized in Table 10.3. Due to structural constraints, regardless of the configuration P_0 or P_1, the motor cannot drive the wheels alone due to the synchronous rotation of engine and motor. That is, there is no pure electric mode.

Once the vehicle speed and vehicle configuration are fixed, the corresponding engine speed and motor speed in each gear can be obtained by the dynamic equation.

By analyzing vehicle configuration, the power balance equation of the transmission system is obtained as follows:

$$\begin{cases} P_{req} = P_e + P_{bat} \\ P_e = T_e n_e / 9550 \\ P_m = T_m n_m \times \eta_m^s / 9550 \end{cases} \tag{10.7}$$

where P_{req} is the power demand, P_e is the engine power, P_{bat} is the battery power. T_e, T_m, n_e, n_m, P_m are the engine torque, motor torque, engine speed, motor speed, motor power, respectively. η_m is the motor efficiency. When $s = -1$, the motor acts as a drive motor, while, it acts as generator when $s = 1$.

TABLE 10.3
The Existing Work Modes of Each Configuration (P_0, P_1, P_2, P_3, P_4)

structure	P_0	P_1	P_2	P_3	P_4
dynamic	$n_e = n_w * i_g i_0, n_m = n_w * i_g i_0$			$n_e = n_w * i_g i_0, n_m = n_w * i_0$	$n_e = n_w * i_g i_0, n_m = n_w$
work modes	working states	mode		working states	mode
		parking/ sliding		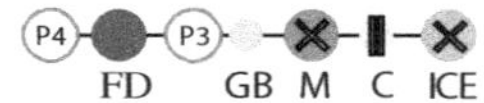	parking/sliding mode
		regenerative braking mode		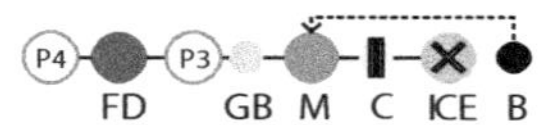	pure electric mode
		engine-only mode			regenerative braking mode
	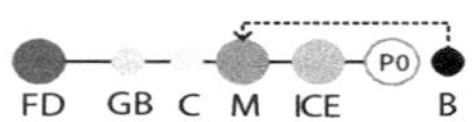	hybrid mode		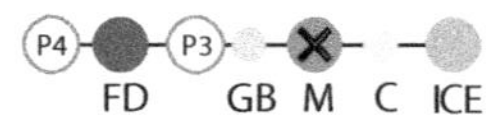	engine-only mode
		driving charging			hybrid mode
					driving charging mode

*× and grey (background) mean that the power component is off. The arrow pointing to the battery indicates that the battery is charging, on the contrary, the arrow pointing to the motor indicates that the battery is discharging. Examples (motor M) represent the P1 configuration and P2 configuration, respectively. *n_e is the engine speed, n_m is the motor speed, n_w is the wheel speed, i_g is the transmission ratio, and i_0 is the final drive ratio.

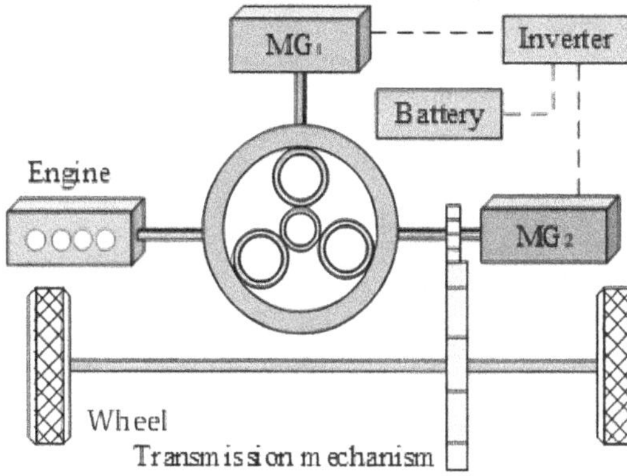

FIGURE 10.6 Powertrain configuration.

2) Hybrid vehicle with planetary gear

The Toyota hybrid powertrain system [72], as a well-known commercial power-split system, is taken as the research object. The structure (series-parallel HEV) is shown in Figure 10.6, which consists of three power sources and a planetary gear set.

The planetary gear set (PG) connects three power sources to implement the power splitting functionality [73]: the engine is connected to planet carrier; MG_1 (motor-generator) and MG_2 are connected to the sun gear and the ring gear (corresponding to the driveline output shaft), respectively. In addition, MG_1 and MG_2 are connected to the battery through an inverter, which is used to convert DC and AC between the battery pack and the motor.

The MG (motor-generator) can operate in charging and discharging mode. By reasonably distributing the state of each gear in the planetary row, the vehicle can be operated in different work modes, which are listed in Table 10.4.

For the planetary gear, the relationship between the speed of sun gear ω_s, the speed of ring gear ω_r and the speed of carrier ω_c is as follows:

$$\omega_c\left(R_r + R_s\right) = \omega_r R_r + \omega_s R_s \tag{10.8}$$

where R_r is gear number of the ring gear (or ring gear radius), R_s is the gear number of the sun gear (or sun gear radius).

Ignoring the energy loss during steady state operation, the torque of sun gear T_s, the torque of ring gear T_r, and the torque of planetary carrier T_c satisfy the following relationship:

$$\begin{cases} T_r = \dfrac{R_r}{R_r + R_s} T_c \\ T_s = \dfrac{R_s}{R_r + R_s} T_c \end{cases} \tag{10.9}$$

TABLE 10.4
Work Modes of Toyota Hybrid Powertrain System

Engine	MG_1	MG_2	Mode
×	×	×	parking/sliding mode
×	×	electric	pure electric mode
×	generate	×	mechanical braking mode
×	generate	electric	start
√	×	×	engine-only mode
√	electric	×	hybrid mode
√	generate	×	stop charging mode
√	×	electric	hybrid mode
√	electric	electric	hybrid mode
√	generate	electric	driving and charging mode

Assuming that the connecting shaft is rigid, the speed between the engine, MG_1, MG_2, and the planetary row meets the following relationship:

$$\begin{cases} \omega_c = \omega_e \\ \omega_s = \omega_{MG1} \\ \omega_r = \omega_{MG2} = vi_0 / R_{wh} \end{cases} \tag{10.10}$$

where i_0 is the final drive ratio, R_{wh} is the wheel radius.

The engine power and battery power are subject to:

$$\begin{cases} P_{req} = P_e + P_{bat} \\ P_e = T_e \omega_e / 9550 \\ P_m = T_{MG1} n_{MG1} \eta_{MG1}^{s} / 9550 + T_{MG2} n_{MG2} \eta_{MG2}^{s} / 9550 \end{cases} \tag{10.11}$$

where $P_e, P_{MG1}, P_{MG2}, P_{bat}$ is the power of engine, MG_1, MG_2, and the battery, respectively. η_{MG1}, η_{MG2} is the efficiency of MG_1 and MG_2.

10.5.1.2 Vehicle Modeling—the Required Power

According to the longitudinal dynamics of the vehicle, the required power P_{req} can be formulated as follows:

$$P_{req} = F_t v \tag{10.12}$$

where F_t is the vehicle traction, v is the vehicle speed, both of them are vectors.

10.5.1.3 Engine Model

To analyze and evaluate fuel economy, the engine model is simplified as a static map to calculate the fuel consumption:

$$M_e = Q_f\left(T_e, n_e\right) \tag{10.13}$$

where n_e is the engine speed, T_e is the engine torque, M_e is the fuel consumption of the engine.

10.5.1.4 Electric Machine Model

The efficiency characteristics of drive motor is expressed as the relationship between motor speed and motor torque, which can be formulated as

$$\eta_m = f\left(T_m, n_m\right) \tag{10.14}$$

where η_m is the motor efficiency, n_m is the motor speed, and T_m is the motor torque, which is defined as positive during propelling and negative during regenerative braking.

Then, the output power of the motor (P_m) can be written as

$$P_m = \begin{cases} T_m\omega_m / 9550\eta_m, & \text{electric motor} \\ T_m\omega_m\eta_m / 9550, & \text{generator} \end{cases} \tag{10.15}$$

10.5.1.5 Power Battery Model

Without the consideration of temperature change and battery aging, a simple but effective internal resistance battery model (static equivalent circuit battery model) is used to calculate the battery power, which is modeled as a voltage source with an open circuit voltage and an internal resistance. Ignoring thermal-temperature effects and battery transients, the state of charge (SOC) can be calculated by Eq. (10.16):

$$SOC_{k+1} = SOC_k - \frac{U_o(k) - \sqrt{U_o^2(k) - 4R_{int}(k)\cdot P_{bat}(k)}}{2R_{int}(k)\cdot C} \tag{10.16}$$

It should be pointed out that the improved battery models are also applicable to the proposed framework. When the SOC value is between 0.2~0.9, there exists a linear relationship between open-circuit voltage and SOC [74, 75]. That is, the open-circuit voltage can be fitted as $U_o = ASOC + B$, where, A and B are the coefficients fitted by experimental data.

10.5.2 The "Kinetic/Potential Energy and Onboard Energy" Conservation Framework

10.5.2.1 The Introduction of Energy Conservation Framework

When the vehicle is driving, from the perspective of energy conservation, energy distribution is essentially the conversion between kinetic energy, potential energy, chemical energy (including electrical energy), and thermal energy, corresponding to vehicle speed, altitude, and works done by the motor (electricity consumption) and the engine (fuel consumption), respectively. To determine the work mode between any reachable state points in the feasible domain, these variables can be organically combined to generate various trigger condition for the powertrain's controllable components, then the unique work mode can be determined.

Based on mathematical analysis, a "kinetic/potential energy and onboard energy" conservation framework is proposed in the physical layer to determine the work modes of the vehicle.

As shown in Figure 10.7, the external factors mainly include the following:

1) Change in kinetic energy (ΔE_k) of the vehicle between the next moment and current moment, which corresponds to vehicle speed.
2) Change in potential energy (ΔE_p) of the vehicle between the next moment and current moment, which corresponds to the altitude.
3) Change in mechanical energy (ΔE) between the next moment and current moment, which corresponds to the sum of ΔE_k and ΔE_p.
4) Vehicle speed.

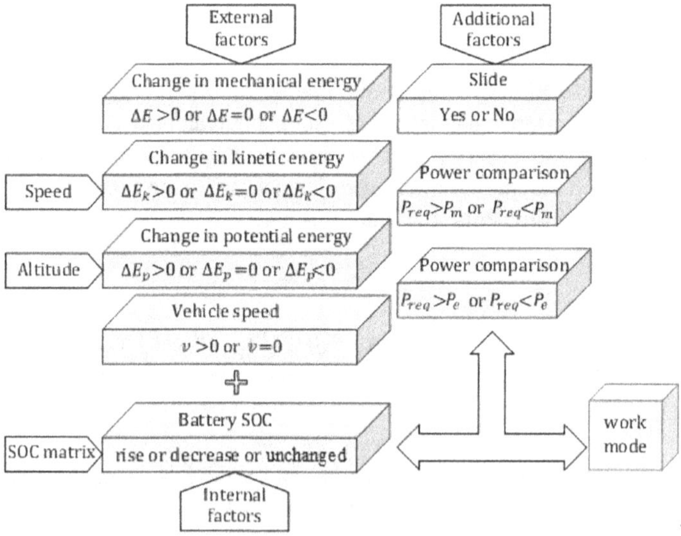

FIGURE 10.7 The schematic diagram of "kinetic/potential energy and onboard energy" conservation framework.

The internal factors mainly refer to the on-board energy, including electricity consumption, which is reflected in the change of battery SOC.

Taking the sliding conditions and power comparison into account, the additional factors mainly include:

1) Whether the vehicle can slide.
2) Comparison between the power demand and the maximum allowable power of the motor or the engine.

The preceding factors can generate various trigger conditions, and each trigger condition corresponds to the unique operation mode of the powertrain's controllable components. It can achieve the one-to-one mapping between work mode and driving condition, which lays a foundation to determine the cost function (i.e., fuel consumption matrix) and control.

It should be noted that the proposed energy conservation framework is applicable to multi-energy source vehicles and single-energy source vehicles. For multi-energy source vehicles, onboard energy involves oil-electric mixing, electric-gas mixing, or electric-hydrogen mixing, and the conversion is between mechanical energy and onboard energy. For single-energy source vehicles, such as pure electric vehicles, the onboard energy refers to electrical energy, and the conservation is between mechanical energy and electric energy.

10.5.2.2 Determination of the Set of Feasible Work Modes

Based on vehicle speed and altitude obtained in the information layer, the change of kinetic energy, potential energy, and total mechanical energy at each moment can be determined.

For the meshed feasible domain (SOC), the state transition between any two reachable state points is determined, which corresponds to the battery power (P_{bat}).

In order to reduce the cumulative errors caused by interpolation problem, the counteracting thought is adopted to determine pure electric mode. When the required power (P_{req}) and the battery power between two reachable state points satisfy $\left|P_{bat} - P_{req}\right| / P_{req} \leq \varepsilon$, the corresponding work mode is regarded as the pure electric mode. Otherwise, the reachable state corresponding to the pure electric mode is determined by interpolation. Therefore, more resulting states in pure electric mode are on the grids to avoid interpolation calculation. To a certain extent, it can reduce cumulative errors caused by the interpolation leakage problem.

For hybrid vehicles with SEM, SOC transition is directly related to the motor power. That is, the working state of the motor can be judged based on the SOC change. If the SOC decreases, the working state of the motor corresponds to electric drive. If the SOC rises, the operating state of the motor corresponds to power generation. Similarly, if the SOC is maintained, the engine is on, while the motor is off. In addition, the comparison between the required power and the maximum allowable motor power at each moment is regarded as an additional condition to distinguish the pure electric mode and the hybrid mode.

Different from the hybrid vehicles with SEM, for hybrid vehicles with planetary gear set (PG), the SOC transition is related to the sum of motor power $\{P_{M1}, P_{M2}\}$. Define the following parameters: $D_0 = min\{P_{M1\_m}, P_{M2\_m}\} = P_{M1\_m}$, $D_1 = max\{P_{M1\_m}, P_{M2\_m}\} = P_{M2\_m}$, $D_2 = P_{M1\_m} + P_{M2\_m}$, $D_3 = P_{em} + P_0$, $D_4 = P_{em} + P_1$, $D_5 = P_{em} + P_{M1\_m} + P_{M2\_m}$. Where, $P_{M1\_m}, P_{M2\_m}, P_{em}$ are the maximum allowable power of MG_1, MG_2 and engine, respectively.

Then, the comparison between the required power with D_0, D_1, D_2, P_{e_max}, D_3, D_4, or D_5 is regarded as an additional condition. For example, in the absence of sliding conditions, when the SOC decreases, if the required power is less than D_2 and $|\left(P_{bat} - P_{req}\right)/P_{req}\langle\varepsilon|$, the vehicle operates in pure electric mode. Then, by comparing the required power with D_0, D_1, and D_2, it can determine whether the MG_1 works alone or the MG_2 works alone or both (work together). Otherwise, the vehicle operates in hybrid mode. By comparing the required power with D_3, D_4, and D_5, it can determine whether the engine and MG_1 work together or the engine and MG_2 work together or all work together.

Based on the "kinetic/potential energy and onboard energy" conservation framework, under each trigger condition, the set of the feasible work mode of the powertrain's controllable components can be determined.

10.6 THE INTERFACE LAYER BETWEEN THE PHYSICAL LAYER AND ENERGY LAYER—DETERMINATION OF STAGE COST AND CONTROLS

For the deterministic trip information, based on the "kinetic/potential energy and onboard energy" conservation framework, the work mode under each trigger condition can be determined. According to the trip information, the required power and ΔSOC, the engine torque, engine speed, motor torque, and motor speed in each gear can be solved based on vehicle dynamics. Correspondingly, the fuel consumption of each gear can be determined according to the engine map.

To improve the calculation efficiency, the fuel consumption under each trigger condition is stored in a three-dimensional matrix, defined as fuel matrix $fuel\left(j,i,k\right)$. Where, i represents the ith state point at the moment k; j represents the jth state point at the moment $\left(k+1\right)$; k represents the moment (or geographical location, the same below). The schematic diagram is shown in Figure 10.8. Where, N_k represents the number of reachable state points at the moment k, N_m represents the maximum number of reachable state points at all moments.

Meanwhile, the corresponding engine power, motor power, engine torque, motor torque, engine speed, motor speed, clutch state, and gear state are stored in the three-dimensional control matrix $P_e\left(j,i,k\right)$, $P_m\left(j,i,k\right)$, $T_e\left(j,i,k\right)$, $T_m\left(j,i,k\right)$, $n_e\left(j,i,k\right)$, $n_m\left(j,i,k\right)$, $clutch\left(j,i,k\right)$, $gear\left(j,i,k\right)$, which lay a foundation for searching the optimal control sequence in the energy layer.

For the stochastic trip information, considering the "curse of dimension," adaptive dynamic programming (ADP) is utilized to improve vehicle economy in the

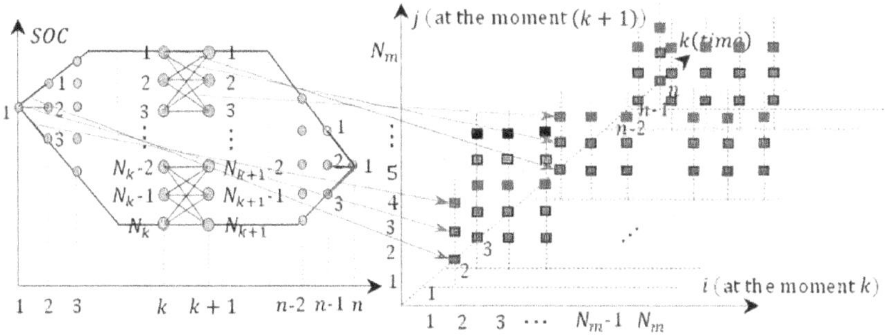

FIGURE 10.8 The schematic diagram of fuel consumption matrix for deterministic information.

* **N_k represents the number of reachable state points at the moment k, N_m represents the maximum number of reachable state points at all moments.**

case of uncertain operating conditions, which includes one action network and two critic networks. Established by a back propagation neural network, the action network is used to approximate the control strategy. The utility function (i.e., stage cost function) is formulated by state variables and control variables.

10.7 THE ENERGY LAYER—OPTIMAL ENERGY DISTRIBUTION

Based on the available trip information and cost function, the optimal energy allocation is completed in the energy level.

10.7.1 The Optimal Solution under the Deterministic Trip Information

For vehicles with a fixed route, the trip information can be acquired in advance. Once the information is determined, a global optimal solution can be obtained by deterministic dynamic programming (DDP).

10.7.1.1 Formulation of Deterministic Dynamic Programming (DDP)

1) State space equation
To ensure the Markov characteristic of the DP model, the gear state must be taken as state variable if the frequent shift problem is considered. Then, a unified state space equation of the DP model for MEVs can be established:

$$x_{k+1} = f\left(x_k, u_k\right) \; k = 1,2,3,\ldots \tag{10.17}$$

where the state variable is $x=[SOC,v,G]$, and the control variable is $u=[T_e,T_m,u_g]$, $f(x_k,u_k)$ is the system dynamics function, k is the moment. v is vehicle speed, G is the gear, T_e is the engine torque, T_m is the motor torque, u_g is the gear change (upshift, downshift or unchanged, corresponding to $1,-1,0$). For the vehicles without transmission, the state space equations are also uniformly expressed by Eq. (10.17) by setting G to a constant value.

2) Constraint conditions

To ensure the safety of power components, the physical constraints on state variables and control variables are considered.

$$\begin{cases} SOC_{min} \le SOC(k) \le SOC_{max} \\ P_{bat\_min} \le P_{bat}(k) \le P_{bat\_max} \\ T_{e\_min} \le T_e(k) \le T_{e\_max} \\ T_{m\_min} \le T_m(k) \le T_{m\_max} \\ u_g \in [0,1,-1] \end{cases} \tag{10.18}$$

where n_e is the engine speed, n_m is the motor speed, P_{bat} is the battery power. The subscripts max and min refer to the maximum and minimum limits of each variable.

In addition, additional constraints are taken into account to avoid frequent gear shifting, frequent engine stopping and starting, and dramatic changes in engine torque. That is,

$$\begin{cases} |T_e(k+1)-T_e(k)| \le M \\ G(k+1)-G(k)=0 \; or \pm 1 \\ \Delta t_G \ge T_G \\ \Delta t_E \ge T_E \end{cases} \tag{10.19}$$

where M represents the maximum allowable transient increment of engine torque, Δt_G expresses the time maintained at a certain gear, T_G is the shortest time of gear maintenance. Δt_E expresses the working time after the engine starts, correspondingly, T_E is the minimum allowable working time of the engine.

3) Cost function

The objectives of DP strategy on MEVs are to find optimal control sequences to obtain the optimal SOC trajectory and minimize fuel consumption over a given driving schedule. Therefore, the optimization objective of DP can be expressed as [17]:

$$J=\sum_{k=1}^{n} L(x_k,u_k) \tag{10.20}$$

where L is the instantaneous cost at each step, n is the total steps (i.e., driving time).

According to the optimality principle [76], the optimal cost function of each step can be regarded as the minimum fuel consumption at this stage. The basic recursive equation of DP can be obtained by

$$J_k^*(x_k) = \min_{u_k}\left[fuel(x_k, u_k) + J_{k+1}^*(x_{k+1})\right] \tag{10.21}$$

$$u^*(k) = \arg\min_{u(k)} J_k(x(k)) \tag{10.22}$$

where J_k^* is the optimal cost-to-go function at state x_k in the *kth* step; $\pi^* = \{u^*(1), u^*(2), \cdots, u^*(N-1)\}$ represents the optimal control sequence. $fuel(\bullet)$ is a function of engine torque and engine speed, which represents the fuel consumption.

10.7.1.2 Update the Fuel Consumption Matrix

For the transient change of engine torque and gear shift, the unreasonable state points can be directly eliminated by updating fuel consumption (redefined as infinite, i.e., inf). However, engine frequent start-stop problem and frequent shift problem are cumulative problems on the time series. When searching for the optimal solution, unreasonable ones should be removed from a global perspective.

According to engine characteristics (transient-state map), when the load is below 30%, the engine does not enter the supercharging zone. There is no limit to the transient increment of engine torque. On the contrary, when the load is greater than 30%, the transient engine torque cannot change dramatically.

Assuming that the consequence of engine torque is $T_e = \{T_e^1, T_e^2, \ldots, T_e^n\}$, the torque increment is $\Delta T_e = \{T_e^2 - T_e^1, T_e^3 - T_e^2, \ldots, T_e^n - T_e^{n-1}\} = \{\Delta T_e^1, \Delta T_e^2, \ldots, \Delta T_e^n\}$, as shown in Figure 10.9 (a). At the moment k, the maximum allowable engine torque (T_{emax}) can be determined by the external characteristic, as shown in Figure 10.9 (b). If the ratio of the required power and the maximum allowable engine power is greater than 30% ($P_{req}^k / P_{emax}^k \geq 0.3$), and the ratio of transient torque increment and the maximum allowable engine torque is greater than 60% ($\Delta T_e^k / T_{emax}^k \geq 0.6$),

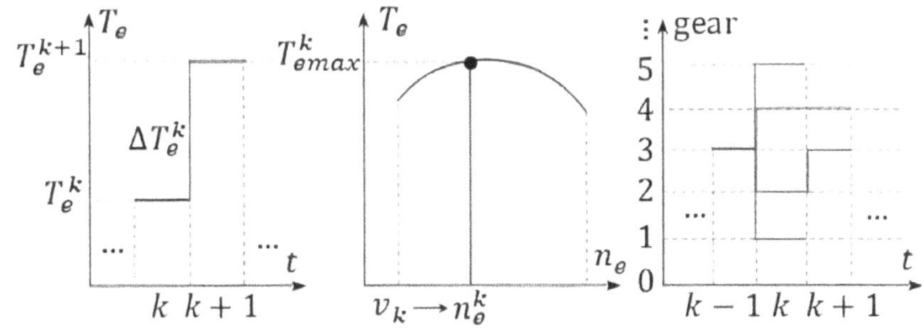

FIGURE 10.9 Schematic diagram of transient torque increment and gear shifting (a) transient torque increment (b) external characteristic curve (c) gear shifting.

the fuel consumption is redefined as infinite (inf). Similarly, as shown in Figure 10.9 (c), if the transient gear shifting exceeds 1, the fuel consumption is redefined as infinite (inf).

Then, the fuel consumption matrix is updated by adding the constraints about transient change of engine torque and gear shifting.

10.7.1.3 The Solution of DDP-Optimal SOC Trajectory Domain

To quickly obtain optimal control, a global domain-searching algorithm is proposed, which outputs all solutions in the form of optimal state domain. The global domain-searching algorithm mainly involves the following steps:

1) Sequentially solve and store the optimal cost of each state point to the starting point
   In the energy layer, by introducing the idea of graph theory, the fuel consumption matrix is transformed into the distance weight between each state point. That is, the optimal energy distribution problem is transformed into the shortest path problem from the starting point (initial SOC) to the ending point (terminal SOC).

   From the starting point, the optimal cost (i.e., shortest distance) of each state point to the starting point is sequentially solved according to the original number, which is stored in a two-dimensional matrix $D(i,j)$, namely, the optimal cost matrix. The optimal value can be solved by:

$$D(j,k+1) = min\{D(i,k) + fuel(j,i,k,g) \mid s_k \in V - \{i\}\} \quad (10.23)$$

   where $fuel(j,i,k,g)$ represents the fuel consumption under the $g-th$ gear from the ith state point at moment k to the jth state point at the moment $(k+1)$. $D(i,k)$ represents the optimal cost from the start point to the ith state point at the moment k.

   To solve the engine frequent start-stop problem and frequent shift problem, the optimal cost matrix is updated by adding constraints on gear maintenance and engine on-off. The specific process is shown in Table 10.5.
2) Sequentially solve and store the optimal state points at the previous moment of each state point

   All state points are renumbered, and each state point at each moment is sequentially numbered. Numbers are stored in a two-dimensional matrix $SOC_{number}(i,k)$, namely, the number matrix, and the corresponding position coincides with SOC matrix.

   Because the same state transition exists in some stages and the solution of DP optimization is a cumulative calculation of the cost function at different stages, there are many equal stage cost functions (fuel consumption) in a certain stage. Therefore, during solving the optimal cost function (i.e., shortest distance) from a certain state point to the starting point, all the optimal state points should be searched.

TABLE 10.5 The Process of Constraint Addition in Calculating Optimal Cost Matrix

Algorithm 1- Add constraints to update the optimal cost matrix

1. Set the time T_G, T_E and initialize $D(1,1) = 0$
2. For $k = 1:1:n$

        For $j = 1:1:num_{k+1}$

      For $i = 1:1:num_k$

        For $g = GM_k : 1 : GN_k$

            $s(i) = D(i,k) + fuel(j,i,k,g)$

            record $\Delta t_G, \nabla t_E$ ← search control matrix: $gear, T_e$

            if ($\Delta t_G \langle T_G || \Delta t_E < T_E$) && $j > max\{T_G, T_E\}$

                $s(i) = inf$

            end

        end

        $D(j, k+1) = min\{s(i)\}$

      end

    end

end

* n is the total driving time, num_k represents the number of state points at the moment k, num_{k+1} represents the number of state points at the moment $k+1$.

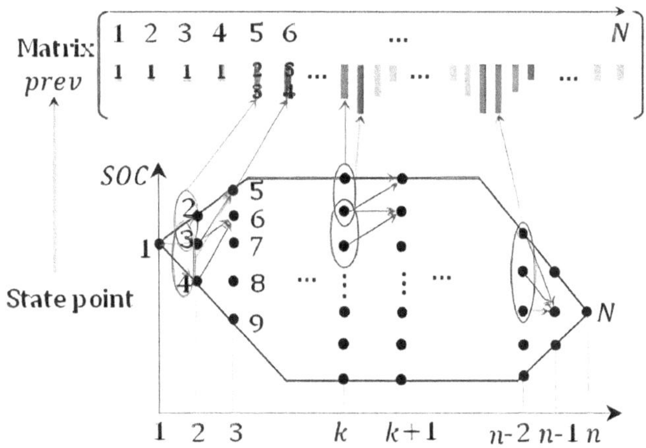

FIGURE 10.10 The schematic diagram of the optimal number matrix.

According to the new number of state points, the serial number of state points at the previous moment on the shortest path (i.e., optimal cost function) are recorded and stored in the corresponding column of a two-dimensional matrix *prev*, namely, the optimal number matrix, as shown in Figure 10.10. The dimension of the matrix *prev* depends on the number of all state points N, which is calculated by $N = \sum_{k=1}^{n} num_k$.

3) Search and store optimal state points at each moment in reverse order
The last state point is the optimal state point of the moment n. Starting from the ending point, the number (Nun) stored in the last column of matrix *prev* can be searched, which is the optimal state point of the moment $(n-1)$ and stored in the $(n-1)th$ column of a two-dimensional matrix $P(i, j)$, namely, the optimal state matrix.

   Then, according to the number Nun, the number stored in $(Nun)th$ column of the matrix *prev* can be searched. The obtained number is the optimal state point of the moment $(n-2)$, which is stored in the $(n-2)th$ column of matrix P. If there exist multiple numbers $\{Nun_1, \ldots, Nun_m\}$ at this moment, the numbers of corresponding columns stored in the matrix *prev* can be searched in turn, which are stored in the $(n-2)th$ column of the matrix P. By analogy, the optimal state points are searched in reverse order until it reaches the starting point. The schematic diagram of the matrix P is shown in Figure 10.11, which stores the serial number of all optimal state points at each moment.

4) Generate optimal SOC trajectory domain
To obtain all the optimal SOC trajectories, the serial number of state point stored in the matrix P is reverted to the original number, which corresponds to the position of each optimal state point in the SOC feasible domain.

   Taking the medium-speed part of WLTP as an example, based on the proposed global domain-searching algorithm, all state points on all optimal SOC trajectories can be searched, as shown in Figure 10.12. The front view and vertical view reflect the position of each optimal

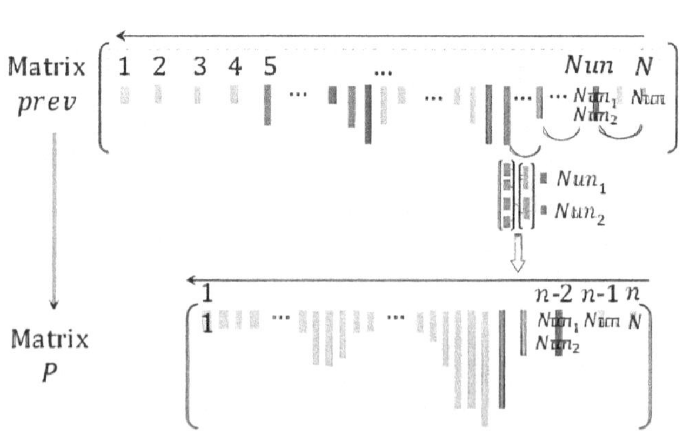

FIGURE 10.11 The schematic diagram of the optimal state matrix.

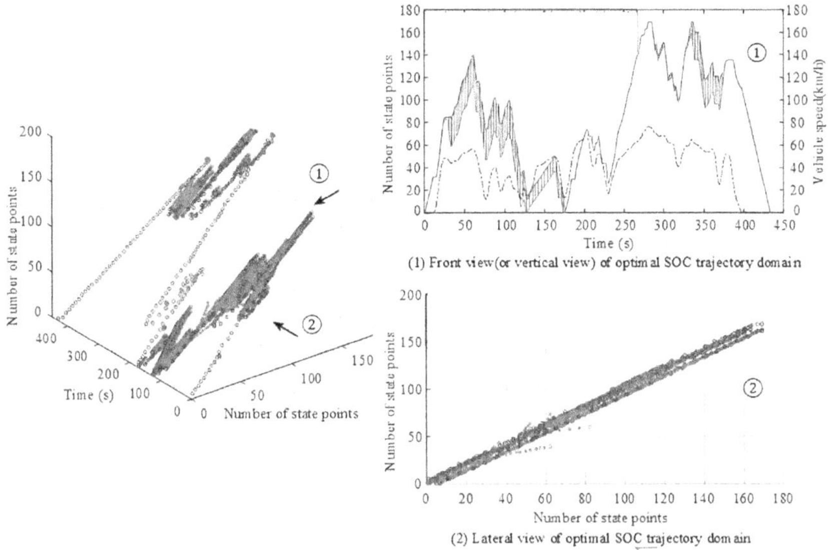

FIGURE 10.12 The optimal SOC trajectory domain under the medium-speed of WLTP.

state point in the SOC feasible domain. It can be seen that all the optimal state points form an optimal trajectory domain. According to the distribution of optimal state points, the equivalent optimal states (at a certain moment) mostly occur in deceleration or the part of multiple continuous rises and falls (within a certain range). The main reason is that the parts correspond to multiple optional working modes; hence, the battery SOC can fluctuate within a certain range, resulting in the same total fuel consumption. Moreover, the position of the optimal state point in the SOC feasible domain reflects the general trend of the optimal SOC trajectory. Thus, the density of the optimal state points in the lateral view can roughly reflect the type of the driving cycle.

Because it is not easy to output all optimal SOC trajectories, all optimal state points are output in the form of optimal domain to improve computational efficiency. That is, the optimal SOC trajectory domain is generated to obtain the optimal results, which lays a foundation for the establishment of fast DP in the subsequent study.

10.7.1.4 The Solution of DDP-Fast DP

To effectively improve the computational efficiency of the global optimization algorithm, the direct method is to reduce the reachable state points as much as possible. Based on the statistical rules of optimal SOC trajectories, a fast DP is developed to improve the real-time applicability of global optimization energy management.

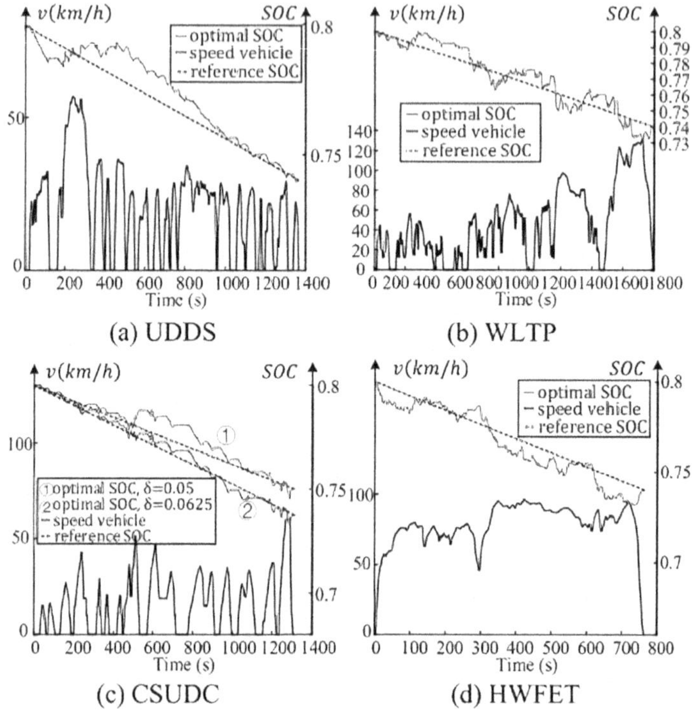

FIGURE 10.13 Simulation results under different standard driving cycle.

1) Generation of reference SOC trajectory
   For a given driving cycle, it can be divided into multiple kinematic segments. As shown in Figure 10.13, by analyzing the trend of the optimal SOC trajectories under different speed distributions with different vehicle models, the following conclusions can be summarized:

   ① If the speed distribution is relatively uniform, the optimal SOC trajectory decreases roughly linearly.
   ② When the speed distribution of a certain segment is significantly higher than other segments, the optimal SOC trajectory shows a linear downward trend as a whole; however, the optimal SOC trajectory corresponding to the protruding speed is roughly V-shaped or inverted V-shaped.
   ③ When the driving time (or mileage) is long enough or the SOC range is wide enough, the optimal SOC trajectory will decrease roughly linearly, regardless of the type of driving cycle.

   It should be pointed out that the first two conclusions are suitable for the situation when the whole driving time is not very long ($n < 2000s$) or the SOC range ($SOC_0 - SOC_f < 0.25$) is not very wide.

TABLE 10.6
The Upper and Lower Deviation between Optimal SOC Trajectories and Reference SOC Trajectory

Cycle	CSUDC	UDDS	NEDC	WLTP	HWEET
upper deviation	0.008	0.0171	0.008	0.02	0.0115
lower deviation-	0.0056	0.0091	0.03	0.021	0.026

Based on the this information, the optimal SOC trajectory obtained by DDP strategy declines linearly as a whole. Thus, we set a reference SOC (SOC_r), defined as:

$$SOC_r(k) = SOC_0 - \left(SOC_0 - SOC_f\right) \cdot k / n \tag{10.24}$$

where SOC_0 is the initial SOC, SOC_f is the terminal SOC, n is the total driving time, k is the moment.

The SOC range is set to 0.06. Aiming for a passenger vehicle with P2 configuration, the upper and lower deviation between the optimal SOC trajectories and the reference SOC trajectory under multiple standard driving cycles is concluded in Table 10.6.

Based on the prior analysis, we can conclude that the maximum deviation between optimal SOC trajectories and reference SOC trajectory is basically no more than 0.03. Particularly, if the speed distribution is relatively uniform and the speed is low ($v < 60$ km/h), the maximum deviation is basically no more than 0.01.

Therefore, by extending a certain allowable deviation on the basis of the reference SOC, a banded-searching region can be formed to narrow the state exploring region. According to the average speed, maximum speed and so forth, the type of speed distribution can roughly divide into three categories: low-speed, middle-speed, and highway (the same below). The corresponding SOC allowable deviation (d_1, d_2, d_3) can be set to 0.01, 0.02 and 0.03, respectively.

2) Suboptimal SOC trajectory domain

To ensure that the theoretical optimal state points do not exceed the boundary of the exploring region, the sub-optimal trajectory domain is proposed to expand the improved state feasible domain to a certain extent. By finding the potential connection between sub-optimal domain and optimal domain, the exploring region of state variables can be narrowed while ensuring the optimality.

Regarding the optimal SOC trajectory domain (with the minimum fuel consumption) as the baseline, the sub-optimal trajectory domain can be determined by setting a tolerance. Considering that the sub-optimal cost can occur at any moment during the whole trip, the terminal moment corresponding to the suboptimal cost needs to be pre-set.

> According to the tolerance, the reverse searching stops until the accumulative cost exceeds the allowable range. The specific process is shown in Table 10.7.

The tolerance is set to 10%. Under the same typical driving cycles, the expansion of the suboptimal domain width relative to the optimal domain width is concluded in Table 10.8.

Based on this analysis, the expansion of the suboptimal domain width relative to the optimal domain width is basically within the range of 10~15%. To make the statistical rule more universal, simulations are performed under different vehicle parameters (same category). The results demonstrate that it needs to be expanded by 5% on the basis of the original rules (10~15%) to make the optimal trajectory domain not exceed the boundary of the exploring region. It should be noted that different categories of the vehicle correspond to the respective statistical rules.

Regarding the reference SOC trajectory as the benchmark, based on the statistical rules, the simplified state feasible domain can be formed for a fast DP. The details are as follows:

Step 1: Determine the maximum deviation of SOC according to speed distribution.

TABLE 10.7
The Process of Sub-Optimal Domain-Searching Algorithm

Algorithm 2—Reverse searching

1. Set the error range ε and the moment s at which the error begins to occur
2. Find the optimal state domain from the moment s to terminal moment
3. For $j = s:-1:1$

    Find the sub-optimal solution aiming to each state point at current moment, and calculate cumulative errors E

    If $E > \varepsilon$

        break;

    End

   End → Obtain the moment d (the moment when the loop stops)
4. Find the optimal state domain from the starting moment to moment d
5. Obtain the suboptimal trajectory domain of SOC

TABLE 10.8
The Expansion of the Suboptimal Domain Width Relative to the Optimal Domain Width

Cycle	CSUDC	UDDS	NEDC	WLTP	HWEET
expansion	10.50%	14.94%	11.43%	14.12%	8.5%

Step 2: Regarding the reference SOC trajectory as the baseline and the deviation width as the radius, a banded-searching region is initially formed.

Step 3: The available trip information is introduced to limit the maximum charging and discharging current, and then the simplified state feasible domain is formed, as shown in Figure 10.14.

Compared with traditional DDP, the proposed fast DP improves computational efficiency while ensuring global optimality. Because the calculation time of the algorithm is related to the hardware configuration of the computing device, it is difficult to objectively evaluate the computational efficiency. Since the number of state points (SOC) has a great impact on the calculation efficiency of DP algorithm, the number of state points required by DP algorithm is utilized to measure the calculation efficiency of the fast DP.

Taking WLTP and UDDS as examples, supposing that speed limits and SOC range are available in advance during the whole trip, the simulation results are listed in Table 10.9.

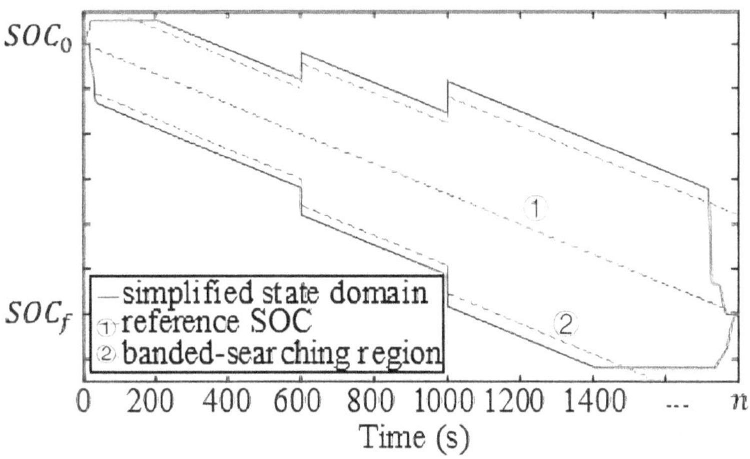

FIGURE 10.14 The schematic diagram of simplified state feasible domain.

TABLE 10.9
The Simulation Results of Fast DP ($d\_SOC = 0.0001, SOC\hat{I}[0.6, 0.8]$)

Cycle	UDDS		WLTP	
algorithm	traditional	fast DP	traditional	fast DP
number (state points)	5001401	244891	6767208	399952
efficiency improvement	95.10%		94.09%	

Under multiple standard driving cycles, the computing efficiency of fast DP can be improved by about 95%. It further illustrates the effectiveness of the proposed method (fast DP).

10.7.2 The Optimal Solution of ADP under the Uncertain Trip Information

In the case of uncertain driving conditions, adaptive dynamic programming (ADP) is regarded as an effective method to improve vehicle economy.

10.7.2.1 Determination of Utility Function

The objective of global optimization is to find a set of optimal control sequences to minimize the cost function (J):

$$J(x_k) = \sum_{i=k}^{\infty} \gamma^{k-i} U(k) \tag{10.25}$$

where $U(k) = fuel(x_k, u_k)$ is the utility function, γ is a discount factor, which reflects the impact of rewards or penalties at different stages (moments) on the overall cost function. The physical constraints on state variables and control variables are consistent with Eqs. (10.18) and (10.19).

Thence, the basic recursive equation of ADP can be obtained by

$$J_k^*(x_k) = \min_{u_k}\left[U(x_k, u_k) + \gamma J_{k+1}^*(x_{k+1})\right] \tag{10.26}$$

$$u_k^* = \arg\min_{u_k}\left[U(x_k, u_k) + \gamma J_{k+1}^*(x_{k+1})\right] \tag{10.27}$$

To ensure optimality, the reference SOC trajectory is generated to restrict the SOC state. Based on the fast DP, the reference SOC at a certain prediction horizon can be roughly determined by the linear decreasing rule. If the reference SOC is not approximate, it means that there exists more braking energy recovery under this scale. Then, the terminal SOC will be re-planned based on CDCS strategy. Specifically, when the vehicle is parking, the reference SOC will remain constant. When the vehicle can only recover energy, the reference SOC is re-planned according to the regenerative braking mode (energy recovery rate is set to 30%).

Within a certain prediction horizon, the speed limits can be acquired or the future driving conditions can be predicted. Aiming to different speed distributions, the fluctuation degree relative to the reference SOC trajectory is different. Hence, the schematic diagram of the state feasible domain is shown in Figure 10.15.

For the speed state, the reachable state points can be determined by introducing constraints about maximum acceleration/deceleration, parking, and braking, as shown in Figure 10.16.

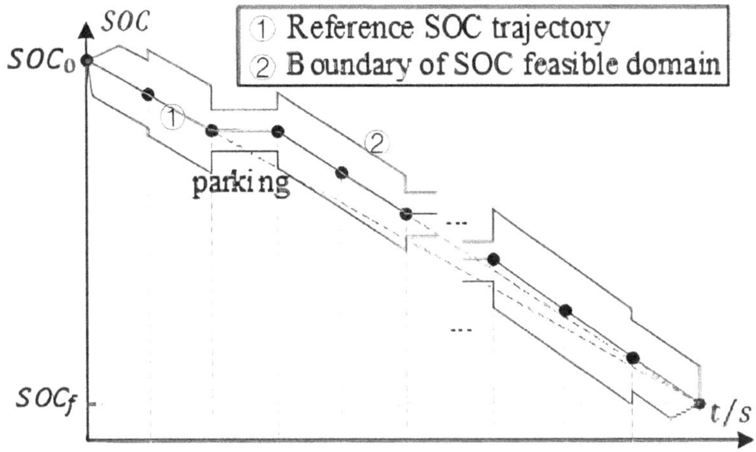

FIGURE 10.15 Schematic diagram of SOC feasible domain.

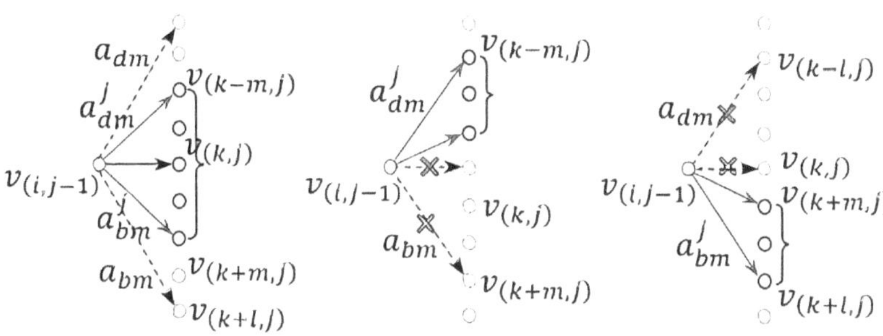

FIGURE 10.16 The schematic diagram of reachable state points (vehicle speed) (a) maximum acceleration (b) acceleration (c) deceleration.

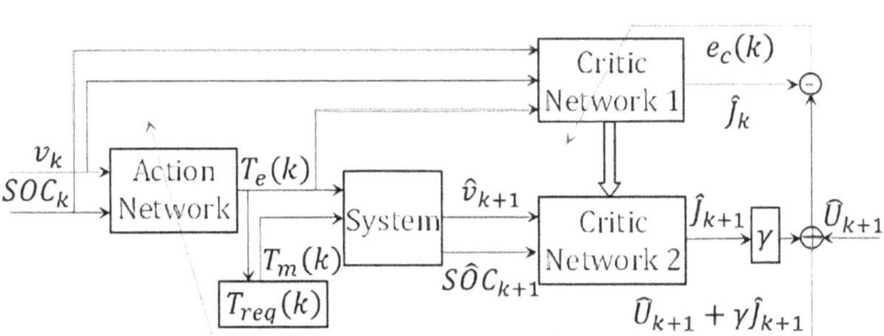

FIGURE 10.17 ADHDP model.

10.7.2.2 Establishment of ADHDP Model

Based on the driving equation, the power demand of the vehicle can be determined. The ADHDP structure is adopted, as shown in Figure 10.17, which includes one action network and two critic networks.

The critic network is used to approximate the optimal value function. The two critic networks show the temporal difference between the moment k and the moment $(k+1)$. Only the weight vector of the critic network at the moment k is updated. The cost can be formularized as:

$$\begin{cases} inputC_k = [u_k, x_k] \\ c_{h1}(k) = inputC_k \times W_{c1}(k) + b_{c1}(k) \\ c_{h2}(k) = \dfrac{1 - e^{-c_{h1}(k)}}{1 + e^{-c_{h1}(k)}} \\ \hat{J}_k = c_{h2}(k) \times W_{c2}(k) + b_{c2}(k) \end{cases} \tag{10.28}$$

where W_{c1}, W_{c2} represent the weight vectors between two adjacent layers; b_{c1}, b_{c2} represent the threshold vectors between two adjacent layers.

The steepest gradient descent method is used to update the weights, formularized as Eq. (10.29). Where, $e_c(k) = \hat{J}_k - (\beta\hat{J}_{k+1} + \hat{U}_{k+1})$ is the backpropagation of critic network, l_c is the learning efficiency of critic network.

$$\begin{cases} \Delta W_{c2}(k) = -l_c \times c_{h2}^T(k) \times e_c(k) \\ \Delta b_{c2}(k) = -l_c \times e_c(k) \\ \Delta W_{c1}(k) = -\dfrac{1}{2} l_c \cdot e_c(k) \cdot x_k^T \cdot \{W_{c2}^T(k) \otimes [1 - c_{h2}(k) \otimes c_{h2}(k)]\} \\ \Delta b_{c1}(k) = -\dfrac{1}{2} l_c \cdot e_c(k) \cdot \{W_{c2}^T(k) \otimes [1 - c_{h2}(k) \otimes c_{h2}(k)]\} \end{cases} \tag{10.29}$$

The action neural network approximates the optimal policy. At each moment, engine torque and motor torque satisfy

$$T_{req}(k) = (T_e(k) + T_m(k)) \cdot i_g(G_k) \cdot i_0 \tag{10.30}$$

where T_{req} is the required torque, i_g is the transmission ratio, i_0 is the final drive ratio.

Thence, the action network only outputs the engine torque, which is formularized as:

$$\begin{cases} a_{h1}(k) = x_k \times W_{a1}(k) + b_{a1}(k) \\ a_{h2}(k) = \dfrac{1 - e^{-a_{h1}(k)}}{1 + e^{-a_{h1}(k)}} \\ a_{h3}(k) = a_{h2}(k) \times W_{a2}(k) + b_{a2}(k) \\ u_k(k) = \dfrac{1}{1 + e^{-a_{h3}(k)}} \end{cases} \tag{10.31}$$

where W_{a1}, W_{a2} represent the weight vectors between two adjacent layers; b_{a1}, b_{a2} represent the threshold vectors between two adjacent layers.

The action network updates weight vectors to minimize the cost J_k. With the gradient descent algorithm, the weight vectors W_{a1} and W_{a2} are updated as Eq. (10.32).

where l_a is the learning efficiency of action network.

$$\begin{cases} \Delta w_{a2}(k) = \frac{1}{2} l_a \cdot a_{h3}^T(k) \times \left\{ W_{c2}^T(k) \otimes \left[1 - c_{h2}(k) \otimes c_{h2}(k)\right] \right\} \times W_{clu}^T(k) \times u_k(1-u_k) \\ \Delta b_{a2}(k) = \frac{1}{2} l_a \cdot \left\{ W_{c2}^T(k) \otimes \left[1 - c_{h2}(k) \otimes c_{h2}(k)\right] \right\} \times W_{clu}^T(k) \\ \Delta w_{a1}(k) = \frac{1}{4} l_a \cdot x_k^T \left\{ \begin{bmatrix} W_{c2}^T(k) \otimes \left[1 - c_{h2}(k) \otimes c_{h2}(k)\right] \\ \times W_{clu}^T(k) u_k (1-u_k) W_{a2}^T(k) \end{bmatrix} \otimes \left[1 - a_{h2}(k) \otimes a_{h2}\right] \right\} \\ \Delta b_{a1}(k) = \frac{1}{2} l_a \left\{ \begin{bmatrix} W_{c2}^T(k) \otimes \left[1 - c_{h2}(k) \otimes c_{h2}(k)\right] \\ W_{clu}^T(k) u_k (1-u_k) W_{a2}^T(k) \end{bmatrix} \otimes \left[1 - a_{h2}(k) \otimes a_{h2}(k)\right] \right\} \end{cases} \tag{10.32}$$

10.7.2.3 The Solution of ADP Under the Information Supported by Historical Data

Based on existing historical data, DP results under different speed distributions can be obtained. The gear shift schedule and reference SOC trajectory under different speed distributions can be extracted by observing DP behaviors. The vehicle speed, acceleration, or power demand are regarded as the basis for rule extraction (map generation).

Within a certain period of time, the vehicle speed conforms to the Markov property. Regarding speed and acceleration as state variables, the state transition probability matrix of vehicle speed can be defined as:

$$T_{ij} = P\left[a_{k+m} = \bar{a}_j \mid V_{k+m-1} = \bar{V}_i\right] \tag{10.33}$$

where, $i \in \{1,2,\ldots,pp\}$ is speed space, $j \in \{1,2,\ldots,qq\}$ is acceleration space; V_{k+m-1} is the speed state at the moment k, a_{k+m} is the acceleration state at the moment $(k+1)$, $m \in \{1,2,\ldots,L_p\}$; L_p is the prediction time.

Regarding the ADHDP model, the vehicle speed for the next 2 seconds (v_{k+1}, v_{k+2}) should be predicted, and then the gear at the current and the next moment (G_k, G_{k+1}) can be determined by the extracted rules. The control process of the ADHDP algorithm is listed in Table 10.10.

10.7.2.4 The Solution of ADP Under the Information with Constraints

Different from the information supported by historical data, there exist multiple reachable speed states. By discretizing vehicle speed at a certain interval (d), the data set of future driving conditions can be formed, that is,

$$\Omega = \begin{bmatrix} \left(v_k^1, v_{k+1}^1\right) & \cdots & \left(v_k^1, v_{k+1}^q\right) \\ \vdots & \ddots & \vdots \\ \left(v_k^p, v_{k+1}^1\right) & \cdots & \left(v_k^p, v_{k+1}^q\right) \end{bmatrix} \tag{10.34}$$

TABLE 10.10
Process of ADHDP Model under the Information Supported by Historical Data

Algorithm 3: Control process of ADHDP algorithm

Initialization: $\gamma, T_c, T_a, l_c, l_a, n, W_{c1}, W_{c2}, b_{c1}, b_{c2}, W_{a1}, W_{a2}, b_{a1}, b_{a2}, x_0$
For $k = 1:1:n$
Step 1: Estimating control and cost function
$[v_k, SOC_k] \rightarrow$ running (14), $T_e(k) \rightarrow T_m(k)$
$[v_k, SOC_k, T_e(k), T_m(k), G_k] \rightarrow$ calculate $[\hat{v}_{k+1}, S\hat{O}C_{k+1}]$
$[v_{k+1}, SOC_{k+1}] \rightarrow$ running (14), $\hat{T}_e(k+1) \rightarrow \hat{T}_m(k+1)$
$[v_k, SOC_k, T_e(k), T_m(k)]_{k,k+1} \rightarrow$ running (11), $\hat{J}_k, \hat{J}_{k+1}$
$\hat{U}_{k+1} = fuel(\hat{T}_e(k+1), \hat{n}_e(k+1))$
$E_c(k) = \frac{1}{2} e_c^2(k), e_c(k) = \hat{J}_k - (\beta \hat{J}_{k+1} + \hat{U}_{k+1})$
$\Delta E_a(k) = |E_a(k) - E_a(k-1)|, E_a(k) = \hat{J}_k$
Step 2: Optimal control
While ($E_c(k) > T_c$) && ($\Delta E_a(k) > T_a$) **do**
Update $W_{c1}(k), W_{c2}(k), b_{c1}(k), b_{c2}(k)$
Update $W_{a1}(k), W_{a2}(k), b_{a1}(k), b_{a2}(k)$
Step 1: Estimating control and cost function $\rightarrow E_c(k), \Delta E_a(k)$
End
End $\rightarrow$ **Output:** $W_{c1}, W_{c2}, W_{a1}, W_{a2}, u_k^* = \hat{u}_k$

where p is the number of reachable speed state at the current moment, q is the number of reachable speed state at the next moment.

The reference SOC can be generated based on the linear decline rule. Because the gear shift schedule cannot be extracted, the optimal gear is determined by three ADHDP models. Similar to Algorithm 3, the control process of the ADHDP model under the information with constraints is shown in Table 10.11.

10.8 CONCLUSIONS AND PROSPECTS

10.8.1 Conclusions

Corresponding to the main problems of DP, a global optimization framework of "information layer—physical layer—energy layer—dynamic programming" (IPE-DP) is proposed, which organically integrates vehicles, information, and energy.

The advantages of the proposed framework are mainly as follows:

1) Standardization
The framework (IPE-DP) organically integrates the information, physical, and energy layers, which can standardize the optimizing process of

TABLE 10.11
Process of ADHDP Model under the Information with Constraints

Algorithm 4: Control process of ADHDP algorithm

Initialization: $\gamma, T_c, T_a, l_c, l_a, n, W_{c1}, W_{c2}, b_{c1}, b_{c2}, W_{a1}, W_{a2}, b_{a1}, b_{a2}, x_0$
For $k = 1:1:n$
  For $i = 1:d:v_p$
    For $j = 1:d:v_q$
      $v_k = i, v_{k+1} = j \rightarrow$ determine $T_{req}(k), T_{req}(k+1)$
      For $l = -1:1:1$
        $u_g^l(k) = l \rightarrow G_k^l = G_{k-1} + u_g^l(k)$
        Step 1: Estimating control and cost $\rightarrow E_c^l(k), \Delta E_a^l(k)$
        Step 2: Optimal control
        While ($E_c^l(k) > T_c$) && (s $\Delta E_a^l(k) > T_a$) **do**
          Update $W_{c1}^l(k), W_{c2}^l(k), b_{c1}^l(k), b_{c2}^l(k)$
          Update $W_{a1}^l(k), W_{a2}^l(k), b_{a1}^l(k), b_{a2}^l(k)$
          Step 1: Estimating control and cost $\rightarrow E_c^l(k), \Delta E_a^l(k)$
        End $\rightarrow$ **Output:** $W_{c1}^l(k), W_{c2}^l(k), W_{a1}^l(k), W_{a2}^l(k), U_k^l(i,j), J_k^l(i,j)$
      End
      $J_k(i,j) = \min\{J_k^l(i,j)\} \rightarrow$ **Output:** $G_k(i,j),\ u_g^k(i,j)$
    End
  End
  $\min\{J_k(i,j)\} \rightarrow$ **Output:** $x_k^* = [v_k^*, SOC_k^*, G_k^*], u_k^* = [T_e^*, T_m^*, u_g^*(k)]$
End

DP strategy and achieve the unifications of different information scenarios, vehicle configurations, and energy conversions. Specifically, for the deterministic trip information, the conservation framework of "kinetic/potential energy and onboard energy" proposed in the physical layer realize the one-to-one mapping between driving condition and feasible work mode; that is, each condition corresponds to a unique work mode.

2) Accuracy
On the one hand, full-factor trip information, including vehicle speed, slope, and slip rate, is acquired in the information layer, which provides a more comprehensive and accurate driving power demand for energy management. On the other hand, the global domain-searching algorithm is developed to obtain all optimal solutions.

3) Guarantee drivability and comfortability
In the energy layer, the engine frequent start-stop problem, frequent shift problem and excessive transient torque response are considered in a DP optimization process to ensure the safety, drivability, and comfort. The consideration of the gear state and engine state ensures that the DP model satisfies the Markov characteristic.

4) Real-time application
The proposed method can effectively reduce the computational burden and improve the real-time performance of global energy management. On the one hand, under deterministic trip information, a fast DP is developed based on the reference SOC trajectory. This approach contributes to quickly obtaining the optimal control and then extracting rules or generating maps for real-time applications. On the other hand, the ADP method is utilized to achieve the online application of global energy management in the case of uncertain operating conditions.

10.8.2 Prospects

In the future, energy management still has a long way to go. In order to apply energy management strategy to a real vehicle, the calculation time of the algorithm should be further reduced; the essence of real-time application is prediction and optimization, but factors such as prediction accuracy and design parameters will affect the control performance of the strategy. Guaranteeing the accuracy of predictions and the resulting optimality of energy management are also issues to consider.

The gradual maturity of technologies such as 5G communication, Internet of Vehicles (IoV), and Big Data has brought new opportunities for energy management research. With the continuous development of intelligent transportation and vehicle networking technology, vehicles driving on the road every day can record a large amount of working condition data through specific devices and methods, and information exchange between vehicles is also possible. Research is no longer limited to fixed vehicles but can focus on all vehicles in a certain area to develop globally optimized regional traffic energy management, implement vehicle scheduling for all vehicles in the region, and alleviate traffic congestion while reducing regional traffic energy consumption and emissions. In addition to considering regional energy consumption, how the car is charged is also important. Multi-energy source vehicles are inextricably linked to the grid. The increase in the number of multi-energy source vehicles has brought about an increase in vehicle electricity consumption and electricity load, which has become one of the important boosters for the growth of electricity load in the future. The contradiction between the growing demand for electricity load and the backward power grid expansion and reconstruction is becoming more and more prominent. Considering the flexibility of vehicles, the time and space of charging can be adjusted. Therefore, charging energy management based on global vehicle time-space load characteristics can be developed, and the load curve of the power grid can be cut and filled to relieve the charging pressure of the power grid.

When it comes to energy conservation and emission reduction, from a narrow vehicle perspective, it is how to reduce energy consumption and emissions during vehicle driving, and from a broad energy perspective, it is necessary to start from the source of vehicle energy. For example, for logistics enterprises, the proportion of energy consumption cost can reach 30% or even higher, and cost reduction and efficiency increase are the long-term concerns of the industry. If the renewable energy power supply system can be combined with global energy management to plan the operation and charging of multi-energy source logistics vehicles, it can simultaneously meet the low-carbon needs of the transportation industry, the consumption needs of regional renewable energy, and the demand for logistics vehicles to reduce costs and increase efficiency.

REFERENCES

1. Sustainable Development Goals [EB/OL], 15–09–25. Accessed [2022–06–13]. www.un.org/sustainabledevelopment/zh/sustainable-development-goals/.
2. Wirasingha SG, Emadi A. Classification and review of control strategies for plug-in hybrid electric vehicles. *IEEE Transactions on Vehicular Technology* 2011; 60(1): 111–122.
3. Energy in China's New Era [EB/OL], 2020–12–21. Accessed [2022–06–13]. www.scio.gov.cn/zfbps/ndhf/42312/Document/1695299/1695299.htm.
4. China Association of Automobile Manufacturers, China Automotive Technology and Research Center Co., Ltd., Toyota. *Annual Report on the Development of China Automotive Industry*. Social Science Literature Press, 2020–02. (In Chinese)
5. The European Parliament, The Council. Regulation (EU) 2019/631[Z], 2019–04–17. https://eur-lex.europa.eu/legal-content/EN/TXT/?uri=CELEX%3A32019R0631.
6. Shi J, Liu B, Zhou W. Introduction to the California zero emission vehicle act. *Auto Industry Research* 2015; 8: 15–16.
7. Sorrentino M, Rizzo G, Arsie I. Analysis of a rule-based control strategy for on-board energy management of series hybrid vehicles. *Control Engineering Practice* 2011; 19(12): 1433–1441.
8. Banvait H, Anwar S, Chen YB. A rule-based energy management strategy for plug-in hybrid electric vehicle (PHEV). *Proceedings of the American Control Conference* 2009: 3938–3943.
9. Zhang BZ, Mi CC, Zhang MY. Charge-depleting control strategies and fuel optimization of blended-mode plug-in hybrid electric vehicles. *IEEE Transactions on Vehicular Technology* 2011; 60(4): 1516–1525.
10. Sciarretta A, Guzzella L. Control of hybrid electric vehicles. *IEEE Control Systems Magazine* 2007; 27(2): 60–70.
11. Bertsekas DP. *Dynamic Programming and Optimal Control* (2 Vols.). Athena Scientific, 1995; 4th edition of Vol. I, published in 2017; 4th edition of Vol. II: Approximate Dynamic Programming, 2012.
12. Song ZY, Hofmann H, Li JQ, Han XB, Ouyang MG. Optimization for a hybrid energy storage system in electric vehicles using dynamic programing approach. *Applied Energy* 2015; 139: 151–162.
13. Borhan H, Vahidi A, Phillips AM, Kuang ML, Kolmanovsky IV, Di Cairano S. MPC-based energy management of a power-split hybrid electric vehicle. *IEEE Transactions on Control Systems Technology* 2012; 20(3): 593–603.
14. Rezaei A, Burl JB, Solouk A, Zhou B, Rezaei M, Shahbakhti M. Catch energy saving opportunity (CESO), an instantaneous optimal energy management strategy for series hybrid electric vehicles. *Applied Energy* 2017; 208: 655–665.
15. Li L, Yang C, Zhang YH, Zhang LP, Song J. Correctional DP-based energy management strategy of plug-in hybrid electric bus for city-bus route. *IEEE Transactions on Vehicular Technology* 2015; 64(7): 2792–2803.

16. Lin CC, Peng H, Grizzle JW, Kang JM. Power management strategy for a parallel hybrid electric truck. *IEEE Transactions on Control Systems Technology* 2003; 11(6): 839–848.
17. Chen BC, Wu YY, Tsai HC. Design and analysis of power management strategy for range extended electric vehicle using dynamic programming. *Applied Energy* 2014; 113: 1764–1774.
18. Gong QM, Li YY; Peng ZR. Trip based power management of plug-in hybrid electric vehicle with two-scale dynamic programming. *Proceedings of IEEE Vehicle Power Propulsion Conference (VPPC)* 2007: 12–19.
19. Zou Y, Liu T, Sun FC, Peng H. Comparative study of dynamic programming and Pontryagin's minimum principle on energy management for a parallel hybrid electric vehicle. *Energies* 2013; 6(4): 2305–2318.
20. Zhang C, Vahidi A. Route preview in energy management of plug-in hybrid vehicles. *IEEE Transactions on Control Systems Technology* 2012; 20(2): 553–546.
21. Wang H, Huang YJ, Khajepour A, Song Q. Model predictive control-based energy management strategy for a series hybrid electric tracked vehicle. *Applied Energy* 2016; 182: 105–114.
22. Wang XM, He HW, Sun FC, Zhang JL. Application study on the dynamic programming algorithm for energy management of plug-in hybrid electric vehicles. *Energies* 2015; 8(4): 3225–3244.
23. Peng JK, He HW, Xiong R. Rule based energy management strategy for a series–parallel plug-in hybrid electric bus optimized by dynamic programming. *Applied Energy* 2017; 185: 1633–1643.
24. He HW, Tang HL, Wang XM. Global optimal energy management strategy research for a plug-in series-parallel hybrid electric bus by using dynamic programming. *Mathematical Problems in Engineering* 2013; 708261.
25. Zhou W, Yang L, Cai YS, Ying TX. Dynamic programming for new energy vehicle based on their work modes part I: Electric vehicles and hybrid electric vehicles. *Journal of Power Sources* 2018; 406: 151–166.
26. Xu N, Kong Y, Chu L, Ju H, Yang ZH, Xu Z, Xu ZQ. Towards a smarter energy management system for hybrid vehicles: A comprehensive review of control strategies. *Applied Sciences-Basel* 2019; 9(10): 2026.
27. Zhang YJ, Guo C, Li G, Liu YG, Chen Z. Cooperative control strategy for plug-in hybrid electric vehicles based on a hierarchical framework with fast calculation. *Journal of Cleaner Production* 2020; 251: 119627.
28. Polimeni S, Meraldi L, Moretti L, Leva S, Manzolini G. Development and experimental validation of hierarchical energy management system based on stochastic model predictive control for off-grid microgrids. *Advances in Applied Energy* 2021; 2: 100028.
29. Zhang S, Xiong R. Adaptive energy management of a plug-in hybrid electric vehicle based on driving pattern recognition and dynamic programming. *Applied Energy* 2015; 155: 68–78.
30. Guo JQ, He HW, Peng JK, Zhou NN. A novel MPC-based adaptive energy management strategy in plug-in hybrid electric vehicles. *Energy* 2019; 175: 378–392.
31. Peng JK, He HW, Xiong R. Study on energy management strategies for series-parallel plug-in hybrid electric buses. *Energy Procedia* 2015; 75: 1926–1931.
32. Liu C, Wang YJ, Wang L, Chen ZH. Load-adaptive real-time energy management strategy for battery/ultracapacitor hybrid energy storage system using dynamic programming optimization. *Journal of Power Sources* 2019; 438: 227024.

33. Li MC, Wang L, Wang YJ, Chen ZH. Sizing optimization and energy management strategy for hybrid energy storage system using multiobjective optimization and random forests. *IEEE Transactions on Power Electronics* 2021; 36(10): 11421–11430.
34. McQueen B, McQueen J. *Intelligent transportation systems architectures*. Norwood, MA: Artech House, 2003.
35. Tian H, Li SE, Wang X, Huang Y, Tian GY. Data-driven hierarchical control for online energy management of plug-in hybrid electric city bus. *Energy* 2018; 142: 55–67.
36. Borozan S, Giannelos S, Strbac G. Strategic network expansion planning with electric vehicle smart charging concepts as investment options. *Advances in Applied Energy* 2021: 100077.
37. Zhang P, Zhou JY. Higher-order nonlinear discrete approximate iteration on the continuing dynamic programming. *IISME* 2012; 459: 571.
38. Vinot E. Time reduction of the dynamic programming computation in the case of hybrid vehicle. *International Journal of Applied Electromagnetics and Mechanics* 2017; 53: 213–227.
39. Zhang YJ, Guo C, Liu YG, Ding F, Chen Z, Hao WM. A novel strategy for power sources management in connected plug-in hybrid electric vehicles based on mobile edge computation framework. *Journal of Power Sources* 2020; 477.
40. Lei ZZ, Qin DT, Hou LL, Peng JY, Liu YG, Chen Z. An adaptive equivalent consumption minimization strategy for plugin hybrid electric vehicles based on traffic information. *Energy* 2020; 190: 116409.
41. Astarita V, Giofre VP, Guido G, Vitale A. A single intersection cooperative-competitive paradigm in real time traffic signal settings based on floating car data. *Energies* 2019; 12(3): 409.
42. Li GP, Zhang JL, He HW. Battery SOC constraint comparison for predictive energy management of plug-in hybrid electric bus. *Applied Energy* 2017; 194: 578–587.
43. Chen Z, Mi CC, Xu J, Gong XZ, You CW. Energy management for a power-split plug-in hybrid electric vehicle based on dynamic programming and neural networks. *IEEE Transactions on Vehicular Technology* 2014; 63(4): 1567–1580.
44. Liu T, Tang XL, Wang H, Yu HL, Hu XS. Adaptive hierarchical energy management design for a plug-in hybrid electric vehicle. *IEEE Transactions on Vehicular Technology* 2019; 68(12): 11513–11522.
45. Liu T, Hu XS, Li SE, Cao DP. Reinforcement learning optimized look-ahead energy management of a parallel hybrid electric vehicle. *IEEE-ASME Transactions on Mechatronics* 2017; 22(4): 1497–1507.
46. Li GQ, Gorges D. Ecological adaptive cruise control and energy management strategy for hybrid electric vehicles based on heuristic dynamic programming. *IEEE Transactions on Intelligent Transportation Systems* 2019; 20(9): 3526–3535.
47. Liu JC, Chen YZ, Zhan JY, Shang F. Heuristic dynamic programming based online energy management strategy for plug-in hybrid electric vehicles. *IEEE Transactions on Vehicular Technology* 2019; 68(5): 4479–4493.
48. Sun HC, Fu ZM, Tao FZ, Zhu LL, Si PJ. Data-driven reinforcement-learning-based hierarchical energy management strategy for fuel cell/battery/ultracapacitor hybrid electric vehicles. *Journal of Power Sources* 2020; 455: 227964.
49. Zhou W, Zhang CN, Li JQ. Analysis and comparison of optimal power management strategies for a series plug-in hybrid school bus via dynamic programming. *International Journal of Vehicle Design* 2015; 69(1–4): 113–131.
50. Agarwal P. Internet GIS: Distributed geographic information services for the Internet and wireless networks. *Professional Geographer* 2004; 56(2): 313–315.

51. Ramadan HS, Becherif M, Claude F. Energy management improvement of hybrid electric vehicles via combined GPS/rule-based methodology. *IEEE Transactions on Automation Science and Engineering* 2017; 14(2): 586–597.
52. Gong QM, Li YY, Peng ZR. Trip-based optimal power management of plug-in hybrid Eelectric vehicles. *IEEE Transactions on Vehicular Technology* 2008; 57(6): 3393–3401.
53. Guo JQ, He HW, Sun C. ARIMA-based road gradient and vehicle velocity prediction for hybrid electric vehicle energy management. *IEEE Transactions on Vehicular Technology* 2019; 68(6): 5309–5320.
54. Zeng XR, Wang JM. A parallel hybrid electric vehicle energy management strategy using stochastic model predictive control with road grade preview. *IEEE Transactions on Control Systems Technology* 2015; 23(6): 2416–2423.
55. Larsson V, Johannesson L, Egardt B. Analytic solutions to the dynamic programming subproblem in hybrid vehicle energy management. *IEEE Transactions on Vehicular Technology* 2015; 64(4): 1458–1467.
56. Chen Z, Wu YT, Guo NY, Shen JW, Xiao RX. Energy management for plug-in hybrid electric vehicles based on quadratic programming with optimized engine on-off sequence. *IEEE Industrial Electronics Society* 2017: 7134–7139.
57. Elbert P, Nusch T, Ritter A, Murgovski N, Guzzella L. Engine ON/OFF control for the energy management of a serial hybrid electric bus via convex optimization. *IEEE Transactions on Vehicular Technology* 2014; 63(8): 3549–3559.
58. Pu J, Yin C. Optimal control of fuel economy in parallel hybrid electric vehicles. *Proceedings of the Institution of Mechanical Engineers, Part D: Journal of Automobile Engineering* 2007; 221(9): 1097–1106.
59. Opila DF, Wang XY, McGee R, Gillespie RB, Cook JA, Grizzle JW. An energy management controller to optimally trade off fuel economy and drivability for hybrid vehicles. *IEEE Transactions on Control Systems Technology* 2012; 20(6): 1490–1505.
60. Ngo DV, Hofman T, Steinbuch M, Serrarens AFA. Analyzing the performance index for a hybrid electric vehicle. *Proceedings of the American Control Conference* 2011: 4592–4597.
61. Fu JT, Song SZ, Fu ZM, Ma JW. Real-time implementation of optimal control considering gear shifting and engine starting for parallel hybrid electric vehicle based on dynamic programming. *Optimal Control Applications and Methods* 2018; 39: 757–773.
62. Miro-Padovani T, Colin G, Ketfi-Cherif A, Chamaillard Y. Implementation of an energy management strategy for hybrid electric vehicles including drivability constraint. *IEEE Transactions on Vehicular Technology* 2016; 65(8): 5918–5929.
63. Fan LK, Zhang YT, Dou HS, Zou RN. Design of an integrated energy management strategy for a plug-in hybrid electric bus. *Journal of Power Sources* 2020; 448: 227391.
64. Xu N, Kong Y, Yan JY, Zhang YJ, Sui Y, Ju H, Liu H, Xu Z. Global optimization energy management for multi-energy source vehicles based on “information layer—physical layer—energy layer—dynamic programming” (IPE-DP). *Applied Energy* 2022; 312: 118668.
65. Kong Y, Xu N, Zhang YJ, Sui Y, Ju H, Liu H, Xu Z. Acquisition of full-factor trip information for global optimization energy management in multi-energy source vehicles and the measure of the amount of information to be transmitted. *Energy* 2021; 236.
66. The State Council of the People's Republic of China. *The law of the People's Republic of China on road traffic safety [Z]*, 2003: Article 43–47. http://www.npc.gov.cn/zgrdw/englishnpc/Law/2007-12/05/content_1381965.htm

67. The State Council of the People's Republic of China. *The regulation on the implementation of the law of the People's Republic of China on road traffic safety [Z]*, 2004: Article 45, 78. http://www.npc.gov.cn/zgrdw/englishnpc/Law/2007-12/05/content_1381965.htm
68. Gazis D. *Traffic theory*. Kluwer Academic Publishers, 2002.
69. Ray L. Nonlinear tire force estimation and road friction identification: Simulation and experiments. *Automatica* 1997; 33(10): 1819–1833.
70. Sun XD, Shi Z, Lei G, Guo YG, Zhu JG. Multi-objective design optimization of an IPMSM based on multilevel strategy. *IEEE Transactions on Industrial Electronics* 2019; 68(1): 139–148.
71. Yang YL, Hu XS, Pei HX, Peng ZY. Comparison of power-split and parallel hybrid powertrain architectures with a single electric machine: Dynamic programming approach. *Applied Energy* 2016; 168:683e90.
72. Liu JM, Peng H. Modeling and control of a power-split hybrid vehicle. *IEEE Transactions on Control Systems Technology* 2008; 6(16): 125–1242.
73. Sun C, Sun FC, He HW. Investigating adaptive-ECMS with velocity forecast ability for hybrid electric vehicles. *Applied Energy* 2017; 185(2): 1644–1653.
74. He HW, Zhang XW, Xiong R, Xu YL, Guo HQ. Online model-based estimation of state-of-charge and open-circuit voltage of lithium-ion batteries in electric vehicles. *Energy* 2012; 39(1): 310–318.
75. Rahimi-Eichi H, Baronti F, Chow MY. Online adaptive parameter identification and state-of-charge coestimation for lithium-polymer battery cells. *IEEE Transactions on Industrial Electronics* 2014; 61(4): 2053–2061.
76. Bellman, R. *Dynamic programming*. Princeton, NJ: Princeton University Press, 1957.

11 Traffic-in-the-Loop Simulation, Optimization, and Evaluation for Electric Vehicles

Fuguo Xu

11.1 INTRODUCTION

The real-time driving emission testing and energy-efficiency improvement strategy design has evolved from finite standard driving cycles to real-world traffic road driving. Since standard driving cycles have lower fidelity in reflecting the traffic conditions that a vehicle would face with during driving time, it is necessary to consider the real-world driving scenario. Moreover, with the utilization of vehicle to vehicle (V2V) and vehicle to infrastructure (V2I) communication, there is a higher probability of energy-efficiency improvement and emission reduction to be further achieved in powertrain control and optimization without making a physical and type change of a vehicle [1].

In the sense of traffic scenario anticipation control scheme design process, to verify the proposed approach, one way is to test it in a real vehicle running on an actual road. Usually, this design process is time-consuming and expensive, increasing design cost [2, 3]. Even worse, since the test version of the control scheme would be inadequate for safety protection for real-world traffic conditions, it may put test drivers in danger situations when testing on an actual traffic road. An alternative way to deal with these problems is to utilize a real-world traffic emulated simulation platform with powertrain equipment or physical-based mathematical model that could capture dynamics accurately, especially during the preliminary stage of control scheme design. By using this kind of traffic-emulated powertrain simulation platform, design cost and timing would decrease substantially.

Researchers have paid attention to this combination of traffic scenario and powertrain, and various attempts have been made towards using traffic scenarios to decrease fuel economy in powertrain optimal control. The most common example is connecting the traffic simulator and powertrain simulator in a sequential

DOI: 10.1201/9781003302827-14

unidirectional structure, where traffic information is only utilized as inputs for powertrain control scheme design. However, the dynamic of ego vehicle in powertrain control would influence traffic participants dynamics, and as a result, the total traffic condition would also change. In a unidirectional traffic platform, this influence could not be simulated, which would make the fuel economy performance under this design control scheme less reliable. To reflect ego vehicle dynamic influence on the traffic scenario, there are bidirectional co-simulation platforms that are set as a closed-loop by using macroscopic traffic simulation, such as simulation of urban mobility (SUMO) and powertrain model [4, 5]. In terms of information interchange, V2V and V2I information is available to the powertrain control. However, that co-simulation framework is actually a velocity tracking sense in the powertrain model to that generated in the traffic platform. Obviously, only tracking vehicle speed information will not capture driver behavior adequately.

11.2 SIMULATION PLATFORM

The framework of vehicles in the connected environment is sketched in Figure 11.1 where the connected vehicles are enabled to communicate with GPS, V2V, and V2I during the trip. For example, the V2I information, such as traffic light phase and timing and the distance to the next intersection can be obtained for the ego vehicle. The distance headway between the ego vehicle and the preceding vehicle should be considered for driving safety, and it is available to the ego vehicle equipped with a distance detecting sensor. The preceding vehicle information, such as distance, speed, and acceleration, is transmitted to the ego vehicle through the V2V. In the road, the maximum speed limit based on traffic rule is also available to the ego vehicle.

A traffic-in-the-loop powertrain simulation system mainly includes a real-world emulated traffic simulation platform and a mathematical physical-based powertrain model. The powertrain model would capture the dynamic progress of components accurately and the energy consumption of each energy supply resource,

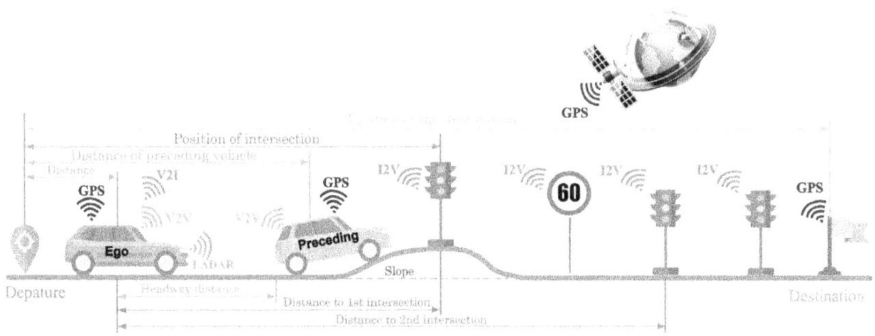

FIGURE 11.1 Electric vehicle running in the connected environment.

such as the fuel consumption of the engine and the electricity consumption of the battery. As depicted in Figure 11.2, the proposed simulation platform consists of IPG CarMaker and MATLAB/Simulink, where the traffic simulation is conducted in CarMaker, and the powertrain simulation is conducted in MATLAB/Simulink, respectively.

11.2.1 Traffic Simulator

A real-world emulated traffic scenario is formulated in IPG CarMaker and its graphical user interface (GUI) is shown in Figure 11.3. In this platform, V2V and V2I communication functions could be simulated. The traffic scenario is edited in advance with road condition, traffic light information, traffic density, and traffic flow generation. Real-world emulated 3D video is displayed in the GUI. Moreover, there are available interfaces with other software, such as MATLAB/Simulink, by which additional powertrain modeling and control scheme verification can be achieved.

In the following part of this section, the sensor of the ego vehicle and communication of V2V and V2I will be shown. For the ego vehicle equipped with a sensor, it is possible to detect the inter distance between the ego vehicle and the preceding vehicle. The real-time signals of detection flag and distance headway are provided in the simulator. The detection flag signals 1 and 0 denote detectable and undetectable, respectively. If the flag signal is 1, the real-time distance headway is available; otherwise, the value of distance headway is given as 0. The profiles of the ego sensor's information in three traffic scenarios are shown in Figure 11.4.

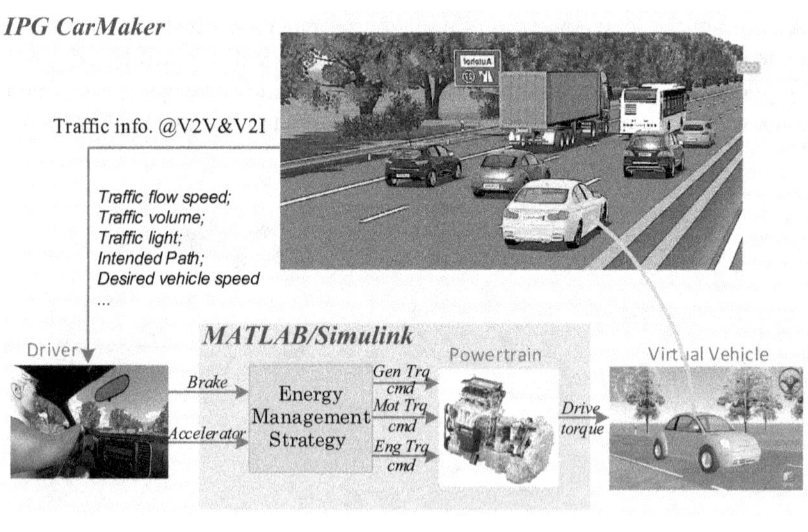

FIGURE 11.2 Framework of the traffic-in-the-loop powertrain simulation platform.

FIGURE 11.3 GUI of traffic simulator.

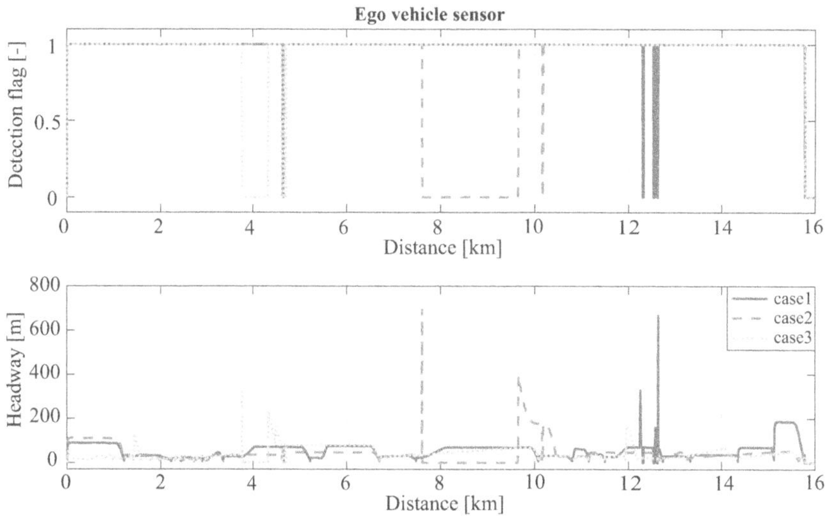

FIGURE 11.4 Ego vehicle sensor information of three traffic cases.

For a vehicle running in the real-world traffic scenario, the other traffic participants in the same scenario should also be considered, consisting of traffic lights and vehicles ahead of the ego vehicle. The real-time position, speed, and acceleration of the preceding vehicle are available to the ego vehicle through V2V communication. The profiles of speed and acceleration of the preceding vehicle in three cases are shown in Figure 11.5.

The ego vehicle should follow the rule of the traffic light. It is noted that only green and red signals are considered in the phase of a traffic light. Thus, it follows the rule that vehicle runs and stops when facing with the green and red signals at

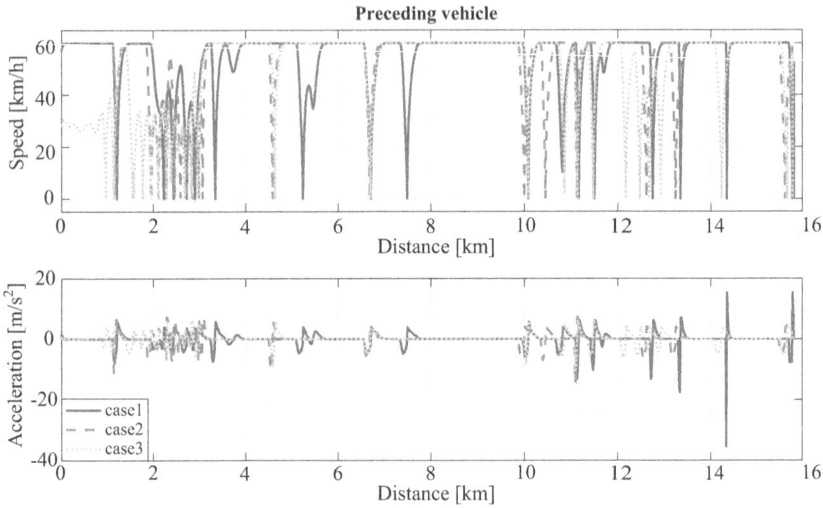

FIGURE 11.5 Speed and acceleration of preceding vehicle of three traffic cases.

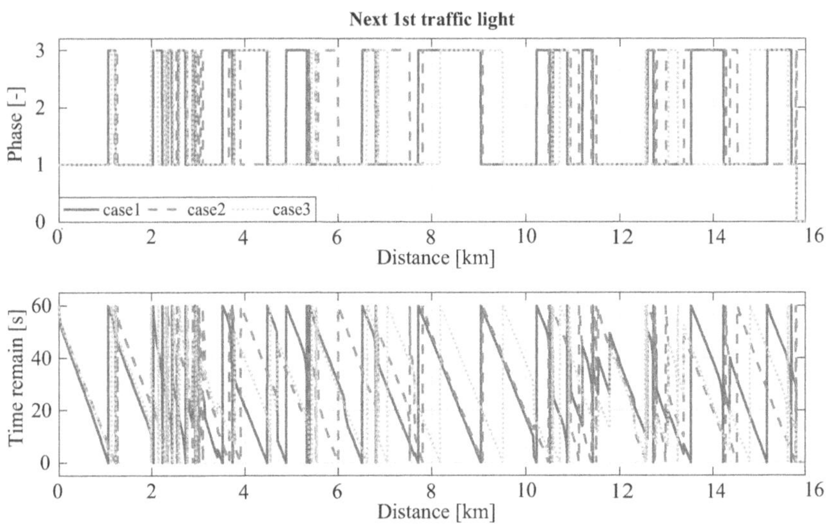

FIGURE 11.6 The traffic lights remain timing and phase of 1st upcoming for three traffic cases.

the intersection, respectively. There is a traffic light at each intersection, and the distance to the intersection (traffic light) is available to the ego vehicle. The phases and remaining timings of upcoming traffic lights in three cases are shown in Figure 11.6. In Figure 11.6, the phase denotes the traffic light information, where 1 and 3 denote the green light and the red light, respectively.

11.2.2 Powertrain Model

The dynamic of the HEV powertrain in Figure 11.7 can be seen as a four-order differential equation with following form:

$$\begin{cases} J_l\dot{\omega}_l = R_r F_1 - T_c \\ J_e\dot{\omega}_e = T_e - R_r F_1 - R_s F_2 \\ J_g\dot{\omega}_g = T_g + R_s F_2 \\ M_E\dot{v} = \alpha T_m + \beta T_c - F(v) - \eta_B F_B \end{cases} \tag{11.1}$$

where F_1 and F_2 are forces working on ring gear and sun gear. T_c is ring gear torque. J_i, T_i, ω_i denote rotational inertia, torque and speed, respectively. $i = [e, m, g, l]$ is equivalent mass with consideration of rotational inertials of powertrain components, including engine, motor and generator, and ring gear. And M_E represent equivalent transmission ratios from motor and planetary gear system to tire. α and β are teeth number of ring gear and sun gear. F_B denotes braking force and η_B is corresponding efficiency. v is vehicle speed and $F(v)$ is resistance force, including rolling force and air resistance, which is written as following:

$$F(v) = \mu mg\cos\theta + 0.5\rho C_d A v^2 \tag{11.2}$$

where $m, \mu, \theta, \rho, C_d, A$, represent vehicle mass, rolling coefficient, slope, gravitational acceleration, air density, drag coefficient, and frontal area, respectively.

For the planetary gear system, speeds of ω_e, ω_g, and ω_l should satisfy the following speed constraint:

$$R_r\omega_l + R_s\omega_g = (R_r + R_s)\omega_e \tag{11.3}$$

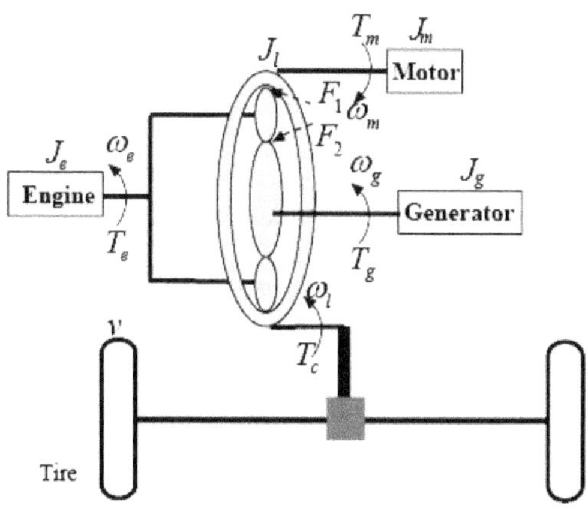

FIGURE 11.7 Physical structure of power-split HEV powertrain.

Forces of F_1 and F_2 are equal. Moreover, with the knowledge of Eq. (11.3), powertrain dynamics shown in Eq. (11.1) could be rewritten as:

$$\begin{cases} J_e\dot{\omega}_e + J_g \dfrac{R_r + R_s}{R_s}\dot{\omega}_g = T_e + \dfrac{R_r + R_s}{R_s}T_g \\ M_E\dot{v} + \beta(J_l\dot{\omega}_l + J_e\dot{\omega}_e + J_g\dot{\omega}_g) = \alpha T_m + \beta(T_e + T_g) - F(v) - \eta_B F_B \end{cases} \tag{11.4}$$

Further, it is assumed that the axis is rigid, speeds of ω_l is equal to ω_m, which has the relationship with v:

$$\omega_l = \frac{Rw}{(R_r + r)R_{tire}} v \tag{11.5}$$

Then, ω_l and ω_g are replaced by ω_e and v, only dynamics of ω_e and v are given and could be summarized as following matrix form:

$$\begin{bmatrix} \dot{\omega}_e \\ \dot{v} \end{bmatrix} = \begin{bmatrix} D_{11} & D_{12} \\ D_{21} & D_{22} \end{bmatrix} \begin{bmatrix} T_e + \dfrac{(R_r + R_s)}{R_s}T_g \\ \beta T_e + \alpha T_m + \beta T_g - F(v) - \eta_B F_B \end{bmatrix} \tag{11.6}$$

where D_{11}, D_{12}, D_{21} and D_{22} are the fixed parameters for transforming Eq. (11.4) into Eq. (11.6).

With determination of engine speed and vehicle speed, speeds of motor and generator would be calculated as following:

$$\begin{cases} \omega_m = \dfrac{R_w}{(R_r + r)R_{tire})} v \\ \omega_g = \dfrac{(R_r + R_s)}{R_s}\omega_e - \dfrac{R_r R_w}{R_s(R_r + r)R_{tire}} v \end{cases} \tag{11.7}$$

where r, R_w, R_{tire} denote teeth numbers before and after differential and tire radius, respectively.

In a traffic scenario simulator, usually there is no powertrain structure for a virtual vehicle. For analyzing traffic information consideration-based energy management strategy design and powertrain dynamic control in optimal sense, a combination of a powertrain model and traffic scenario simulator would be conducted. Dynamics of a vehicle in longitudinal scenes could be seen as a 1D differential equation; however, it is a 2D dynamic function with engine speed and vehicle speed as states in the HEV powertrain model. The torque matching problem should be solved for the co-simulation platform. This chapter mainly investigates

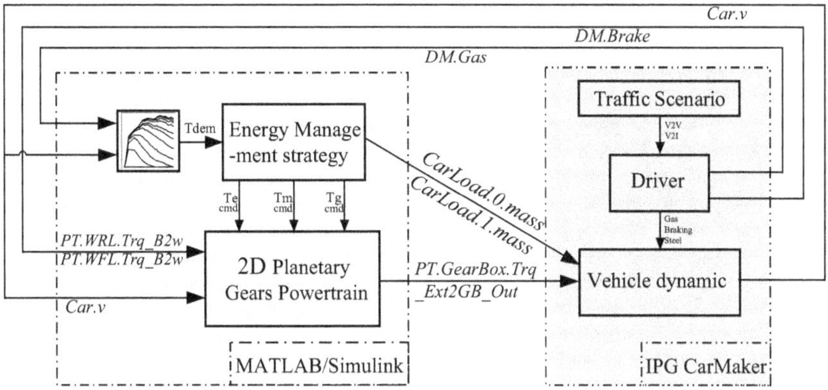

FIGURE 11.8 Signal exchange between CarMaker and MATLAB.

an integration approach for mechanical 2D dynamics of the powertrain with the limited 1D vehicle dynamics in the CarMaker traffic simulator. The signal exchange between the traffic simulator and HEV powertrain simulator is depicted in Figure 11.8.

11.3 OPTIMIZATION

11.3.1 Optimal Powertrain Control

For the optimal powertrain control of electric vehicles, the main target is to achieve efficiency improvement to reduce energy consumption. Thus, the cost function L in the optimal control problem represents energy consumption. The goal of the optimal control problem is to minimize this energy consumption by deriving an optimal control solution u^* for the cost functional with dynamics model constraints and inequality constraints, described as follows:

$$u^* = \arg\min_u \int_{t_0}^{t_f} L(x,u)dt, \tag{11.8}$$

$$\text{s.t.}\begin{cases} \dfrac{d_x}{d_s} = f(x,u,w), \\ h(x,u) \geq 0, \\ u_{min} \leq u \leq u_{max}, \\ x(s_0) = x_0, \end{cases}$$

where t_0 and t_f denote the initial time and the terminal time of the predictive horizon; x, u, and w denote the state variables, control inputs, and outside disturbance, respectively. There are inequality constraints h that the optimal solutions have to

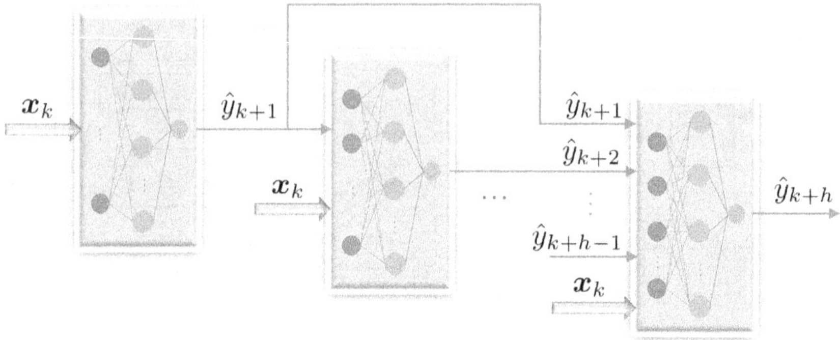

FIGURE 11.9 Structure of the chain predictor.

satisfy. Moreover, there are the physical limitations of the control input u, defining as u_{min} and u_{max}. The initial state $x(t_0)$ is determined as x_0. Also, the outside disturbance in the current time t_0 is available; however, this disturbance w in future time of predictive horizon is unknown. How to deal with this problem to further improve the performance will be discussed in the following subsection.

11.3.2 Traffic Information–Based Prediction

With the rich historical and real-time traffic participants data, it is possible to employ the advanced data-based learning approach to predict the disturbance w that is mentioned in the previous optimal powertrain problem. In this part, an example of prediction using traffic information is introduced. Specially, the controlled vehicle is equipped with an automated driving technology, and the preceding vehicle speed is seen as the disturbance when the headway distance between the preceding vehicle and the ego vehicle are considered. Since there are multiple disturbance signals to be predicted within the optimization horizon between t_0 and t_f, the chain predictor is introduced, shown in Figure 11.9. The detail information and output information of the predictor are listed in Table 11.1. For the prediction, Gaussian process (GP) is employed.

11.4 EVALUATION

In this section, the prediction results and optimization results are given, respectively.

11.4.1 Prediction Results

To show the performance of prediction, Figure 11.10 depicts the preceding vehicle speed v_p prediction results in four cases: (a) maximum speed, (b) decreasing speed, (c) minimum speed, and (d) increasing speed. It can be seen that for the maximum and minimum speeds, the prediction accuracy is high. Even though an error of

TABLE 11.1
Meaning of Input and Output Signals for Multi-Step-Ahead Δq Predictor

Signal	Variable	Description
	$\rho(k)$	Traffic density at kth step
Input	$d_{next}(k)$	Distance to next junction at kth step
	$d_{past}(k)$	Distance to past junction at kth step
	$TL_{ph}(k)$	Traffic light phase at kth step
	$TL_{time}(k)$	Traffic light timing at kth step
	$\Delta q(k)$	Vehicle increment number at kth step
Output	$\Delta q(k+1)$	Vehicle increment number at k+1th step
	$\Delta q(k+2)$	Vehicle increment number at k+2th step
	$\Delta q(k+3)$	Vehicle increment number at k+3th step
	$\Delta q(k+4)$	Vehicle increment number at k+4th step
	$\Delta q(k+5)$	Vehicle increment number at k+5th step

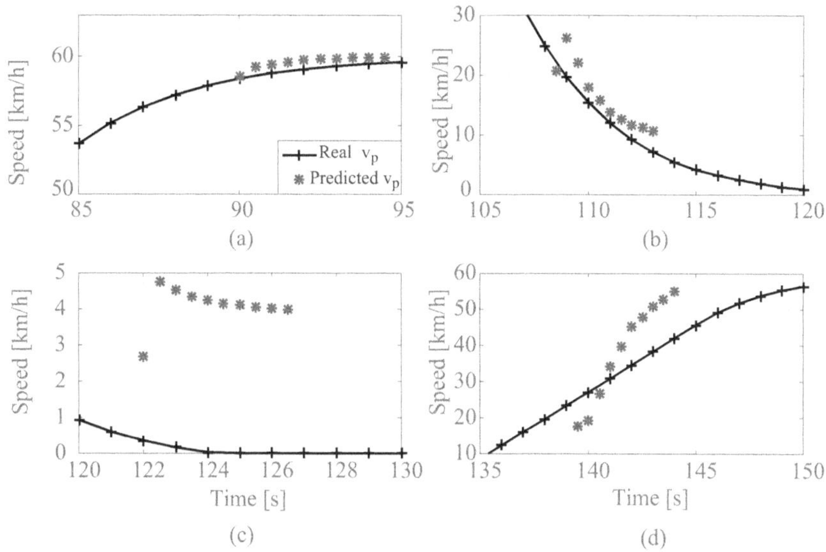

FIGURE 11.10 Online preceding vehicle speed comparison between current and prediction for different cases.

approximately 4 km/h is observed when vehicle speed decreases to approximately zero, the proposed control scheme would also plan a zero ego vehicle speed. For the increasing and decreasing speed speed scenarios, the prediction accuracies are not as promising as in the case of stable vehicle speeds. For a receding horizon optimal control scheme, only the first control input is applied. The prediction of the preceding vehicle motion trend is therefore sufficient for real-time optimization.

11.4.2 Optimization Results

Figure 11.11 depicts the simulation result comparison for different initial conditions, which are $s_0 = 10[m]$, $s_0 = 28[m]$ and $s_0 = 35[m]$. The black solid lines denote the rules of maximum and minimum $s - hv$ that the car-following scenario must abide by. *W/I GP-Predictor* and *W/O GP-Predictor* represent the performances of the proposed control scheme with the chain GP-based predictor

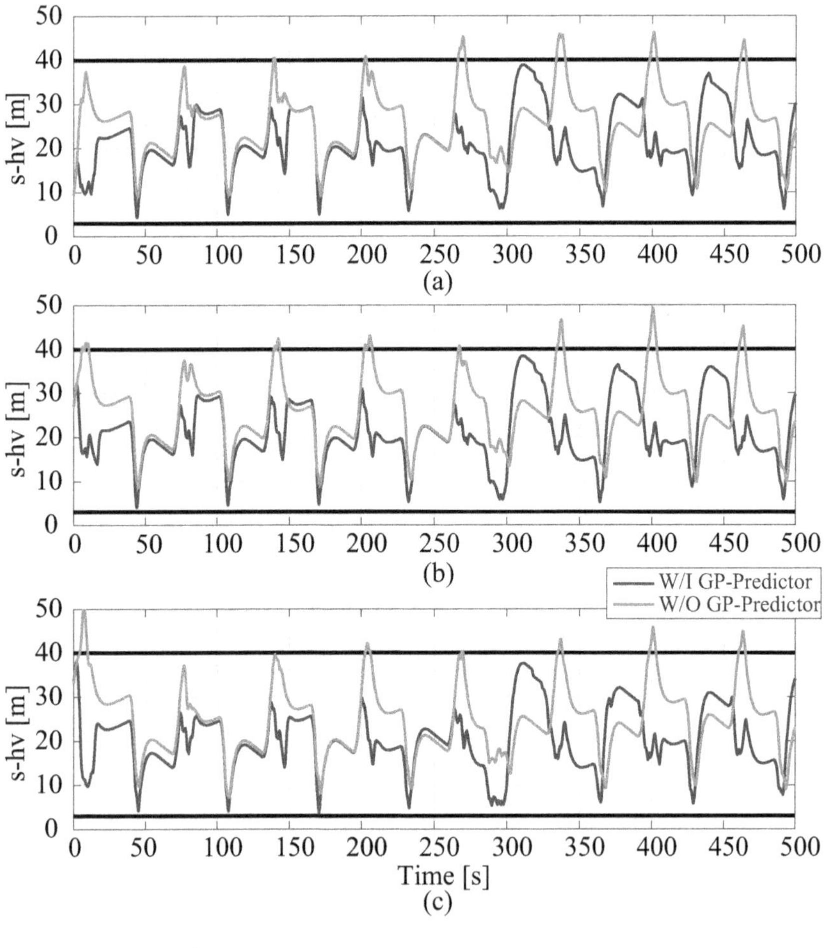

FIGURE 11.11 Simulation comparison of inter-vehicle distance for different initial distance headways between the ego vehicle and preceding vehicle.

TABLE 11.2
Performance Comparisons under Different Control Schemes for Different Cases

	Expense [¥/km]			
Scenario	**W/I GP**	**W/O GP**	**DP**	**DP W/I Dis.**
Case 1	16.94	18.03	15.16	22.45
Case 2	17.29	20.08	15.52	21.01
Case 3	16.97	20.49	15.57	22.54
Case 4	16.83	20.74	15.43	24.77
Case 5	15.13	15.74	13.17	22.43

and the proposed control scheme without the chain GP-based predictor. In the *W/O GP-Predictor* case, the future preceding speeds $v_p(k+j)$, $j = 1, \ldots, 5$ are observed to be equal to the current $v_p(k)$ obtained through V2V. It is observed that there is no constraint violation scenario with the chain GP-based control scheme under different initial conditions. However, without the predictor, the constraint violation scenario occurs. This is caused when the preceding vehicle speed starts to increase. Since when the preceding vehicle starts to accelerate, the further preceding speed without the GP-based predictor is seen to be the same as the current speed, which is approximately zero, and the distance headway increases.

Finally, Table 11.2 shows the fuel economy of the proposed optimal control scheme for five different traffic scenarios. In Table 11.2, *W/I GP* and *W/O GP* denote the proposed receding horizon-based optimal control schemes with the GP-based predictor and without the predictor, respectively. Moreover, a dynamic programming (DP) based optimal control scheme is also developed for performance comparison, where *DP* and *DP W/I Dis.* denote the DP algorithm with all disturbance input available in advance and only part of it available, respectively. In Case 3 and Case 4, the traffic densities are set higher. Therefore, the receding horizon-based control scheme without the GP-based predictor exhibits less ability for fuel economy improvement than in the other cases. Moreover, the traffic density is low in Case 5. The difference in fuel economy performance improvement between the proposed control scheme with the predictor and without the predictor is small in Case 5. With less traffic density, there are fewer vehicles in the traffic scenario, so the traffic look-ahead horizon is more stable, which would exert less influence on the fuel economy of the ego vehicle. It is observed that the proposed control scheme with the predictor can achieve higher fuel economy, but it is still less than that of the DP-based control scheme with all information pre-known. However, it is better than that of DP with less pre-known disturbance input information.

11.5 CONCLUSION AND OUTLOOK

In this chapter, the simulation, optimization, and evaluation for an electric vehicle in a simulation platform that combines traffic simulator and a powertrain simulator are introduced. In the traffic simulator, the communication between vehicle, infrastructure, and ego vehicle can be simulated. However, the high-fidelity powertrain model is built to emulate the powertrain dynamic. Prediction and optimization results of a typical example in optimal powertrain control to minimize the energy consumption are given to show the potential application prospects of this platform.

In the proposed simulation platform, currently only simulated traffic scenario and simulated powertrain are integrated. In the future work, how to integrate the real powertrain test bench, such as an engine, into this platform still needs more exploration. Moreover, real-world traffic information through fast communication will be available to this platform, and the digital twin technology may be the possible approach to deal with this problem. However, the rapid prototyping platform should be considered for running the virtual model when real-world data is introduced to this platform for real-time application.

REFERENCES

1. Guanetti, J., Kim, Y. and Borrelli, F. (2018). Control of connected and automated vehicles: State of the art and future challenges. *Annual Reviews in Control*, *45*, 18–40.
2. Zhao, D. and Peng, H. (2017). *From the lab to the street: Solving the challenge of accelerating automated vehicle testing*. arXiv preprint arXiv:1707.04792.
3. Palcu, P. and Bauman, J. (2018). Whole-day driving prediction control strategy: Analysis on real-world drive cycles. *IEEE Transactions on Transportation Electrification*, *4*(1), 172–183.
4. Rajendran, A. V., Hegde, B., Ahmed, Q. and Rizzoni, G. (2017). *Design and development of traffic-in-loop powertrain simulation*. 1st IEEE Conference on Control Technology and Applications, Hawai'i, August 27–30, 261–266.
5. Macedo, J., Kokkinogenis, Z. et al. (2013). *A HLA-based multi-resolution approach to simulating electric vehicles in Simulink and SUMO*. IEEE Conference on Intelligent Transportation Systems, Hague, The Netherlands, October 6–9, 2367–2372.

12 Driving Control and Traffic Modeling for Automated, Connected, and Electrified Vehicles (ACEVs)

Yinglong He

12.1 CHALLENGES FOR ROAD TRANSPORT

Road transport, connecting and providing access to people, goods, and services in societies, is fundamental to economic and social activities worldwide [1]. In the coming decades, passenger and freight mobility demand is expected to rise significantly because of continuing population growth and urbanization [2].

As shown in Figure 12.1 (a), the International Transport Forum (ITF)has projected global passenger transport demand to increase over twofold between 2020 and 2050, from 55 trillion to 121 trillion passenger kilometers (pkm). Private vehicles are the preferred mode of travel, which will account for 50–52% over the next 30 years. Public transport ridership will increase at an annual rate of 3.2% through 2020 and 2050, covering 35% road trips by 2050. Two- and three-wheelers were only responsible for 5% of worldwide road passenger transport in 2000, but that proportion is likely to reach 13% in 2050. According to the data in Figure 12.1 (b), the current global freight transport demand of 15 trillion tonne kilometres (tkm) is expected to increase by 4 trillion tkm (29%) by 2030 and 17 trillion tkm (117%) by 2050, with heavy-duty vehicles maintaining a dominant share (over 75%) of road freight movements.

The rising demand for road transport poses unprecedented environmental, economic, and social challenges, particularly with the increasing urgency to save energy, reduce carbon emissions and air pollutants, avoid crashes, and relieve congestion [3]. For example, as reported by the International Energy Agency (IEA), over 53% of global primary oil consumption in 2010 was used to meet 94% of the total transport energy demand [4]. The Intergovernmental Panel on Climate Change (IPCC) reported that, in 2014, the global transportation sector was responsible for almost one-quarter of greenhouse gas (GHG) emissions, with

DOI: 10.1201/9781003302827-15

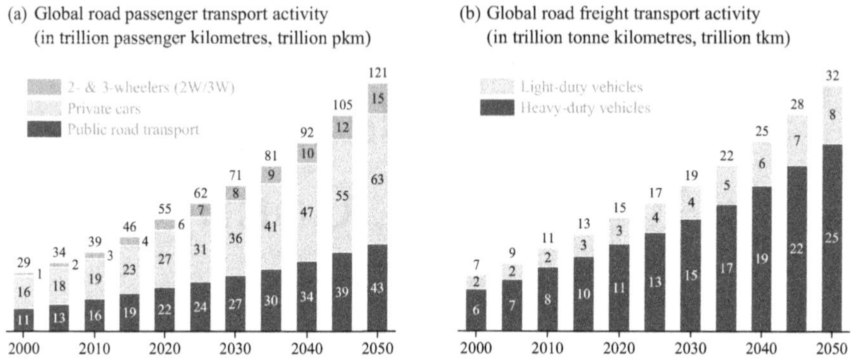

FIGURE 12.1 Global road transport activity evolution (2000–2050) for (a) passenger (in trillion passenger kilometres, trillion pkm) and (b) freight (in trillion tonne kilometres, trillion tkm).

Sources: Adapted from International Council on Clean Transportation (ICCT) "Global transportation roadmap" [10], International Transport Forum (ITF) "ITF transport outlook 2019" [11] and Khalili et al. [12].

Note: Figures may not sum, because of rounding.

about 72% thereof from road transport [5, 6]. In the 2018 Global Status Report on Road Safety [7], the World Health Organization (WHO) highlighted that the number of annual road traffic fatalities continued to increase steadily since 2000, reaching 1.35 million in 2016. In addition, according to INRIX Roadway Analytics in 2017, over the next 10 years, across all 25 most-congested cities of the United States (US), for example, Los Angeles, New York, and San Francisco, traffic jams are estimated to cost the drivers $480 billion due to lost time, wasted fuel, and emitted carbon [8, 9].

12.2 AUTOMATED, CONNECTED, AND ELECTRIFIED VEHICLES (ACEVS)

Overcoming these problems, such as carbon emissions, air pollution, nonrenewable energy consumption, deaths, injuries, wasted time, and congestion, requires rethinking the entire road transport system. Therefore, we must bring together technologies, systems design methods, and business models to supply sustainable mobility solutions at a low cost to consumers and to societies [1]. Although much uncertainty remains about how, exactly, future sustainable mobility will unfold, many of the pivotal building blocks and their potential are becoming clear. As the key to these developments, three trends, namely, automated, connected, and electrified vehicles (usually represented by the acronym ACEVs), are radically altering the way people and goods move [13]. Summaries of recent advances in these three directions are provided as follows.

12.2.1 Automated Vehicles

Commercially available automated vehicles are mainly supported by advanced driver assistance systems (ADAS), which are intended to help drivers in their driving activities [14]. The market for ADAS is expected to show strong momentum in the coming decades, fuelled largely by regulatory and consumer interest in safety solutions that protect drivers and prevent accidents [15]. For example, in 2019, some 40 countries, led by the European Union (EU) and Japan, have agreed on a draft United Nations (UN) Regulation making autonomous emergency braking (AEB) systems mandatory for all new vehicles [16]. In addition, current ADAS technologies will ultimately pave the way for fully autonomous vehicles, which are now a major focus of research and development (R&D), public interest, and press coverage [17]. As summarized in Figure 12.2, ADAS features can be categorized into three broad groups: aiding the driver, warning the driver, and assisting the driver in performing certain basic driving functions [18].

12.2.1.1 ADAS Aiding Features

These features, enabled by sensors such as mono-vision cameras, infrared lights, and lasers, provide additional display or illumination, improving improve the

	ADAS FEATURES			PARTIAL AUTOMATION
	Aiding	Warning	Assisting	
Definition	Aiding features can improve the driver's visibility by providing additional display or illumination.	Warning features alert the driver to potential danger through sensory cues (auditory, visual or haptic).	Assisting features can engage in steering, acceleration, and/or brake systems if necessary.	Automated features allow the vehicle to be driven/stopped without driver intervention in certain scenarios.
Feature (year)	Night vision (NV, 2000) Rear view camera (2002) Adaptive front lighting system (AFS, 2006) Surround view system (SVS, 2007)	Park assist (2002) Forward collision warning (FCW, 2003) Lane departure warning (LDW, 2005) Blind spot detection / rear cross traffic alert (BSD/RCTA, 2006) Driver monitoring system (DMS, 2006)	Adaptive cruise control (ACC, 2007) Lane keeping assist (LKA, 2010) Automatic parking (2006) Autonomous emergency braking (AEB, 2008) Pedestrian avoidance (2014) Intelligent speed adaptation (ISA, 2018)	Single lane highway autopilot (2016) Autonomous valet parking (2017) Traffic jam autopilot (2017) Highway autopilot with lane changing (2018) Urban autopilot (2022)
Technology	Mono cameras Infrared (night vision) Laser lights	Mono and stereo cameras Radar (short) Steering inertia Ultrasonic	Mono and stereo cameras Radar (short and long) Lidar Ultrasonic	Mono and stereo cameras Radar (short and long) Lidar GPS/mapping Ultrasonic
Others			Drivers begin to share control	System integration

FIGURE 12.2 Categories of ADAS features.

Source: Adapted from Boston Consulting Group (BCG) "A roadmap to safer driving through advanced driver assistance systems" [18].

driver's visibility. For instance, night vision (NV) systems utilize the infrared camera to capture the front view in adverse circumstances, such as darkness and poor weather [19]. The rear view camera can alleviate the rear blind spot to facilitate parking and reversing [20]. The adaptive front lighting system (AFS) optimizes the distribution of light from the headlights according to vehicle speed and steering input [21]. The surround view system (SVS) gives drivers a 360-degree surround view to ensure easy, safe, and comfortable parking and maneuvering [22].

Aiding features, most of which were introduced before 2008, are not newcomers in the automotive field. Their cost to customers is decreasing at an annual rate of 4% to 9% [18].

12.2.1.2 ADAS Warning Features

Warning features can alert the driver to potential dangers through sensory cues such as audio, visual, and haptic signals. For example, the park assist system uses proximity sensors to measure the distances to nearby objects while parking and alerts the driver of obstacles via a beeping noise [23]. The forward collision warning (FCW) system monitors traffic conditions ahead and provides alerts to the driver when possible forward collisions are detected [24]. The lane departure warning (LDW) system visually tracks lane markers and alerts the driver if the vehicle unintentionally drifts out of the travelling lane [25].

Other warning features include blind spot detection (BSD), rear cross traffic alert (RCTA), and driver monitoring systems (DMS), all of which have been commercially available since 2006.

12.2.1.3 ADAS Assisting Features

Assisting features can actively engage in driving tasks to ensure safety and comfort, such as longitudinal control (e.g., acceleration, braking, and gear shifting) and lateral control (e.g., steering) of the vehicle movement [26]. Such features are usually enabled by advanced digital technologies and intelligent tools, for example, mono- and stereo-vision cameras, short- and long-range radars, and light detecting and ranging (lidar), using reflected light signals to assess the driving environment [18]. ADAS products with assisting features have hit the automotive market in the last decade, some of which are elaborated as follows.

1) **Adaptive cruise control (ACC):** Vehicles equipped with ACC systems are now reaching mainstream production [27]. These systems utilize radar and control the throttle and brake pedals of the vehicle, aiming to automatically maintain a safe inter-vehicle spacing in congested traffic or a constant speed (user-specified) in free-flow traffic, which potentially improve safety, fuel economy, and road capacity [28].
2) **Intelligent speed adaptation (ISA):** It uses the onboard global positioning system (GPS) and a digital map to improve the driver's compliance with speed limits [29]. If the vehicle speed exceeds a safe or legally enforced speed, the ISA system can provide early warnings of safety issues or directly control the brake and throttle pedals to prevent speeding [30].

3) **Lane keeping assist (LKA):** It is an extended version of the lane departure warning (LDW) system. Instead of alerting the driver to the unintended lane departure, LKA can intervene in the actual driving task by providing steering torque and therefore helping the vehicle return to the center of the lane [31].
4) **Automatic parking:** It uses ultrasonic sensors and/or cameras to scan for a suitable parking space and then moves the vehicle from a traffic lane into the parking spot [31]. The automatic parking system aims to enhance the comfort and safety of driving in constrained environments where much attention and experience is required [32].
5) **Autonomous emergency braking (AEB):** It is enabled by sensors (e.g., cameras, radars, and/or lidars) to detect the presence of possible hazards in front of the vehicle [33]. In an emergency, the system can automatically enhance braking effort to prevent an accident or mitigate its severity [34].

12.2.2 Connected Vehicles

Connected vehicles are another key enabler for future mobility. As digital disruption is transforming the automotive and transport sectors, more and more relevant applications and services in various domains will follow, such as infotainment, telematics, driver assistance, and autonomous driving [3]. Vehicles should, therefore, be capable of exchanging information not only with other vehicles, but also with entities, like pedestrians, roadside equipment, and the internet. These innovative connectivity features are collectively called vehicle-to-everything (V2X), which will lend support to a more active role of the vehicle in safety and mobility applications, for example, avoiding collisions, monitoring intersections, gathering real-time traffic information, and exchanging incident alerts. The resulting market is expected to increase across the globe and reach around $100 billion in 2025 at a compound annual growth rate (CAGR) of 18% between 2018 and 2025 [35].

According to the type of peer that the vehicle communicates with, V2X communication is a term including many different acronyms, such as

1) **Vehicle-to-infrastructure (V2I):** Communicating with roadside units (RSUs), such s traffic lights, speed limits, and traffic cameras.
2) **Vehicle-to-vehicle (V2V):** Communicating between vehicles, without having to access the telecom network.
3) **Vehicle-to-pedestrian (V2P):** Scanning for and communicating with pedestrians or cyclists within close proximity.
4) **Vehicle-to-device (V2D):** Communicating with electronic devices such as smartphones, smartwatches, and smart keys.
5) **Vehicle-to-grid (V2G):** Allowing electrified vehicles (EVs) to communicate with the power grid and to store and discharge electricity, which enables innovative demand management on power-limited grids.
6) **Vehicle-to-network (V2N):** Operating in the licensed cellular spectrum (e.g., 5G), which is good for infotainment.

V2X applications often need extreme connectivity performances, for example, ultra-low latency (<1ms) for real-time applications, highly secure and reliable connectivity for safety-critical tasks, and high-speed broadband (>1 Gb/s >) to support augmented reality (AR) [35]. Different V2X communication standards available on the market can be categorized into two types depending on the underlying technology: WLAN-based (IEEE 802.11p) and cellular-based (C-V2X) [36].

12.2.3 Electrified Vehicles

The International Energy Agency (IEA) reported in 2020 that 17 countries have announced 100% zero-emission vehicle targets or the phase-out of internal combustion engine vehicles (ICEVs) by 2050 [37]. These decisions are linked to the Paris Climate Agreement signed by more than 190 nations in 2016 [38]. Electrified vehicles, therefore, feature prominently in current and planned policy frameworks to achieve greenhouse gas (GHG) emissions reduction targets [39].

To address the global push towards more environmentally friendly and energy-efficient transport, electrified propulsion systems are evolving with the automotive industry. Table 12.1 overviews five types of commercially available powertrain technologies, which are categorized by their energy sources, and provides an evaluation of how they stack up against key environmental, performance, and economic dimensions [40].

Internal combustion engine vehicles (ICEVs) are almost exclusively powered by gasoline or diesel and well-established in global road transport because of their low cost, high power density, proven durability, fuel use flexibility, and a widespread network of refuelling stations [41]. Over the past decades, advanced engine technologies, such as downsizing, turbocharging, and exhaust after-treatment, have contributed to reducing tailpipe emissions and energy losses; however, it is widely recognized that little room is left for significant improvements [40].

Hybrid electric vehicles (HEVs) are developed to optimize the use of the internal combustion engine (ICE) through its interplay with electric components such as the low-voltage (LV) battery and the electric motor (EM) [42]. As a result, the ICE can be better managed to avoid low-efficiency and high-emission operations, such as idling, cold start, and strong acceleration [43]. Moreover, the EM can serve as a generator in regenerative braking to convert the otherwise wasted kinetic energy into electric energy, which is then used to charge the battery and extend the driving range [44].

Plug-in hybrid electric vehicles (PHEVs) have a similar powertrain architecture to HEVs, however, are usually fitted with a more powerful EM and a much larger battery [45]. The battery can be recharged by plugging it into an external power source (utility grid or renewable energy sources such as solar cells), by the onboard ICE-powered generator, or through regenerative braking [46]. Despite the ICE being used occasionally to assist the propulsion, PHEVs can provide a significant share of the all-electric range (AER), the value of which typically lies between 30 and 60 km today and possibly between 60 and 80 km in the near future [40].

Battery electric vehicles (BEVs), operating only on stored electricity, consist mainly of the high-voltage (HV) battery and the EM. Compared with the ICE,

TABLE 12.1
Pros and Cons of Electrified Vehicles Categorized by Energy Sources

			Electrified vehicles			
		ICEV	**HEV**	**PHEV**	**BEV**	**FCEV**
Environment	T2W emissions [a]	D	C	B	A	A
	W2W emissions [b]	D	D	B	B	B
	Recycling	B	B	A	D	C
Performance	Range Refueling time [c]	A	A	A	C	B
	Acceleration	A	A	C	D	B
	Top speed	B	B	A	A	A
		A	A	B	C	C
Economics	TCO today [d]	A	A	B	C	D
	Price today	A	B	C	D	D
	Infrastructure costs	A	A	B	D	C
Key characteristics, indicative	ICE power, kW	50–400	50–400	50–400	—	—
	Electric power, kW	—	<25	<100	>100	>100
	Battery capacity, kWh	—	<2	<30	>40	<10
	T2W CO_2 savings, % CO_2	—	10–20	50–80[e]	100	100

[a] Tank-to-wheel (T2W) emissions, i.e., tailpipe emissions that a vehicle produces locally via the combustion of fossil fuels; these emissions are subject to current regulations globally.
[b] Well-to-wheel (W2W) emissions, i.e., emissions related to the fuel cycle or generation of electricity, the production of the vehicle and battery, and the use of the vehicle; largely dependent on a country's energy mix.
[c] Considering only the time needed to refuel/charge the vehicle, not infrastructure availability.
[d] Total cost of ownership (TCO), strongly depending on region and vehicle segment.
[e] Estimated CO2 savings considered for certification tests.

Source: Adapted from McKinsey Center for Future Mobility—"Reboost: A comprehensive view on the changing powertrain component market and how suppliers can succeed" [40].

Note: A = excellent; B = good; C = moderate; D = challenged; ICE(V) = internal combustion engine (vehicle); HEV = hybrid electric vehicle; PHEV = plug-in hybrid electric vehicle; BEV = battery electric vehicles; FCEV = fuel cell electric vehicles.

which produces peak torque within a limited range of speed, the EM is able to deliver high starting torque across a broad range of speed [47]. Therefore, many parts, such as the reduction gearbox, the increasingly complex engine management system, and a wide variety of fluids (e.g., engine oil and transmission fluid), are not required for BEVs, meaning lower maintenance costs [48]. Furthermore, as battery prices continue to drop, BEVs are expected to offer lower total cost of ownership (TCO) to customers than ICE and hybrid vehicles [40].

Fuel cell electric vehicles (FCEVs) use fuel cell (FC) stacks to convert the chemical energy of onboard gaseous hydrogen (H_2) into electricity, which is then stored in a battery to drive the vehicle's EM. The compressed hydrogen demonstrates significantly higher gravimetric and volumetric energy densities than batteries [49]. The refuelling time of a few minutes makes FCEVs well-suited for applications with high-power and long-distance requirements, such as trucks and commercial vehicles [40]. In addition, from the perspective of life cycle emissions and environmental impacts, FCEVs are more sustainable than BEVs, ICEVs, and hybrids, with even more room for improvement as technologies of hydrogen production and delivery mature [50].

12.3 ECO-COOPERATIVE AUTOMATED DRIVING SYSTEMS (ADS)

As one of the most important ADAS features (shown in Figure 12.2), adaptive cruise control (ACC) systems have reached the consumer market for over two decades [51, 52]. As presented in Figure 12.3 (a) and (b), these systems are enabled by radar sensors and used to automatically maintain a desired inter-vehicle spacing in congested traffic or a constant speed in free-flow traffic, which potentially improves the safety, capacity, and fuel economy of road transport [53, 54].

Figure 12.3 (c) illustrates the longitudinal kinematics for ACC implementation, which is important for the control problem formulation and can be described by

$$\begin{cases} a_n(t) = \dot{v}_n(t) = \ddot{x}_n(t), \\ v_{n-1}(t) = \dot{x}_{n-1}(t), \\ a_{n-1}(t) = \dot{v}_{n-1}(t), \\ s_n(t) = x_{n-1}(t) - x_n(t) - l_{n-1}, \\ \Delta x_n(t) = x_{n-1}(t) - x_n(t), \\ \Delta s_n(t) = s_n(t) - s_{n,des}(t), \\ \Delta v_n(t) = v_{n-1}(t) - v_n(t), \end{cases} \tag{12.1}$$

where subscripts n and $n-1$ represent the ego and the preceding vehicles, respectively; x, v, and a are the vehicle's longitudinal position (m), speed (m/s), and acceleration (m/s^2), respectively; t denotes time (s); l means the vehicle length (m); s_n is the spacing (m) between the front bumper of the ego vehicle and the back bumper of the preceding one; $s_{n,des}$ is the desired spacing determined by the spacing policy inherent in the ACC system; Δx_n is the headway (m); Δs_n represents the spacing error (m) between the actual and desired values; Δv_n is the relative speed (m/s) between the preceding and ego vehicles.

In the remainder of the section, we revisit the literature on spacing policies and control algorithms of ACC systems, followed by a review of studies on cooperative ACC (CACC) and ecological ACC (EcoACC), both of which are more sophisticated variants of ACC.

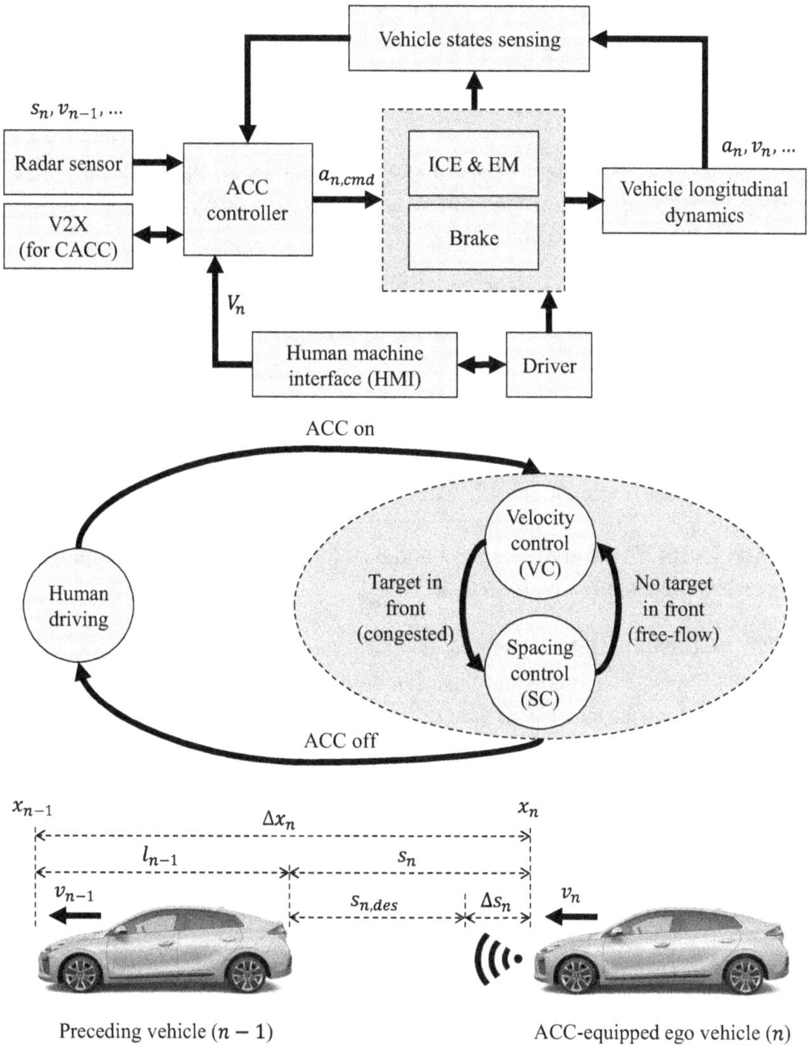

FIGURE 12.3 Overview of an adaptive cruise control (ACC) system. (a) Control architecture of an ACC system (b) Operation modes of an ACC system (c) Longitudinal kinematics for ACC implementation.

Source: Adapted from Bu et al. [55], Li et al. [56], Wang [57], Bernsteiner [58]
Note: V2X = vehicle-to-everything; ACC = adaptive cruise control; CACC = cooperative adaptive cruise control; ICE = internal combustion engine; EM = electric motor

12.3.1 Spacing Policies of ADS

The spacing policy determines the desired spacing ($s_{n,des}$) in Eq. (12.1) and plays a crucial role in driving safety, traffic throughput, and string stability [59]. String stability means disturbances (e.g., spacing error Δs_n and relative speed Δv_n) of an individual vehicle in a platoon (or vehicle string) do not amplify when they propagate upstream [60].

In previous studies, prevailing spacing policies can be grouped into three types including constant distance (CD), constant time headway (CTH), and nonlinear distance (NLD), as detailed next [56].

12.3.1.1 Constant Distance (CD)

In the CD policy, the desired inter-vehicle spacing ($s_{n,des}$) is independent of the driving environment [61, 62],

$$s_{n,des}(t) = s_0, \tag{12.2}$$

where s_0 is a positive constant (m). Although potentially increasing traffic capacity [63], this policy cannot guarantee string stability unless the leading vehicle broadcasts its information (e.g., speed and acceleration) via V2V communication to all other vehicles in the platoon [64].

12.3.1.2 Constant Time Headway (CTH)

In the CTH policy, which is widely used in commercial ACC systems, the desired spacing ($s_{n,des}$) varies as a linear function of the ego vehicle speed (v_n) [65],

$$s_{n,des}(t) = s_0 + t_h v_n(t), \tag{12.3}$$

where t_h is the time headway (s) and s_0 is the minimum (or standstill) spacing. This policy is to some extent similar to that of human driver's behavior [56] and has been used to develop ACC controllers that can guarantee string stability [66]. The major drawback of the CTH policy, however, is poor robustness against traffic flow fluctuation [67, 68].

12.3.1.3 Nonlinear Distance (NLD)

In the NLD policy, the desired spacing ($s_{n,des}$) is a nonlinear function of one or multiple state variables (e.g., v_n, v_{n-1}, x_n, and x_{n-1}) associating with the driving environment, the generic mathematical formulation of which can be represented as

$$s_{n,des}(t) = g\left(v_n(t), v_{n-1}(t), x_n(t), x_{n-1}(t), \ldots\right), \tag{12.4}$$

NLD spacing policies are expected to outperform CD and CTH policies in enhancing both traffic flow stability and traffic capacity [69]. For instance, an NLD policy is developed based on Greenshield's speed-density relationship and demonstrates superior stability and safety properties [67], as given by

$$s_{n,des}(t) = \frac{1}{\rho_{max}\left(1 - \frac{v_n(t)}{v_{free}}\right)} - l_{n-1}, \tag{12.5}$$

where ρ_{max} and v_{free} denote the jam density and free-flow speed of the traffic, respectively [59]. Another NLD spacing policy, which includes a quadratic term of ego vehicle speed (v_n), is proposed and optimized to improve both string stability and traffic flow stability [68], as described by

$$s_{n,des}(t) = 3 + 0.0019v_n(t) + 0.0448v_n(t)^2. \tag{12.6}$$

More complicated NLD policies have been discussed in the literature. For example, a safe distance policy is reported to take into account the vehicles' braking capability [70, 71], which can be mathematically expressed as

$$s_{n,des}(t) = s_0 + \tau v_n(t) - \left(\frac{v_n(t)^2}{2\alpha_{min}} - \frac{v_{n-1}(t)^2}{2\hat{\alpha}_{min}}\right), \tag{12.7}$$

where s_0 is the minimum (or standstill) spacing (m); τ is the time constant (s); both α_{min} and $\hat{\alpha}_{min}$ are negative constants (m/s^2), representing the braking capabilities of the ego (n) and preceding ($n-1$) vehicles, respectively.

12.3.2 Adaptive Cruise Control (ACC) Algorithms

As shown in Figure 12.3 (b), the ACC system usually operates in two modes, that is, spacing control (SC, in congested traffic) and velocity control (VC, in free-flow traffic) [57]. The transition between these two modes can be determined by a fixed spacing threshold such as the maximum unambiguous range (MUR) of the radar [72, 73],

$$a_{n,cmd}(t) = \begin{cases} a_{n,cmd}^{VC}(t), s_n(t) > s_{th,} \\ a_{n,cmd}^{SC}(t), s_n(t) \le s_{th,} \end{cases} \tag{12.8}$$

where $s_{th,}$ is the spacing threshold (m); $a_{n,cmd}$ is the final acceleration command (m/s2) adopted by the ACC system; while $a_{n,cmd}^{VC}$ and $a_{n,cmd}^{SC}$ are acceleration commands (m/s2) generated in the VC and SC modes, respectively. Alternatively, most previous studies implicitly implement the mode transition by comparing acceleration commands of two modes, in which the more restrictive choice is adopted [74, 75],

$$a_{n,cmd}(t) = \min\left(a_{n,cmd}^{VC}(t), a_{n,cmd}^{SC}(t)\right). \tag{12.9}$$

In free-flow traffic, the ACC velocity control (VC) mode regulates the throttle and brake pedals to follow the set speed, which is specified in advance by the driver via human-machine interface (HMI) [52]. This mode is usually accomplished by a linear proportional controller [76, 74, 73], as given by

$$a_{n,cmd}^{VC}(t) = k_0\left(V_n - v_n(t)\right), \tag{12.10}$$

where k_0 is a positive coefficient (s^{-1}) and V_n is the set speed (m/s) of the ACC system.

In congested traffic, the ACC spacing control (SC) mode aims to maintain a desired inter-vehicle spacing ($s_{n,des}$), which is determined by the spacing policy as previously discussed. In the literature, the development of ACC algorithms is mainly focused on this control mode since congested traffic is a much more challenging environment for ACC operation [57]. Typical ACC spacing control algorithms will be presented as follows.

12.3.2.1 Linear Controller

The linear controller is probably the most widely used ACC spacing control (SC) method in either academia or industry because it can facilitate theoretical analyses and hardware implementations [57, 77, 56]. A generic form of linear ACC controller is described by

$$a_{n,cmd}^{SC}(t) = k_s \Delta s_n(t) + k_v \Delta v_n(t), \tag{12.11}$$

where Δs_n and Δv_n are the spacing error (between the actual and desired spacing values) and the relative speed (between the preceding and ego vehicles), respectively, as given by Eq. (12.1); while k_s and k_v are positive coefficients (s^{-2} and s^{-1}, respectively). The linear ACC control law using simple spacing policies, such as constant distance (CD) or constant time headway (CTH), cannot guarantee collision-free at safety-critical (e.g., approaching a standstill vehicle at high speeds) or dense traffic conditions [72, 78, 79]. In these conditions, the ACC system is switched off in practice [57].

12.3.2.2 Model Predictive Control (MPC)

Advanced ACC systems usually have multiple design objectives, for example, minimizing tracking error (also known as spacing regulation error), preserving string stability, increasing ride comfort, and improving fuel economy, some of which are contradictory [80]. In addition, there are often multiple constraints imposed on ACC design solutions, such as actuator (the brake and throttle pedals) limit and safety limit [55]. MPC is a control framework that can optimize multiple objectives under different design constraints; therefore, the MPC-based ACC is usually formulated as a constrained optimization problem [81]. However, existing MPC algorithms suffer from a high computational load [80]. In the literature, a hybrid MPC approach is used to design ACC systems that can improve safety and tracking capability [82, 83]. An MPC-based ACC system is proposed to explicitly include acceleration constraints to meet comfort and safety needs [84]. In addition, a multi-objective MPC-based ACC system is developed to simultaneously address multiple critical issues, including tracking capability, fuel economy, and driver desired response, in which limits on both acceleration and jerk are applied to guarantee ride comfort of the vehicle [80].

12.3.2.3 Sliding Mode Control (SMC)

SMC is a variable structure control method, which can switch back and forth between two continuous control laws based on the current position in the state space [85]. This switching approach makes SMC robust to nonlinear vehicle dynamics, actuator constraints, and external disturbances [86]. However, the main obstacle for its implementation is the chattering problem (the phenomenon of finite-frequency, finite-amplitude oscillations) [87]. There are some studies employing SMC algorithms to design ACC systems. For example, an ACC system based on the SMC algorithm and the CTH spacing policy is proposed, which can detect the presence of possible hazards in front of the vehicle and then switch between two spacing control modes, adaptive cruise mode and collision avoidance mode [88]. Similar SMC-based ACC systems are developed to overcome bounded disturbances under different spacing policies [89].

12.3.2.4 Fuzzy Logic (FL) Controller

FL-based ACC systems are proposed in several studies. For instance, an FL-based ACC system adopting the CTH spacing policy is reported to guarantee string stability [90]. An adaptive FL method is used to develop an ACC system that can emulate human drivers' behavior [91]. Another fuzzy controller is developed for both high-level (to maintain safe spacing) and low-level (to adjust the throttle and brake pedals) control tasks, exhibiting a good performance at a broad range of speeds [92].

12.3.3 Cooperative ACC (CACC)

Using V2V and V2I communications, CACC extends ACC to achieve cooperative maneuvers [74]. A series of CACC studies, in terms of controller design, stability analysis, and real-world driving experiments, have been conducted in the California Partners for Advanced Transit and Highways (PATH) program [93, 94]. With the connectivity between vehicles, CACC allows the vehicle to maintain smaller headway compared to ACC [95].

By sharing vehicle operating states (e.g., position, speed, and emergency braking) in a distributed manner, CACC vehicles within a certain geographic area can cooperate with each other, which can maximize the benefits of automated driving in the following ways [96]. First, driving safety is enhanced because the downstream traffic information can be broadcasted to upstream vehicles in advance, significantly reducing the actuation delay [57]. Second, road capacity is increased owing to the reduced time (or distance) headway between consecutive vehicles [52]. Finally, energy economy and pollutant emissions are improved because both unnecessary speed changes and aerodynamic drag can be reduced [96].

Based on CACC and ACC, the platoon control techniques have attracted extensive interest for decades. Typical research topics of platoon control include spacing policies [97, 68], string stability [98, 99], dynamics heterogeneity [100, 101], time delays [102, 103], communication topologies (e.g., bidirectional) [104, 56], etc.

12.3.4 Ecological ACC (EcoACC)

Increasing concerns over energy consumption and GHG emissions are giving rise to the development of EcoACC [105], which regulates the individual vehicle's speed by minimizing not only spacing error (Δs_n) and relative speed (Δv_n) but also fuel consumption. Most studies on EcoACC focus on internal combustion engine vehicles (ICEVs). For example, using road grade information provided by a high-definition (HD) map, predictive EcoACC systems are proposed to control the inter-vehicle spacing in an energy-efficient manner [106–108].

In addition, a few efforts are being made to develop EcoACC systems for HEVs and PHEVs, which can leverage the synergistic benefits of ACC and EMS in improving energy efficiency and exhaust emissions. For instance, a multi-objective EcoACC system is proposed to simultaneously improve fuel economy and ride comfort for an HEV, which adopts a rule-based (RB) energy management strategy (EMS) [109]. Another EcoACC system designed for PHEVs can use the traffic light information to predict the future trajectory of the preceding vehicle and then provide optimal speed and power control signals, which can minimize the fuel consumption and satisfy the constraints associated with the vehicle's safety and comfort [110]. By combining an ACC based on action dependent heuristic dynamic programming (ADHDP) and an adaptive EMS, an EcoACC system is developed to achieve near-optimal fuel economy and comfortable driving [111].

12.4 MICROSCOPIC TRAFFIC MODELING

Unlike ACC models, which are discussed in Section 12.3 and developed to control the vehicle's longitudinal driving behavior through mechanical (e.g., ICE, EM, and brake system) and electrical (e.g., radar and camera) components, microscopic car-following (CF) models focus on describing the resulting vehicle behavior and then estimating its impact on traffic flow dynamics in traffic simulations [112]. Traffic flow theory and microscopic traffic simulation are indispensable tools to maximize the benefits of ACEVs to the road network [113].

From the perspective of motion description, existing CF models can be categorized into two groups, namely, kinematics-based (or behavioral) and dynamics-based [114]. The former is simple and concerns only variables derived from trajectories of vehicles, such as position, spacing, speed, and acceleration. The latter, however, can take into account variables underlying the vehicle movement, like forces, torque, and energy [115]. It is worth noting that both types of CF models should be able to strike a balance between simplicity and accuracy and therefore are feasible for large-scale traffic simulation [116].

12.4.1 Kinematics-Based Traffic Models

Kinematics-based (or behavioral) CF models describe the movement of the following vehicle as a function of its kinematic relationship to the preceding vehicle, as depicted in Figure 12.3 (c), and, in the last seven decades, have been studied intensively by model simulation, experimental campaign, and traffic observation [117].

In Table 12.2, typical kinematics-based CF models are listed chronologically with the deterministic acceleration or speed equation and the associated parameters given in the last two columns [118]. In the early 1950s, Reuschel [119] and Pipes [120] did pioneering work on the development of behavioral CF models. Their model formulations only consider either the inter-vehicle spacing (s_n) or the relative velocity (Δv_n , between the preceding and ego vehicles), and, therefore, have significant limitations in describing the vehicle behavior [121]. The Gazis–Herman–Rothery (GHR) model, proposed by Gazis et al. in 1961 [122], defines a nonlinear acceleration equation that considers reaction time (τ) [113]. However, it is built upon strong assumptions, leading to critical drawbacks that are frequently reported by researchers, for example, the model overestimates the vehicle's ability to perceive small changes in the relative speed (Δv_n) and the headway (Δx_n) [123, 124]. Newell's model assumes that the following vehicle's response directly depends on the headway (Δx_n) [125], which, however, might result in unrealistic acceleration behavior [126].

As a major milestone towards the development of safety distance (or collision avoidance) CF models [126], Gipps model [127] developed in 1981 has been extensively studied [128]. It describes the vehicle speed in a way assuming that the driver leaves enough safe distance in front and thus can safely stop the vehicle in case the preceding vehicle commences an emergency brake [129]. Similar to ACC models given in Eq. (12.9), Gipps model consists of two driving regimes, free-flow and congested, and chooses the more restrictive one from the resulting speeds of these two regimes [130].

The optimal velocity (OV) mode, proposed by Bando et al. [131], assumes that the vehicle attempts to follow an optimal (or safe) speed (V_{opt}) that depends on the headway (Δx_n). This model has received considerable attention because of its ability to accurately describe many traffic flow characteristics in the real world, such as traffic instability, traffic congestion evolution, and the formation of stop-and-go waves. However, a comparison between the simulated and observed vehicle trajectories indicates that the OV model exhibits significantly high acceleration and unrealistic deceleration [132]. To overcome this limitation, Helbing et al. adopted the basic concept of the OV model, utilized negative velocity difference, and developed the generalized force model (GFM), which demonstrates a good agreement with empirical trajectory data [133] but is poor in predicting the delay time of vehicle movement [132]. Consequently, Jiang et al. [134] modified GFM in 2001 by including the negative and positive velocity differences and then proposed the full velocity difference model (FVDM) [132].

The intelligent driver model (IDM), proposed by Treiber et al. [138] in 2000, is a breakthrough in the development of desired measure CF models, which usually

TABLE 12.2
Typical Kinematics-Based Car-Following (CF) Models

Model	Year	Acceleration or speed equations	Parameters
Reuschel [119]	1950	$\upsilon_n(t) = c \cdot s_n(t)$,	c
Pipes [120]	1953	$a_n(t) = c \cdot \Delta\upsilon_n(t)$,	c
Gazis et al. [122] GHR	1961	$a_n(t+\tau) = c \cdot \upsilon_n(t+\tau)^m \cdot \upsilon_n(t) / (\Delta x_n(t))^l$,	τ, c, m, l
Newell [125]	1961	$a_n(t) = c \cdot \upsilon_n(\Delta x_n(t))^l$, or $a_n(t) = \frac{1}{\tau}[V_{opt}(t) - \upsilon_{n-1}(t)]$, where $V_{opt}(t) = V_n \cdot \left[1 - e^{-\frac{e}{V_n}(\Delta x_n(t) - s_0)}\right]$,	c, l or c, s_0
Bierley [135]	1963	$a_n(t) = \alpha \cdot \Delta\upsilon_n(t) + \beta \cdot (\Delta x_n(t))^l$,	α, β, l
Gipps [127]	1981	$\upsilon_n(t+\tau) = \min\left\{\upsilon_n(t) + 2.5\alpha_{max}\tau \cdot \left(1 - \frac{\upsilon_n(t)}{V_n}\right)\left(0.025 + \frac{\upsilon_n(t)}{V_n}\right)^{0.5}, \alpha_{max}\tau + \sqrt{\alpha_{min}{}^2\tau^2 - \alpha_{min} \cdot \left[2(\Delta x_n(t) - s_0) - \tau\upsilon_n(t) - \frac{\upsilon_{n-1}(t)^2}{\alpha_{min}}\right]}\right\}$	$\tau, s_0, a_{max}, a_{min}, \hat{a}_{min}$
Leutzbach et al. [136]	1986	$a_n(t+\tau) = \frac{(\Delta\upsilon_n(t))^2}{2(\Delta x_n(t) - s_{safe})} + a_{n-1}(t)$,	τ, s_{safe}
Sultan et al. [137]	2004	$a_n(t+\tau) = c \cdot \upsilon_{n-1}(t)^m \cdot \frac{\Delta\upsilon_n(t)}{(\Delta x_n(t))^l} + k_1 a_{n-1}(t) + k_2 a_n(t)$,	τ, c, m, k_1, k_2

Model	Year	Acceleration or speed equations	Parameters
Bando et al. [131] OV	1995	$a_n(t) = c \cdot \left[V_{opt}(t) - v_n(t)\right]$, where $V_{opt}(t) = \frac{V_n}{2}\left[\tanh\left(\Delta x_n(t) - s_0\right) + \tanh(s_0)\right]$	c, s_0
Helbing et al. [133] GFM a	1998	$a_n(t) = c \cdot \left[V_{opt}(t) - v_n(t)\right] + \lambda \cdot H\left(-\Delta v_n(t)\right) \cdot \Delta v_n(t)$, where $V_{opt}(t) = V_1 + V_2 \tanh\left[k_1 s_n(t) - k_2\right]$,	$c, \lambda, V_1,$ V_2, k_1, k_1
Treiber et al. [138] IDM	2000	$a_n(t) = a_{\max} \cdot \left[1 - \left(\frac{v_n(t)}{V_n}\right)^{\delta} - \left(\frac{s_{n,des}(t)}{s_n(t)}\right)^2\right]$, where $s_{n,des}(t) = s_0 + \max \left[0, t_h v_n(t) - \frac{v_n(t)\Delta v_n(t)}{\sqrt[2]{-a_{max} \cdot a_{min}}}\right]$,	$\delta, t_h, s_0,$ a_{max}, a_{min}
Jiang et al. [134] FVDM	2001	$a_n(t) = c \cdot \left[V_{opt}(t) - v_n(t)\right] + \lambda \Delta v_n(t)$,	c, λ

a $H(\cdot)$ is Heaviside function.

Source: Adapted from Chen et al. [118, 139], Reuschel [119], and Gipps [127]

Note: Subscripts n and $\boldsymbol{n-1}$ indicate the ego and preceding vehicles, respectively; $\Delta \boldsymbol{x}_n(\boldsymbol{t}) = \boldsymbol{x}_{n-1}(\boldsymbol{t}) - \boldsymbol{x}_n(\boldsymbol{t})$ is the headway (m); $\Delta \boldsymbol{V}_n(\boldsymbol{t}) = \boldsymbol{V}_{n-1}(\boldsymbol{t}) - \boldsymbol{V}_n(\boldsymbol{t})$ is the relative speed (m/s); $\boldsymbol{s}_n(\boldsymbol{t}) = \boldsymbol{x}_{n-1}(\boldsymbol{t}) - \boldsymbol{x}_n(\boldsymbol{t}) - \boldsymbol{l}_{n-1}$ is the spacing (or gap, m); $\boldsymbol{V}_n$ is the free-flow (or desired) speed; GHR = Gazis-Herman-Rothery model; OV = optimal velocity model; GFM = generalized force model; IDM = intelligent driver model; FVDM = full velocity difference model.

TABLE 12.3
Development of Dynamics-Based Car-Following (CF) Models

Model	Year	Full load engine power[a]	Driveline and resistances	Gearshift behavior[b]	Driving behavior[c]	Collision avoidance[d]	Traffic stability[d]	Model output
Searle et al. [141]	1999	X p_{ice}^{max}	X	X	X	X	X	Maximum acceleration
Rakha et al. [142]	2001	X p_{ice}^{max}	✓	X	X	X	X	
Rakha et al. [143]	2002	X $\beta(\upsilon_n)\cdot p_{ice}^{max}$	✓	X	X	X	X	
Rakha et al. [144]	2004	X $\beta(\upsilon_n)\cdot p_{ice}^{max}$	✓	X	✓ ω_{gb}	X	X	
Rakha et al. [145] RPA	2009	X $\beta(\upsilon_n)\cdot p_{ice}^{max}$	✓	X	✓ t_{gb}	✓ $\upsilon_{ts}^{lim}(V_n, s_n)$	✓ $\upsilon_{ts}^{lim}(V_n, s_n)$	
Rakha et al. [146]	2012	✓ $p_{ice}^{FL}(\upsilon_n, \varphi_g)$	✓	✓ ω_{gb}	✓ t_{gb}	X	X	
Fadhloun et al. [147]	2015	X p_{ice}^{max}	✓	X	✓ $t_{gb}(\upsilon_n, V_n)$	X	X	Typical
Fadhloun et al. [148] FR	2019	X $\beta(\upsilon_n)\cdot p_{ice}^{max}$	✓	X	✓ $\omega_{gb}(\upsilon_n, \upsilon_{n-1}, V_n, s_n)$			
Makridis et al. [116] MFC	2019	✓ $p_{ice}^{FL}(\upsilon_n, \varphi_g)$	✓	✓ ω_{gb} (GS)	✓ $\omega_{gb}(\upsilon_n, V_n, DS)$	X	X	
He et al. [115] Electric MFC	2020	✓ $p_{em}^{FL}(\upsilon_n, \varphi_g)$	✓	✓ ω_{gb} (GS)	✓ $\omega_{gb}(\upsilon_n, V_n, DS)$	X	X	

[a] The ICE full load power (P_{ice}^{FL}) across the entire operating speed range can be calculated in three ways: 1) p_{ice}^{max}, which is a constant and equal to the peak power that the ICE can produce; 2) $\beta(\upsilon_n)\cdot P_{ice}^{max}$, which is a percentage of the ICE peak power; or 3) $P_{ice}^{FL}(\upsilon_n, \varphi_g)$, which is a function of the vehicle speed υ_n and the engaged gear ratio φ_g.
[b] The gear shifting points within the vehicle speed range are defined by ω_{gb} for each driver.
[c] The driver's typical driving behavior can be represented as a percentage of the vehicle's full load capabilities in two ways: 1) to multiply the full load ICE power curve by a power reduction (or throttle opening) factor, t_{db}; or 2) to multiply the vehicle's acceleration potential curve by an acceleration reduction factor, ω_{gb}. Both of these factors can be a constant or a function (the symbol with arguments in parentheses) for each driver.
[d] Collision avoidance and traffic stability, both of which belong to the interaction term (instead of the free-flow term) of the CF model, can be either formulated as speed limit functions (i.e., v^{lim} and υ_{ts}^{lim}, respectively) or directly incorporated into the driving behavior function ω_{db}.
Note: β is the full load power adjustment factor for the ICE; υ_n and υ_{n-1} are speeds of the ego (n) and preceding ($n-1$) vehicles, respectively; V_n is the desired or free-flow speed; s_n is the inter-vehicle spacing (or gap); GS and DS are calibratable parameters capturing gearshift style and acceleration style, respectively; ICE = internal combustion engine; RPA = Rakha-Pasumarthy-Adjerid model; FR = Fadhloun-Rakha model; MFC = microsimulation free-flow acceleration model.

assume that vehicles aim to simultaneously reach both the desired (or free-flow) speed (V_n) and the desired spacing ($s_{n,des}$). Its acceleration equation can ensure a smooth transition between the free-flow and congested driving regime [126]. Nevertheless, IDM does not provide a lower bound for the acceleration and, therefore, may lead to unrealistically large deceleration when the inter-vehicle spacing (s_n) drops significantly (e.g., in cut-in maneuvers) [140].

12.4.2 Dynamics-Based Traffic Models

Dynamics-based CF models take into account features such as force, torque, and power underlying the vehicle movement [115]. As summarized in Table 12.3, it has been a growing field across several disciplines in recent years. Relevant publications remain few, almost all of which focus on conventional ICEVs powered by a gasoline or diesel ICE with a multi-ratio transmission. The pioneering work of Searle et al. [141] is crucial to our wider understanding of this area. In their model, however, the full load (FL) engine power (P_{ice}^{FL}) is assumed to be constant and equal to the peak engine power (P_{ice}^{max}) across the entire engine speed range. Also, the dynamics of the internal driveline (e.g., transmission and engine accessories) and external resistances (e.g., aerodynamic, rolling, and grade) are not considered; instead, their effects on the power losses are all incorporated into a constant, acceleration efficiency ((η_a)). Moreover, driving behavior, that is, how the driver utilizes the vehicle's full load capabilities, is not indicated, thus, the model's output is the maximum acceleration characteristics. In the subsequent study by Rakha et al. [142], the enhanced model accounts for detailed dynamics of the driveline and the resistances. Although the full load engine power (P_{ice}^{FL}) is still constant, the tractive force is upper bounded by the maximum force that can be sustained between the vehicle tires and the road surface. Rakha et al. [143] then developed a variable power dynamics model in 2002 by introducing a power adjustment factor (β), which dealt with the full load power reduction caused by the effect of successive gear shifting at low speeds. The factor β is a ramp function (with a positive intercept) of vehicle speed (v_n), when $v_n \leq v_p$ (v_p is the speed at which maximum power occurs).

Rakha et al. [144] are the first of many who have attempted to capture typical driving behavior and vehicle dynamics at the same time. The typical acceleration output is achieved by introducing an acceleration reduction factor (w_{db}), which denotes the ratio of the driver's actual acceleration to the vehicle's theoretical capability. In contrast, the model proposed in 2009 by Rakha et al. [145] adopts a power reduction factor (t_{db}) aiming to represent the percentage of the throttle opening; therefore, the power delivered from the engine can be assumed to be equal to a constant percentage of the full load engine power. Additionally, to capture the interaction of vehicles in the traffic flow, the model considers two speed constraints: v_{ca}^{lim} (v_{n-1}, s_n) to avoid any collisions with the preceding vehicle and v_{ts}^{lim} (V_n, s_n) to maintain cruising in the steady-state traffic stream. Rakha et al. [146] developed the first model that explicitly captures the dynamics of engine power and gearshift. This research is critical, given that the full load engine power (P_{ice}^{FL})

during driving cannot be assumed to be constant over the entire vehicle speed range nor to be adjusted by a ramp function (β) at the low-speed region. In fact, it is well known that the engine power at full load conditions varies as a function of the current vehicle speed (sv_n) and the engaged gear ratio (φ_g). Moreover, the gearshift behavior (ω_{gb}) in this model is designed as per a fixed engine speed threshold.

These three typical acceleration models reported by Rakha et al., however, reproduce the driving behavior (ω_{gb} or t_{db}) and the gearshift behavior (ω_{gb}) using constant factors, thus they provide limited information in relation to each driver's unique pattern of driving. Recognizing this deficiency, an enhancement proposed in 2015 by Fadhloun et al. [147] made the model representative of different driving patterns. The basic idea behind their research is to employ a variable throttle opening (t_{db}) as a function of the vehicle speed (v_n) and the driver's desired speed (V_n). In 2019, Fadhloun et al. [148] further improved the model by incorporating the two constraints regarding vehicle interactions (i.e., collision avoidance and steady-state traffic stream) into the driving behavior function $\omega_{db}(v_n, v_{n1}, V_n, s_n)$, which serves as an acceleration reduction factor. Makridis et al. [116] demonstrated the first study to account for different drivers' driving behavior and gearshift behavior simultaneously, by introducing two calibratable parameters, driving style (*DS*) and gearshift style (*GS*). Built on this work, He et al. [115] proposed a CF model (i.e., electric MFC) that marks the first attempt to account for the acceleration and deceleration dynamics of electrified vehicles.

12.5 CONCLUSIONS AND OUTLOOK

In this chapter, we have explored the challenges and advancements in driving control and traffic modeling for automated, connected, and electrified vehicles (ACEVs). The integration of automation, connectivity, and electrification has the potential to revolutionize the road transport system, improving safety, efficiency, and environmental sustainability. However, several challenges need to be addressed to fully realize the benefits of ACEVs and ensure their successful deployment.

The first challenge lies in the complex nature of road transport. The road environment is dynamic and unpredictable, with various elements and stakeholders involved. ACEVs must navigate through diverse traffic conditions, interact with human-driven vehicles, and adapt to infrastructure limitations. These challenges require robust and reliable driving control systems that can handle complex scenarios and ensure safe and efficient operation.

Automated, connected, and electrified vehicles (ACEVs) play a crucial role in transforming the transportation landscape. Automation enables vehicles to take over driving tasks, reducing human errors and enhancing safety. Connectivity allows vehicles to communicate with each other and with the infrastructure, enabling cooperative systems and advanced functionalities such as platooning and intersection management. Electrification, likewise, promotes cleaner and more sustainable transportation by replacing conventional internal combustion engines with electric powertrains. The integration of these technologies opens up new possibilities for ACEVs, but it also introduces new challenges in terms of system integration, interoperability, and infrastructure requirements.

Eco-cooperative automated driving systems (ADS) represent a promising approach to optimize traffic flow and reduce energy consumption. By enabling vehicles to cooperate and coordinate their movements, eco-cooperative ADS can minimize congestion, reduce travel times, and improve fuel efficiency. These systems rely on advanced algorithms, communication protocols, and sensor technologies to facilitate smooth and efficient traffic operations. However, the deployment of eco-cooperative ADS requires addressing technical, regulatory, and societal challenges, such as data privacy, liability, and acceptance by users and stakeholders.

Microscopic traffic modeling is an essential tool for understanding and predicting the behavior of ACEVs in traffic scenarios. Car-following (CF) models, which describe the behavior of vehicles following each other, play a crucial role in simulating traffic flow dynamics. Kinematics-based CF models focus on variables derived from vehicle trajectories, while dynamics-based CF models consider the underlying forces and power dynamics. These models aim to strike a balance between simplicity and accuracy, enabling large-scale traffic simulations. However, further research is needed to improve the realism and applicability of CF models, especially in the context of ACEVs and their unique characteristics.

This chapter proposes several directions for future exploration. Initially, the advancement and implementation of ACEVs necessitate a holistic methodology that integrates knowledge from diverse fields including automotive engineering, computer science, transportation planning, and policy development. It is imperative to foster collaboration among researchers, industry participants, and policymakers to effectively tackle the technical, regulatory, and societal obstacles linked to ACEVs. Additionally, research endeavors should prioritize improving the safety, dependability, and energy efficiency of ACEVs while also striving to create comprehensive and validated traffic models that accurately depict the intricate interactions between ACEVs and the surrounding road infrastructure.

In conclusion, the integration of automation, connectivity, and electrification in road transport presents immense opportunities for improving safety, efficiency, and sustainability. However, several challenges need to be overcome to realize the full potential of ACEVs. By addressing these challenges and advancing research in driving control, eco-cooperative ADS, and traffic modeling, we can pave the way for a future where ACEVs play a central role in creating a safer, more efficient, and environmentally friendly transport system.

REFERENCES

1. Lawrence D Burns. A vision of our transport future. *Nature*, 497(7448):181–182, 2013.
2. Michel Noussan, Manfred Hafner, and Simone Tagliapietra. *The Future of Transport Between Digitalization and Decarbonization: Trends, Strategies and Effects on Energy Consumption*. Springer Nature, 2020.

3. SE Underwood, S Marshall, and J Niles. *Automated, Connected, and Electric Vehicles: An Assessment of Emerging Transportation Technologies and a Policy Roadmap for More Sustainable Transportation*. University of Michigan, Graham Institute for Sustainability, 2014.
4. IEA. *World Energy Outlook 2012*. International Energy Agency (IEA), 2012.
5. IPCC. *Climate Change 2014: Synthesis Report*. Intergovernmental Panel on Climate Change (IPCC), 2014.
6. Jonn Axsen, Patrick Plötz, and Michael Wolinetz. Crafting strong, integrated policy mixes for deep CO2 mitigation in road transport. *Nature Climate Change*, 10(9):809–818, 2020.
7. WHO. *Global Status Report on Road Safety 2018: Summary*. World Health Organization (WHO), 2018.
8. Bob Pishue. *US Traffic Hot Spots: Measuring the Impact of Congestion in the United States*. INRIX Research, 2017.
9. Tanzina Afrin and Nita Yodo. A survey of road traffic congestion measures towards a sustainable and resilient transportation system. *Sustainability*, 12(11):4660, 2020.
10. International Council on Clean Transportation (ICCT) Global Transportation Roadmap. https://theicct.org/publication/global-transportation-energy-and-climate-roadmap/.
11. ITF. *ITF Transport Outlook 2019*. International Transport Forum (ITF), 2019.
12. Siavash Khalili, Eetu Rantanen, Dmitrii Bogdanov, and Christian Breyer. Global transportation demand development with impacts on the energy demand and greenhouse gas emissions in a climate-constrained world. *Energies*, 12(20):3870, 2019.
13. Timo Möller, Asutosh Padhi, Dickon Pinner, and Andreas Tschiesner. *The Future of Mobility Is at Our Doorstep*. McKinsey Center for Future Mobility, 2019.
14. Jean-Paul A Barthès and Philippe Bonnifait. Multi-agent active collaboration between drivers and assistance systems. In *Advances in Artificial Transportation Systems and Simulation*, pages 163–180. Elsevier, 2015.
15. Seunghyuk Choi, Florian Thalmayr, Dominik Wee, and Florian Weig. *Advanced Driver-Assistance Systems: Challenges and Opportunities Ahead*. McKinsey & Company, pages 1–11, 2016.
16. UN Regulation on Advanced Emergency Braking Systems for Cars to Significantly Reduce Crashes. https://unece.org/sustainable-development/press/un-regulation-advanced-emergency-braking-systems-cars-significantly.
17. Vipin Kumar Kukkala, Jordan Tunnell, Sudeep Pasricha, and Thomas Bradley. Advanced driver-assistance systems: A path toward autonomous vehicles. *IEEE Consumer Electronics Magazine*, 7(5):18–25, 2018.
18. Xavier Mosquet, Michelle Andersen, and Aakash Arora. A roadmap to safer driving through advanced driver assistance systems. *Auto Tech Review*, 5(7):20–25, 2016.
19. Sangyong Lee and Yeonghun Yu. Study of the night vision system in vehicle. In *2012 IEEE Vehicle Power and Propulsion Conference*, pages 1312–1316. IEEE, 2012.
20. Wikipedia. *Backup Camera*. https://en.wikipedia.org/wiki/Backup_camera.
21. Zheng Tuo. Research on adaptive front-lighting system algorithm based on machine vision. In *Proceedings of the 2020 Artificial Intelligence and Complex Systems Conference*, pages 21–26. IEEE, 2020.
22. Yi Gao, Chunyu Lin, Yao Zhao, Xin Wang, Shikui Wei, and Qi Huang. 3D surround view for advanced driver assistance systems. *IEEE Transactions on Intelligent Transportation Systems*, 19(1):320–328, 2017.

23. Pär Degerman. *Automatic Parallel Parking: Park Assist Systems for Passenger Vehicles*. PhD thesis, Linköpings Universitet, 2007.
24. Meixin Zhu, Xuesong Wang, and Jingyun Hu. Impact on car following behavior of a forward collision warning system with headway monitoring. *Transportation Research Part C: Emerging Technologies*, 111:226–244, 2020.
25. Ankita Kamble and Sandhya Potadar. Lane departure warning system for advanced drivers assistance. In *2018 Second International Conference on Intelligent Computing and Control Systems (ICICCS)*, pages 1775–1778. IEEE, 2018.
26. AAA. *Advanced Driver Assistance Technology Names: AAA's Recommendation for Common Naming of Advanced Safety Systems*. American Automobile Association (AAA), 2019.
27. Mark Peplow. Traffic flow: Cruising through congestion. *Nature*, 429(6993), 2004.
28. Joel Vander Werf, Steven E Shladover, Mark A Miller, and Natalia Kourjanskaia. Effects of adaptive cruise control systems on highway traffic flow capacity. *Transportation Research Record*, 1800(1):78–84, 2002.
29. Jeremy J Blum, Azim Eskandarian, and Stephen A Arhin. Intelligent speed adaptation (ISA). In *Handbook of Intelligent Vehicles*, pages 582–602. Springer, 2012.
30. Wikipedia. *Intelligent Speed Adaptation*. https://en.wikipedia.org/wiki/Intelligent_speed_adaptation.
31. JP Thalen. *ADAS for the Car of the Future*. B.S. thesis, University of Twente, 2006.
32. Wikipedia. *Automatic Parking*. https://en.wikipedia.org/wiki/Automatic_parking.
33. Ping-Min Hsu, Ming Hung Li, and Kuo-Ching Chang. *Noise Filtering in Autonomous Emergency Braking Systems with Sensor Fusions*. Technical Report, SAE Technical Paper, SAE, 2015.
34. Brian Fildes, Michael Keall, Niels Bos, Anders Lie, Yves Page, Claus Pastor, Lucia Pennisi, Matteo Rizzi, Pete Thomas, and Claes Tingvall. Effectiveness of low speed autonomous emergency braking in real-world rear-end crashes. *Accident Analysis & Prevention*, 81:24–29, 2015.
35. Thierry Avot. *5G on the Highway to V2X*. Altran, 2018.
36. Wikipedia. *Vehicle-to-Everything*. https://en.wikipedia.org/wiki/Vehicle-to-everything.
37. IEA. *Global EV Outlook 2020*. IEA, 2020.
38. Wikipedia. *Paris Agreement*. https://en.wikipedia.org/wiki/Paris_Agreement.
39. Florian Knobloch, Steef V Hanssen, Aileen Lam, Hector Pollitt, Pablo Salas, Unnada Chewpreecha, Mark AJ Huijbregts, and Jean-Francois Mercure. Net emission reductions from electric cars and heat pumps in 59 world regions over time. *Nature Sustainability*, 3(6):437–447, 2020.
40. A Cornet, R Hensley, C Hirschberg, P Schaufuss, A Tschiesner, A Venus, and J Werra. Reboost: A comprehensive view on the changing powertrain component market and how suppliers can succeed. *Online: Mckinsey.com*, 2019.
41. Gautam Kalghatgi. Is it really the end of internal combustion engines and petroleum in transport? *Applied Energy*, 225:965–974, 2018.
42. Chris Mi and M Abul Masrur. *Hybrid Electric Vehicles: Principles and Applications with Practical Perspectives*. John Wiley & Sons, 2017.
43. Guoqi Ma, Masood Ghasemi, and Xingyong Song. Integrated powertrain energy management and vehicle coordination for multiple connected hybrid electric vehicles. *IEEE Transactions on Vehicular Technology*, 67(4):2893–2899, 2017.
44. Liang Li, Xiangyu Wang, Rui Xiong, Kai He, and Xujian Li. AMT downshifting strategy design of HEV during regenerative braking process for energy conservation. *Applied Energy*, 183:914–925, 2016.

45. Ji Li, Quan Zhou, Yinglong He, Bin Shuai, Ziyang Li, Huw Williams, and Hongming Xu. Dual-loop online intelligent programming for driver-oriented predict energy management of plug-in hybrid electric vehicles. *Applied Energy*, 253:113617, 2019.
46. Yinglong He, Chongming Wang, Quan Zhou, Ji Li, Michail Makridis, Huw Williams, Guoxiang Lu, and Hongming Xu. Multiobjective component sizing of a hybrid ethanol-electric vehicle propulsion system. *Applied Energy*, 266:114843, 2020.
47. Zhen Tian, Chengning Zhang, and Shuo Zhang. Analytical calculation of magnetic field distribution and stator iron losses for surface-mounted permanent magnet synchronous machines. *Energies*, 10(3):320, 2017.
48. Paul Nieuwenhuis, Liana Cipcigan, and Hasan Berkem Sonder. The electric vehicle revolution. In *Future Energy*, pages 227–243. Elsevier, 2020.
49. Asel Sartbaeva, VL Kuznetsov, SA Wells, and PP Edwards. Hydrogen nexus in a sustainable energy future. *Energy & Environmental Science*, 1(1):79–85, 2008.
50. Deloitte China. Fueling the future of mobility hydrogen and fuel cell solutions for transportation. *Financial Advisory*, 1(1), 2019.
51. Michail Makridis, Konstantinos Mattas, Caterina Mogno, Biagio Ciuffo, and Georgios Fontaras. The impact of automation and connectivity on traffic flow and CO2 emissions: A detailed microsimulation study. *Atmospheric Environment*, 226:117399, 2020.
52. Steven E Shladover, Dongyan Su, and Xiaoyun Lu. Impacts of cooperative adaptive cruise control on freeway traffic flow. *Transportation Research Record*, 2324(1):63–70, 2012.
53. Konstantinos Mattas, Michalis Makridis, Pauliana Hallac, María Alonso Raposo, Christian Thiel, Tomer Toledo, and Biagio Ciuffo. Simulating deployment of connectivity and automation on the antwerp ring road. *IET Intelligent Transport Systems*, 12(9):1036–1044, 2018.
54. Arne Kesting, Martin Treiber, Martin Schönhof, Florian Kranke, and Dirk Helbing. Jam-avoiding adaptive cruise control (ACC) and its impact on traffic dynamics. In *Traffic and Granular Flow'05*, pages 633–643. Springer, 2007.
55. Fanping Bu and Ching-Yao Chan. Adaptive and cooperative cruise control. In *Handbook of Intelligent Vehicles*, pages 191–207. Springer, 2012.
56. Shengbo Eben Li, Yang Zheng, Keqiang Li, Yujia Wu, J Karl Hedrick, Feng Gao, and Hongwei Zhang. Dynamical modeling and distributed control of connected and automated vehicles: Challenges and opportunities. *IEEE Intelligent Transportation Systems Magazine*, 9(3):46–58, 2017.
57. Meng Wang. *Generic Model Predictive Control Framework for Advanced Driver Assistance Systems*. PhD thesis, Delft University of Technology, 2014.
58. Stefan Bernsteiner. *Integration of Advanced Driver Assistance Systems on Full-Vehicle Level-Parametrization of an Adaptive Cruise Control System Based on Test Drives*. PhD thesis, Technical University Graz, 2016.
59. Cunxue Wu, Zhongming Xu, Yang Liu, Chunyun Fu, Kuining Li, and Minghui Hu. Spacing policies for adaptive cruise control: A survey. *IEEE Access*, 8:50149–50162, 2020.
60. C Lei, EM Van Eenennaam, W Klein Wolterink, G Karagiannis, Geert Heijenk, and J Ploeg. Impact of packet loss on CACC string stability performance. In *2011 11th International Conference on ITS Telecommunications*, pages 381–386. IEEE, 2011.
61. Steven E Shladover. Longitudinal control of automated guideway transit vehicles within platoons. *Journal of Dynamic Systems, Measurement, and Control*, 320–310, 1978.

62. Hossein Chehardoli and Mohammad R Homaeinezhad. Third-order leaderfollowing consensus protocol of traffic flow formed by cooperative vehicular platoons by considering time delay: Constant spacing strategy. *Proceedings of the Institution of Mechanical Engineers, Part I: Journal of Systems and Control Engineering*, 232(3):285–298, 2018.
63. S Darbha, KR Rajagopal, and Vipin Tyagi. A review of mathematical models for the flow of traffic and some recent results. *Nonlinear Analysis: Theory, Methods & Applications*, 69(3):950–970, 2008.
64. Arash Farnam and Alain Sarlette. About string stability of a vehicle chain with unidirectional controller. *European Journal of Control*, 50:138–144, 2019.
65. Darbha Swaroop and KR Rajagopal. A review of constant time headway policy for automatic vehicle following. In *ITSC 2001: 2001 IEEE Intelligent Transportation Systems. Proceedings (Cat. No. 01TH8585)*, pages 65–69. IEEE, 2001.
66. Rajesh Rajamani and Chunyu Zhu. Semi-autonomous adaptive cruise control systems. *IEEE Transactions on Vehicular Technology*, 51(5):1186–1192, 2002.
67. Junmin Wang and Rajesh Rajamani. Should adaptive cruise-control systems be designed to maintain a constant time gap between vehicles? *IEEE Transactions on Vehicular Technology*, 53(5):1480–1490, 2004.
68. Jing Zhou and Huei Peng. Range policy of adaptive cruise control vehicles for improved flow stability and string stability. *IEEE Transactions on Intelligent Transportation Systems*, 6(2):229–237, 2005.
69. Kumaragovindhan Santhanakrishnan and Rajesh Rajamani. On spacing policies for highway vehicle automation. *IEEE Transactions on Intelligent Transportation Systems*, 4(4):198–204, 2003.
70. F Broqua. Cooperative driving: Basic concepts and a first assessment of intelligent cruise control strategies. In *DRIVE Conference (1991: Brussels, Belgium). Advanced Telematics in Road Transport*. Vol. II. Elsevier, 1991.
71. Youping Zhang, B Kosmatopoulos, Petros A Ioannou, and CC Chien. Autonomous intelligent cruise control using front and back information for tight vehicle following maneuvers. *IEEE Transactions on Vehicular Technology*, 48(1):319–328, 1999.
72. Datta N Godbole, Natalia Kourjanskaia, Raja Sengupta, and Marco Zandonadi. Breaking the highway capacity barrier: Adaptive cruise control-based concept. *Transportation Research Record*, 1679(1):148–157, 1999.
73. Joel VanderWerf, Steven Shladover, Natalia Kourjanskaia, Mark Miller, and Hariharan Krishnan. Modeling effects of driver control assistance systems on traffic. *Transportation Research Record*, 1748(1):167–174, 2001.
74. Bart Van Arem, Cornelie JG Van Driel, and Ruben Visser. The impact of cooperative adaptive cruise control on traffic-flow characteristics. *IEEE Transactions on Intelligent Transportation Systems*, 7(4):429–436, 2006.
75. Bart Van Arem, Hans Driever, Philippus Feenstra, Jeroen Ploeg, Gerdien Klunder, Isabel Wilmink, Allard Zoutendijk, Zoltan Papp, and Bart Netten. *Design and Evaluation of an Integrated Full-Range Speed Assistant*. Technical Report. TNO, 2007.
76. Yinglong He, Quan Zhou, Michail Makridis, Konstantinos Mattas, Ji Li, Huw Williams, and Hongming Xu. Multiobjective co-optimization of cooperative adaptive cruise control and energy management strategy for PHEVs. *IEEE Transactions on Transportation Electrification*, 6(1):346–355, 2020.
77. Yang Zheng, Shengbo Eben Li, Jianqiang Wang, Dongpu Cao, and Keqiang Li. Stability and scalability of homogeneous vehicular platoon: Study on the influence of information flow topologies. *IEEE Transactions on Intelligent Transportation Systems*, 17(1):14–26, 2015.

78. Francesco Viti, Serge P Hoogendoorn, Tom P Alkim, and Gerben Bootsma. Driving behavior interaction with ACC: Results from a field operational test in the Netherlands. In *2008 IEEE Intelligent Vehicles Symposium*, pages 745–750. IEEE, 2008.
79. Gerdien Klunder, Minwei Li, and Michiel Minderhoud. Traffic flow impacts of adaptive cruise control deactivation and (re)activation with cooperative driver behavior. *Transportation Research Record*, 2129(1):145–151, 2009.
80. Shengbo Li, Keqiang Li, Rajesh Rajamani, and Jianqiang Wang. Model predictive multi-objective vehicular adaptive cruise control. *IEEE Transactions on Control Systems Technology*, 19(3):556–566, 2010.
81. ang Zheng, Shengbo Eben Li, Keqiang Li, Francesco Borrelli, and J Karl Hedrick. Distributed model predictive control for heterogeneous vehicle platoons under unidirectional topologies. *IEEE Transactions on Control Systems Technology*, 25(3):899–910, 2016.
82. Daniele Corona, Mircea Lazar, Bart De Schutter, and Maurice Heemels. A hybrid MPC approach to the design of a smart adaptive cruise controller. In *IEEE International Conference on Control Applications*, pages 231–236. IEEE, 2006.
83. Daniele Corona and Bart De Schutter. Adaptive cruise control for a smart car: A comparison benchmark for MPC-PWA control methods. *IEEE Transactions on Control Systems Technology*, 16(2):365–372, 2008.
84. Vibhor L Bageshwar, William L Garrard, and Rajesh Rajamani. Model predictive control of transitional maneuvers for adaptive cruise control vehicles. *IEEE Transactions on Vehicular Technology*, 53(5):1573–1585, 2004.
85. Wikipedia. *Sliding Mode Control*. https://en.wikipedia.org/wiki/Sliding_mode_control.
86. Clara Marina Martinez and Dongpu Cao. *iHorizon-Enabled Energy Management for Electrified Vehicles*. Butterworth-Heinemann, 2018.
87. J Guldner and VI Utkin. The chattering problem in sliding mode systems. In *Fourteenth International Symposium of Mathematical Theory of Networks and Systems, MTNS2000*. IEEE, 2000.
88. Antonella Ferrara and Claudio Vecchio. Second order sliding mode control of vehicles with distributed collision avoidance capabilities. *Mechatronics*, 19(4):471–477, 2009.
89. Xianggui Guo, Jianliang Wang, Fang Liao, and Rodney Swee Huat Teo. Distributed adaptive integrated-sliding-mode controller synthesis for string stability of vehicle platoons. *IEEE Transactions on Intelligent Transportation Systems*, 17(9):2419–2429, 2016.
90. Sang-Jin Ko and Ju-Jang Lee. Fuzzy logic based adaptive cruise control with guaranteed string stability. In *2007 International Conference on Control, Automation and Systems*, pages 15–20. IEEE, 2007.
91. José Eugenio Naranjo, Carlos González, Jesús Reviejo, Ricardo García, and Teresa De Pedro. Adaptive fuzzy control for inter-vehicle gap keeping. *IEEE Transactions on Intelligent Transportation Systems*, 4(3):132–142, 2003.
92. José Eugenio Naranjo, Carlos González, Ricardo García, and Teresa De Pedro. ACC + Stop&go maneuvers with throttle and brake fuzzy control. *IEEE Transactions on Intelligent Transportation Systems*, 7(2):213–225, 2006.
93. Kakan C Dey, Li Yan, Xujie Wang, Yue Wang, Haiying Shen, Mashrur Chowdhury, Lei Yu, Chenxi Qiu, and Vivekgautham Soundararaj. A review of communication, driver characteristics, and controls aspects of cooperative adaptive cruise control (CACC). *IEEE Transactions on Intelligent Transportation Systems*, 17(2):491–509, 2015.
94. teven E Shladover. Path at 20—history and major milestones. *IEEE Transactions on Intelligent Transportation Systems*, 8(4):584–592, 2007.
95. Qing Xu and Raja Sengupta. Simulation, analysis, and comparison of ACC and CACC in highway merging control. In *IEEE IV2003 Intelligent Vehicles Symposium. Proceedings (Cat. No. 03TH8683)*, pages 237–242. IEEE, 2003.

96. Ziran Wang, Guoyuan Wu, and Matthew J Barth. A review on cooperative adaptive cruise control (CACC) systems: Architectures, controls, and applications. In *2018 21st International Conference on Intelligent Transportation Systems (ITSC)*, pages 2884–2891. IEEE, 2018.
97. Gábor Rödönyi. An adaptive spacing policy guaranteeing string stability in multi-brand ad hoc platoons. *IEEE Transactions on Intelligent Transportation Systems*, 19(6):1902–1912, 2017.
98. Carlos Flores and Vicente Milanés. Fractional-order-based ACC/CACC algorithm for improving string stability. *Transportation Research Part C: Emerging Technologies*, 95:381–393, 2018.
99. Jeroen Ploeg, Nathan Van De Wouw, and Henk Nijmeijer. Lp string stability of cascaded systems: Application to vehicle platooning. *IEEE Transactions on Control Systems Technology*, 22(2):786–793, 2013.
100. Meng Wang. Infrastructure assisted adaptive driving to stabilise heterogeneous vehicle strings. *Transportation Research Part C: Emerging Technologies*, 91:276–295, 2018.
101. Yougang Bian, Yang Zheng, Wei Ren, Shengbo Eben Li, Jianqiang Wang, and Keqiang Li. Reducing time headway for platooning of connected vehicles via V2V communication. *Transportation Research Part C: Emerging Technologies*, 102:87–105, 2019.
102. I Ge Jin and Gábor Orosz. Optimal control of connected vehicle systems with communication delay and driver reaction time. *IEEE Transactions on Intelligent Transportation Systems*, 18(8):2056–2070, 2016.
103. Linjun Zhang and Gábor Orosz. Motif-based design for connected vehicle systems in presence of heterogeneous connectivity structures and time delays. *IEEE Transactions on Intelligent Transportation Systems*, 17(6):1638–1651, 2016.
104. Ji-Wook Kwon and Dongkyoung Chwa. Adaptive bidirectional platoon control using a coupled sliding mode control method. *IEEE Transactions on Intelligent Transportation Systems*, 15(5):2040–2048, 2014.
105. Bijan Sakhdari. *Real-time Autonomous Cruise Control of Connected Plugin Hybrid Electric Vehicles Under Uncertainty*. PhD thesis, University of Waterloo, 2018.
106. Sangjun Park, Hesham A Rakha, Kyoungho Ahn, Kevin Moran, Bart Saerens, and Eric Van den Bulck. Predictive ecocruise control system: Model logic and preliminary testing. *Transportation Research Record*, 2270(1):113–123, 2012.
107. Shengbo Eben Li, Qiangqiang Guo, Shaobing Xu, Jingliang Duan, Shen Li, Chengjun Li, and Kuifeng Su. Performance enhanced predictive control for adaptive cruise control system considering road elevation information. *IEEE Transactions on Intelligent Vehicles*, 2(3):150–160, 2017.
108. Shaobing Xu, Shengbo Eben Li, Bo Cheng, and Keqiang Li. Instantaneous feedback control for a fuel-prioritized vehicle cruising system on highways with a varying slope. *IEEE Transactions on Intelligent Transportation Systems*, 18(5):1210–1220, 2016.
109. Yugong Luo, Tao Chen, and Keqiang Li. Multi-objective decoupling algorithm for active distance control of intelligent hybrid electric vehicle. *Mechanical Systems and Signal Processing*, 64:29–45, 2015.
110. Bijan Sakhdari, Mahyar Vajedi, and Nasser L Azad. Ecological adaptive cruise control of a plug-in hybrid electric vehicle for urban driving. In *2016 IEEE 19th International Conference on Intelligent Transportation Systems (ITSC)*, pages 1739–1744. IEEE, 2016.
111. Guoqiang Li and Daniel Görges. Ecological adaptive cruise control and energy management strategy for hybrid electric vehicles based on heuristic dynamic programming. *IEEE Transactions on Intelligent Transportation Systems*, 20(9):3526–3535, 2018.

112. Haiyang Yu, Rui Jiang, Zhengbing He, Zuoduo Zheng, Li Li, Runkun Liu, and Xiqun Chen. Automated vehicle-involved traffic flow studies: A survey of assumptions, models, speculations, and perspectives. *Transportation Research Part C: Emerging Technologies*, 127:103101, 2021.
113. Lily Elefteriadou et al. *An Introduction to Traffic Flow Theory*. Vol. 84. Springer, 2014.
114. Wikipedia. *Equations of Motion*. https://en.wikipedia.org/wiki/Equations_of_motion.
115. Yinglong He, Michail Makridis, Konstantinos Mattas, Georgios Fontaras, Biagio Ciuffo, and Hongming Xu. Introducing electrified vehicle dynamics in traffic simulation. *Transportation Research Record*, 2674(9):776–791, 2020.
116. Michail Makridis, Georgios Fontaras, Biagio Ciuffo, and Konstantinos Mattas. MFC free-flow model: Introducing vehicle dynamics in microsimulation. *Transportation Research Record*, 2673(4):762–777, 2019.
117. Mark Brackstone and Mike McDonald. Car-following: A historical review. *Transportation Research Part F: Traffic Psychology and Behaviour*, 2(4):181–196, 1999.
118. Xiqun Chen. *Modeling Traffic Flow Dynamic and Stochastic Evolution*. PhD thesis, Tsinghua University, 2012.
119. A. Reuschel. Fahrzeugbewegungen in der kolonne. *Osterreichisches Ingenieur Archiv*, 4:193–215, 1950.
120. Louis A Pipes. An operational analysis of traffic dynamics. *Journal of Applied Physics*, 24(3):274–281, 1953.
121. Yong Zhang, Ping Ni, Minwei Li, Hao Liu, and Baocai Yin. A new carfollowing model considering driving characteristics and preceding vehicle's acceleration. *Journal of Advanced Transportation*, 2017(1), 2017.
122. Denos C Gazis, Robert Herman, and Richard W Rothery. Nonlinear followthe-leader models of traffic flow. *Operations Research*, 9(4):545–567, 1961.
123. Saidi Siuhi and Mohamed S. Kaseko. Parametric study of stimulus-response behavior for car-following models. In *Proceedings of the Transportation Research Board 89th Annual Meeting, USA*. IEEE, 2010.
124. Saidi Siuhi. *Parametric Study of Stimulus-Response Behavior Incorporating Vehicle Heterogeneity in Car-Following Models*. PhD thesis, University of Nevada, 2009.
125. Gordon Frank Newell. Nonlinear effects in the dynamics of car following. *Operations Research*, 9(2):209–229, 1961.
126. Mohammad Saifuzzaman. *Incorporating Risk Taking and Driver Errors in Carfollowing Models*. PhD thesis, Queensland University of Technology, 2016.
127. Peter G Gipps. A behavioural car-following model for computer simulation. *Transportation Research Part B: Methodological*, 15(2):105–111, 1981.
128. Biagio Ciuffo, Vincenzo Punzo, and Marcello Montanino. Thirty years of Gipps' car-following model: Applications, developments, and new features. *Transportation Research Record*, 2315(1):89–99, 2012.
129. Vincenzo Punzo and Fulvio Simonelli. Analysis and comparison of microscopic traffic flow models with real traffic microscopic data. *Transportation Research Record*, 1934(1):53–63, 2005.
130. Vincenzo Punzo and Antonino Tripodi. Steady-state solutions and multiclass calibration of Gipps microscopic traffic flow model. *Transportation Research Record*, 1999(1):104–114, 2007.
131. Masako Bando, Katsuya Hasebe, Akihiro Nakayama, Akihiro Shibata, and Yuki Sugiyama. Dynamical model of traffic congestion and numerical simulation. *Physical Review E*, 51(2):1035, 1995.
132. Hongxing Zhao, Ruichun He, and Changxi Ma. An extended car-following model at signalised intersections. *Journal of Advanced Transportation*, 2018(1), 2018.

133. Dirk Helbing and Benno Tilch. Generalized force model of traffic dynamics. *Physical Review E*, 58(1):133, 1998.
134. Rui Jiang, Qingsong Wu, and Zuojin Zhu. Full velocity difference model for a car-following theory. *Physical Review E*, 64(1):017101, 2001.
135. Robert L Bierley. Investigation of an intervehicle spacing display. *Highway Research Record*, 25, 1963.
136. W Leutzbach and R Wiedemann. Development and applications of traffic simulation models at the karlsruhe institut für verkehrswesen. *Traffic Engineering & Control*, 27(5):270–278, 1986.
137. Beshr Sultan, Mark Brackstone, and Mike McDonald. Drivers' use of deceleration and acceleration information in car-following process. *Transportation Research Record*, 1883(1):31–39, 2004.
138. Martin Treiber, Ansgar Hennecke, and Dirk Helbing. Congested traffic states in empirical observations and microscopic simulations. *Physical Review E*, 62(2):1805, 2000.
139. Xiqun Chen et al. *Stochastic Evolutions of Dynamic Traffic Flow*. Springer, 2015.
140. Arne Kesting, Martin Treiber, and Dirk Helbing. Enhanced intelligent driver model to access the impact of driving strategies on traffic capacity. *Philosophical Transactions of the Royal Society A: Mathematical, Physical and Engineering Sciences*, 368(1928):4585–4605, 2010.
141. John Searle. *Equations for Speed, Time and Distance for Vehicles Under Maximum Acceleration*. Technical Report. SAE Technical Paper. SAE, 1999.
142. Hesham Rakha, Ivana Lucic, Sergio Henrique Demarchi, José Reynaldo Setti, and Michel Van Aerde. Vehicle dynamics model for predicting maximum truck acceleration levels. *Journal of Transportation Engineering*, 127(5):418–425, 2001.
143. Hesham Rakha and Ivana Lucic. Variable power vehicle dynamics model for estimating truck accelerations. *Journal of Transportation Engineering*, 128(5):412–419, 2002.
144. Hesham Rakha, Matthew Snare, and Francois Dion. Vehicle dynamics model for estimating maximum light-duty vehicle acceleration levels. *Transportation Research Record*, 1883(1):40–49, 2004.
145. Hesham Rakha. Validation of Van Aerde's simplified steadystate car-following and traffic stream model. *Transportation Letters*, 1(3):227–244, 2009.
146. Hesham A Rakha, Kyoungho Ahn, Waleed Faris, and Kevin S Moran. Simple vehicle powertrain model for modeling intelligent vehicle applications. *IEEE Transactions on Intelligent Transportation Systems*, 13(2):770–780, 2012.
147. Karim Fadhloun, Hesham Rakha, Amara Loulizi, and Abdessattar Abdelkefi. Vehicle dynamics model for estimating typical vehicle accelerations. *Transportation Research Record*, 2491(1):61–71, 2015.
148. Karim Fadhloun and Hesham Rakha. A novel vehicle dynamics and human behavior car-following model: Model development and preliminary testing. *International Journal of Transportation Science and Technology*, 9(1):14–28, 2020.

Index

Note: **Boldface** page references indicate tables. *Italic* references indicate figures.

www.ingramcontent.com/pod-product-compliance
Lightning Source LLC
LaVergne TN
LVHW011116300926
849160LV00008B/402

* 9 7 8 1 0 3 2 2 9 9 4 0 2 *